Innovative Micropropagation of Horticultural and Medicinal Plants

Innovative Micropropagation of Horticultural and Medicinal Plants

Guest Editors

Biljana K. Filipović
Milana Trifunović Momčilov

MDPI

Basel • Beijing • Wuhan • Barcelona • Belgrade • Novi Sad • Cluj • Manchester

Guest Editors
Biljana K. Filipović
Department of Plant Physiology
University of Belgrade
Belgrade
Serbia

Milana Trifunović Momčilov
Department of Plant Physiology
University of Belgrade
Belgrade
Serbia

Editorial Office
MDPI AG
Grosspeteranlage 5
4052 Basel, Switzerland

This is a reprint of the Special Issue, published open access by the journal *Horticulturae* (ISSN 2311-7524), freely accessible at: https://www.mdpi.com/journal/horticulturae/special_issues/H502260E4P.

For citation purposes, cite each article independently as indicated on the article page online and as indicated below:

Lastname, A.A.; Lastname, B.B. Article Title. *Journal Name* **Year**, *Volume Number*, Page Range.

ISBN 978-3-7258-5547-6 (Hbk)
ISBN 978-3-7258-5548-3 (PDF)
https://doi.org/10.3390/books978-3-7258-5548-3

Contents

About the Editors

Biljana K. Filipović

Biljana K. Filipović is a senior research associate in the Department of Plant Physiology at the Institute for Biological Research "Siniša Stanković" – National Institute of the Republic of Serbia, University of Belgrade. Her research focuses on plant biotechnology, molecular biology, phytochemistry, and stress physiology with special interest in in vitro morphogenesis and propagation of medicinal, rare, and endangered plants with a spectrum of health-related specialized metabolites. She has participated in several national and EU research projects. She has published her research results in indexed international journals, also collaborating as a reviewer.

Milana Trifunović-Momčilov

Milana Trifunović-Momčilov is a principal research fellow at the Department for Plant Physiology, Institute for Biological Research "Siniša Stanković" – National Institute of the Republic of Serbia, University of Belgrade. In general, her main research interests include physiology, molecular biology, and plant biotechnology. She has expertise in numerous research methods, including in vitro culture, as well as histological and molecular biology techniques. Additionally, she also conducts experimental studies in plant stress physiology, including the evaluation of morpho-physiological, biochemical, and molecular alterations, as a result of abiotic factors, primarily salt stress and drought.

horticulturae

MDPI

Article

Saffron In Vitro Propagation: An Innovative Method by Temporary Immersion System (TIS), Integrated with Machine Learning Analysis

Waed Tarraf [1,*], Tolga İzgü [1], Özhan Şimşek [2], Nunzia Cicco [3] and Carla Benelli [1]

[1] Institute for BioEconomy (IBE), National Research Council (CNR), 50019 Sesto Fiorentino, Florence, Italy; tolga.izgu@ibe.cnr.it (T.İ.); carla.benelli@ibe.cnr.it (C.B.)
[2] Horticulture Department, Agriculture Faculty, Erciyes University, 38030 Kayseri, Türkiye; ozhan12@gmail.com
[3] Institute of Methodologies for Environmental Analysis, National Research Council, 85050 Tito Scalo, Potenza, Italy; nunzia.cicco@imaa.cnr.it
* Correspondence: waed.tarraf@ibe.cnr.it

Abstract: The propagation of *Crocus sativus* L. relies exclusively on corm multiplication. As underground storage organs, corms are susceptible to a wide range of pathogens, environmental stresses, and diseases, making traditional propagation methods often ineffective with the loss of valuable material. In vitro propagation offers an alternative for the saffron culture under controlled conditions. In particular, the innovative application of the Temporary Immersion System (TIS) represents a technological advancement for enhancing biomass production with a reduction in operational costs. The current study utilized the Plantform™ bioreactor to propagate in vitro saffron corms from the 'Abruzzo' region (Italy), integrating machine learning models to assess its performance. The evaluation of saffron explants after 30, 60, and 90 days of culture showed a marked improvement in growth and microcorm production compared to conventional in vitro culture on semisolid medium, supported by the machine learning analysis. Indeed, the Random Forest algorithm revealed a predictive accuracy with an R^2 value of 0.81 for microcorm number, showcasing the capability of machine learning models to forecast propagation outcomes effectively. These results confirm that applying TIS in saffron culture could lead to economically viable, large biomass production within a controlled environment, irrespective of seasonality. This study represents the first endeavor to use TIS technology to enhance the in vitro propagation of saffron in conjunction with machine learning, suggesting an innovative approach for cultivating high-value crops like saffron.

Keywords: *Crocus sativus* L.; micropropagation; liquid culture; Plantform™ bioreactor; machine learning

1. Introduction

Saffron (*Crocus sativus* L.) is a perennial plant growing widely in Iran, Morocco, Greece, Türkiye, China, and many other countries [1]. It is an ancient plant; the first mention of its cultivation dates around 2300 B.C. by King Sargon, and it was identified in a fresco painting in the Knossos palace (Minos, Crete in Greek) dated 1700–1600 B.C. [2]. It is a plant with traditional importance, indeed it was used in local cultural, intellectual, physical, and spiritual areas [3]. The commercial part of the saffron plant consists of dry red stigma or red powder, also called 'red gold', used as a spice to obtain dye and scents, and as medicinal drugs to human health since ancient times.

The saffron plant is vegetatively propagated through corm multiplication, particularly by forming daughter corms from the mother plant. Corms, which are underground organs, are susceptible to a wide range of pathogens and diseases that can lead to significant losses in valuable production. Additionally, saffron propagation is labor-intensive, as the corms must be manually removed and replanted [4].

Therefore, different propagation pathways, such as in vitro culture, have been investigated. In vitro propagation has already been successfully applied to saffron from different geographical origins, i.e., Iran [5], India [6], Morocco [7], Türkiye [8], China [9], and Spain [10], allowing its culture in sterile, controlled conditions and without season dependency. However, the Temporary Immersion System (TIS) approach can be considered for large-scale in vitro propagation and rapid acclimatization. This system is based on alternating cycles of temporary immersion of the cultured explants into the liquid medium followed by a dry period. The alternating contact between the plants and the liquid medium can encourage the development and adaptation of plants to the next stage of acclimatization [11–13]. Moreover, there is the possibility of reducing hand labor and the cost of production markedly. Indeed, careful positioning of explants in the agarized medium is not necessary, and the replacement of the new fresh liquid medium is easily performed. TIS allows us to achieve many other advantages [14], and it was applied with success on several ornamental, fruit, and woody species [15–24]. TIS application requires the use of plant bioreactors that can be of different types and sometimes are equipped for gas exchange by ventilation phases [14,25]. In this study, the Plantform^{TM} bioreactor was used; it consists of a single container made of polycarbonate transparent and is autoclavable at 120 °C (http://www.plantform.se: accessed on 8 April 2024). Inside, there is a basket containing plant material. Filters and silicon tubes are connected to timers and air pumps to regulate the immersion and ventilation times. The containers can be placed above each other, saving space in the climate chamber. The Plantform^{TM} bioreactor was applied to propagate saffron corms coming from the geographical origin 'Abruzzo region' (Italy).

Many researchers frequently find it challenging to use conventional statistical methods to assess big and complex datasets in the context of in vitro micropropagation, a composite biological process impacted by genotypes, culture medium, and environment [26]. Recently, new technology based on artificial intelligence, such as machine learning, is developing quickly in several scientific and industrial domains [27] but its integration within the plant and agricultural sciences remains relatively emergent. Artificial Neural Networks (ANNs), described as a class of nonlinear computational strategies, are employed for various tasks encompassing data clustering, generating forecasts, and classifying complex systems. Machine learning algorithms emerge as potent and predictive tools for decision-making in the sector of in vitro plant propagation, due to their proficiency in clarifying and defining the complexity of processes that involve a multitude of factors. Currently, these models have been applied in various in vitro culture investigations, including micropropagation, regeneration and in vitro organogenesis, stress physiology, and salt stress [28–33].

Therefore, this study aimed to evaluate the response of the Italian saffron ecotype from Abruzzo to the culture in the Plantform^{TM} bioreactor in terms of growth and microcorm production compared to the conventional in vitro propagation on semisolid medium. To support the research scope, the artificial neural network (ANN) analysis and machine learning (ML) algorithms (Gaussian process, GP; random forest, RF; support vector machine; SVM) were integrated to model and predict the effects of two culture systems on different growth parameters.

2. Materials and Methods

2.1. Plant Material, Sterilization, and In Vitro Establishment

For in vitro culture establishment, healthy saffron corms were collected from Navelli (Protected Designation of Origin) situated in the Abruzzo region (42°14′19″ N 13°43′46″ E). The corms were used as source of explants (Figure 1a). For cleaning and disinfection of corms, the external tunic was removed, and the corms were treated with 0.1% fungicide (Enovit Metil FL; SIPCAM—Soc, Milan, Italy) for 20 min and then sterilized with 80% (v/v) ethanol for 1 min, followed by dipping in 0.2% (w/v) $HgCl_2$ for 20 min. Following that, the corms were rinsed three times with distilled sterile water. The apical and the lateral buds from the corms were excised (in total 250 buds) at the end of October and used for in vitro culture (Figure 1b). The buds were cultured initially in glass tubes (Figure 1c) and then transferred in 500 mL glass jars on Murashige and Skoog (MS, Sigma-Aldrich, St. Louis, MO, USA) medium [34] containing 1 mg L^{-1} 6-benzyladenine; (BAP; Sigma-Aldrich, St. Louis, MO, USA), 1 mg L^{-1} 1-Naphthaleneacetic acid (NAA; Sigma-Aldrich, St. Louis, MO), 100 mg L^{-1} Ascorbic acid (Carlo Erba, Cornaredo, MI, USA), 30 g L^{-1} sucrose, and 3 g L^{-1} Gelrite™ (Sigma-Aldrich, St. Louis, MO, USA), mentioned as MS1. The pH of the media was adjusted to 5.8. The cultures were kept at 24 °C, under a 16 h photoperiod with 60 μmol m^{-2} s^{1} of photosynthetically active radiation provided by cool-white fluorescent lamps and subcultured every 30 days. After three subcultures, new, well-developed in vitro corms were used to start the TIS and semisolid cultures.

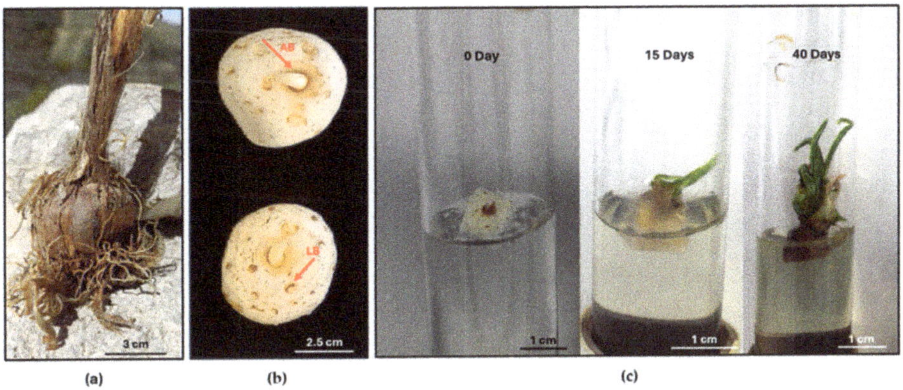

Figure 1. Corms of saffron from the Abruzzo region (**a**); corms with apical buds (AB) and lateral buds (LB) used for in vitro culture (**b**); development of initial explants in in vitro culture (**c**).

2.2. *Preparation of In Vitro Cultures in Temporary Immersion System and Semisolid Medium*

In the Plantform^{TM} bioreactor, 500 mL of liquid MS1 medium was used. A total of 6 explants were assessed. The immersion and ventilation periods of the TIS culture were controlled using air pumps and a timer placed in the climate chamber; the immersion frequency was 5 min every 4 h, with 15 min aeration every 4 h. The experiment was repeated twice. In the semisolid culture, each jar was prepared with 100 mL of MS1 medium, and 6 jars were set up. One corm was placed per jar.

The cultures were kept for 30, 60, and 90 days under the same culture conditions as stated above. In the TIS culture, at each evaluation period, only the MS1 was replaced without removing the corms, while in the semisolid medium, the explants were transferred to a fresh MS1 as they stood. The number of new microcorms, number of developed shoots, and root formation were evaluated for each period. Moreover, corms weight was recorded to estimate the Relative Growth Rate (RGR) index of the cultures after 30, 60, and 90 days using the following formula: [ln FW final − ln FW initial] × 100/days of culture (ln = natural logarithm; FW: fresh weight, [22]). RGR index is based on the initial and final fresh weights of the plant material and the time of culture.

2.3. *Statistical Analysis*

The present study employed the *t*-test method using R-programming to perform a rigorous statistical analysis to examine the differences between two culture systems concerning several attributes. The *t*-test is a statistical tool commonly used, that allowed the comparison of the mean values of each feature between the TIS and semisolid medium, offering important information about any differences influenced by the experimental settings. The experiment was repeated twice and consisted of three replicates with two explants used either for TIS or semisolid per each replicate.

2.4. Modeling Procedure

In this study, we used three Machine Learning (ML) algorithms—Gaussian process (GP), Random Forest (RF), and Support Vector Machine (SVM), as well as the well-known ANN-based Multilayer Perceptron (MLP) to model and predict the micropropagation and rooting efficiency of saffron through TIS and the semisolid medium. To thoroughly evaluate the anticipated performances of the MLP and ML models, we partitioned the dataset into training and testing subsets, and we used a 10-fold cross-validation technique.

TIS and semisolid medium are the input variables. On the other hand, the microcorm, shoot, and root numbers were the target (output) variables. R-programming was used with the Caret package (Classification and Regression Training) to implement coding. Several metrics were used to assess and compare the precision and accuracy of the MLP and ML models. These metrics included the root means square error (RMSE), which shows how closely the regression line matches the observed data points. The coefficient of determination (R^2) shows the degree of relationship between the model and dependent variable, and the mean absolute error (MAE) computes the average error between the predicted and observed values (Equations (1)–(3)).

$$\mathrm{MAE} = \frac{1}{n}\sum_{i=1}^{n} |Y_i - \hat{Y}_i| \tag{1}$$

$$\mathrm{RMSE} = \sqrt{\frac{\left(\sum_{i=1}^{n} (Y_i - \hat{Y}_i)^2\right)}{n}} \tag{2}$$

$$R^2 = 1 - \frac{\sum_{i=1}^{n} (Y_i - \hat{Y}_i)^2}{\sum_{i=1}^{n} (Y_i - \tilde{Y})^2} \tag{3}$$

where Y_i is the actual value; $\hat{Y}_i$ is the predicted value; $\tilde{Y}$ is the mean of the actual values; and n is the sample count.

2.4.1. Multilayer Perceptron

With an input layer, an output layer, and one or more hidden layers, the Multilayer Perceptron (MLP) is a popular example of an Artificial Neural Network (ANN). The input and output variables from the training set were used to train the MLP using a supervised training technique. Until the desired value in Equation (4) was attained, the training procedure was repeated [26].

$$E = \frac{1}{n}\sum_{n=1}^{n} (y_s - \hat{y}_s) \tag{4}$$

where n is the number of observations; y_s is the sth observation variable; and $\hat{y}_s$ is the sth of the predicted variable.

2.4.2. Gaussian Process

To better comprehend the spread of random variables, the Gaussian Process (GP) model, which extends the Gaussian probability distribution, is a useful tool for supervised learning. This model works very well for solving regression and classification issues. It is a non-parametric classifier that calculates the likelihood that input samples belong to particular classes, particularly for binary datasets. One of its main features is its capacity to operate effectively with tiny datasets and produce reliable, accurate, and computationally efficient results [35]. Equation (5) presents the mathematical reasoning for each input (x) and matching output (y).

$$y_i = f(x_i) + \varepsilon \tag{5}$$

2.4.3. Random Forest

The RF approach is well-known for its effectiveness and simplicity of usage, and it has proven successful in both regression and classification problems. Prior studies have demonstrated that the RF model has several noteworthy qualities, such as its capacity to avoid overfitting, adeptness at managing noise, and effective handling of a high number of data [36]. The trained tree makes the final decision in the RF technique, which uses bagging, also referred to as bootstrap aggregation. The basic idea of the RF model is shown in Equation (6).

$$y = \sum_{i=1}^{n} (a_i - a_i^*)k(x, x_i) + b \tag{6}$$

2.4.4. Support Vector Machines

Even with relatively small datasets, the SVM approach demonstrates efficiency; however, a significant amount of training data is often required for effective learning. Additionally, the SVM framework lessens issues with other methods, such as overfitting, slow convergence rates, and becoming trapped in local minima. Due to their higher resistance to the challenges sometimes associated with classic artificial intelligence approaches, SVMs are distinguished by this special quality [37]. The SVM algorithm, which helps identify which class has the longest separator plane, is declared in Equation (7).

$$f(x) = w\varphi(x) + b \tag{7}$$

3. Results and Discussion

3.1. *In Vitro Propagation of Saffron*

Many papers that have reported on in vitro saffron propagation come from several countries [5,7,38–41] and some reviews [42–44] have collected all the information on this fundamental method for propagating the *Crocus* spp. [2], especially in *C. sativus*, to develop callus or new shoot buds and obtain new microcorms formation. The in vitro culture system is a biotechnological tool that enhances saffron propagation by increasing biomass production and plant quality in a sterile and controlled environment, independent of seasonal availability. Indeed, the low multiplication rate and the bulb rot are the major issues in the traditional propagation of saffron, considering that about four or five corms per mother corm are produced in one year, and soil infestation reduces productivity.

Different types of explants were used to obtain embryogenic callus, microcorms, or shoot regeneration, but corms or small parts of corms, including one bud, were generally better explants for the production of shoots and new cormlets [45]. In our research, the initial explants were a small portion of corms with apical or lateral buds, excised in October, from the corms. The best period for introducing saffron buds in vitro culture seems to be September–December [46,47].

Corms collected from fields are contaminated with soil and contaminants, thus their sterilization is a crucial step towards the success of the in vitro culture, and the optimization of the disinfection protocol is essential for ensuring large-scale corm production. Different disinfection processes have been applied to saffron corms [48]. In our study, a sequence of treatments involving ETOH, a fungicide, $HgCl_2$, and sterilized water was identified as the best combination to obtain an 87% survival rate of explants, in agreement with findings by Taheri-Dehkordi et al. [5]. Furthermore, as reported by Yasmin et al. [47], the in vitro development of saffron buds is season-dependent, with the highest success rates observed from September to October (95%) and followed by November to December (85%). Therefore, employing an effective set of sterilization steps combined with choosing the best period for initiating in vitro cultures led to a high survival rate of explants.

Different media were used for the in vitro culture of saffron as reported in the detailed review of Salami et al. [45]. Our preliminary results showed that MS medium containing 1 mg L^{-1} BA and 1 mg L^{-1} NAA was the best combination for the in vitro culture of the Italian ecotype and was consequently applied in the current study.

The conventional in vitro propagation using the semisolid medium is proposed as a system that increases the number of corms produced [49–51], but it is often a cost-labor method. Consequently, the introduction of semi-automation by plant bioreactors for in vitro propagation has been proposed as a strategy to reduce production costs [52–54]. In this study, applying the Plantform^{TM} bioreactor on the Abruzzo saffron ecotype improved in vitro propagation efficiency. Table 1 and Figure 2 present the *t*-test comparisons between the two distinct groups, TIS and semisolid medium, across different growth parameters (microcorm number, shoot number, and root number). For each parameter, significant differences in mean values between the two culture systems were noted; the *p*-values indicate statistically significant differences. Notably, TIS outperforms the semisolid medium in the number of microcorms, with a significantly higher mean value of 107.5, which is further supported by a p-value < 0.001. Likewise, a variation in the shoot number was observed, and the TIS yielded a mean value of 11.26, which was greater than the semisolid medium's mean of 3.07. The statistical significance is confirmed by a corresponding p-value < 0.001, which suggests a significant difference in the number of shoots between the culture systems.

Table 1. The *t*-test analysis of growth parameters between TIS and semisolid medium.

Parameter	System	Mean	df	*p*-Value
Microcorm number	TIS SM	107.50 9.81	72.28	<0.001
Shoot number	TIS SM	11.26 3.07	74.36	<0.001
Root number	TIS SM	4.13 1.38	128.7	<0.001

TIS: Temporary Immersion System; SM: semisolid medium; df: Degree of Freedom.

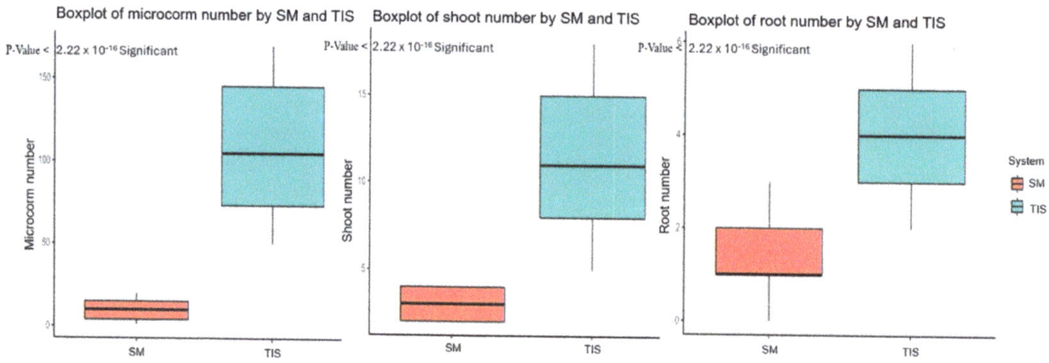

Figure 2. The *t*-test boxplot of growth parameters (microcorm, shoot, and root number) in TIS and semisolid medium (SM) cultures.

There is also a significant difference revealed by the root number, where the temporary immersion has a mean value of 4.13, with respect to the mean of 1.38 in semisolid medium; a p-value < 0.001, showing a clear variation in root number between the two systems. Overall, the statistical analysis significantly highlighted the critical role that culture systems play in influencing the investigated parameters (Figure 2).

RGR is an analytical tool used to characterize plant growth. It can be calculated from measurements taken on the same sample at two different times without the loss of vegetal material. This index was also applied in in vitro cultures [22,24,55–57]. In the current study, the RGR index demonstrated an improvement in the growth of saffron in TIS than in the semisolid medium since the first evaluation period (30 days). The same trend was confirmed after 60 and 90 days of continuous culture, as indicated in Table 2. These results are in agreement with Gatti et al. [22] on the efficiency of Plantform^{TM} in increasing biomass production of *Quercus robur* L. over semisolid culture in a microbox. Our findings showed the highest RGR of 1.27 after 90 days of culture in Plantform^{TM} versus 0.87 in the solid medium. A similar RGR of 1.12 and 1.15 was observed at 30 and 60 days, respectively, indicating a considerable increase in biomass obtained during TIS cultures as recorded by Benelli and De Carlo, [24] using the same type of Plantform^{TM} bioreactor for *Olea europaea* L.

Table 2. Relative growth rate (RGR index) of *C. sativus* in Plantform^{TM} and semisolid medium after 30, 60, and 90 days of the beginning of in vitro culture.

Culture Systems	RGR Index Days		
	30	**60**	**90**
TIS (Plantform^{TM})	1.12	1.15	1.27
Semisolid medium	0.60	0.66	0.87

In saffron, the application of the TIS was reported in only one study by Blázquez et al. [58], which utilized embryogenic calli as explants to explore and highlight the effects of polyamine metabolism during the somatic embryogenesis process.

The Rita® bioreactor was used and in this system, the production of embryogenic calli increased four-fold compared to the semisolid medium. This finding is in accordance with our study. Indeed, the explants in Plantform^{TM} yielded a higher number of microcorms (~eleven-fold) with respect to the semisolid medium system (Figure 3b–e). Moreover, it was possible to maintain the corms in vitro culture for 90 days, equivalent to three subcultures, without handling the plant material, requiring only the replacement of the culture medium. The TIS has been applied to several bulbous plants to achieve rapid propagation, particularly ornamental bulbous plants. For instance, in *Hippeastrum* × *chmielii* Chm., the use of twin-flasks as a temporary culture bioreactor was effective in obtaining a greater number of new bulbs (6.3–6.5) than the agar-solidified medium (3.9) within 4 weeks [59].

Increased bulb induction in *Leucojum aestivum* L. was achieved using the Rita® bioreactor (VITROPIC, St-Mathieu de Tréviers, France)with an immersion period of 5 min every 2 h. Regardless of the constituents of the culture medium, a bulb rate production of 74.6% was obtained in the TIS system, while in the solid culture media, it was 71.7% [60]. Moreover, the study reported that the bulbs derived from TIS had a higher fresh weight (about 1.6 times) than the bulbs in the solid medium, and the acclimatization step was better in plants obtained from Rita® systems.

Similarly, Ruffoni et al. [61] showed the efficiency of Rita® vessels in producing an increased number of *Gladiolus* corms, which could be easily transplanted into the field, in contrast to cultures in a semisolid medium. In TIS *Gladiolus* propagation, an immersion time of 3 min every 3 h was applied for 30, 45, and 90 days of culture.

Several studies have investigated the *Lilium* spp. propagation in bioreactors [62]. For example, in three oriental lilies [63], bioreactor culture with an immersion time of 5 min every 2 h led to an increase in bulb mass production by 10–13 times. Barberini et al. [64] reported the application of Rita® for multiplying *L. bulbiferum*, with good results in terms of material quality to be easily transferred in the greenhouse. Lian et al. [65] propagated the *Lilium* oriental hybrid 'Casablanca' in an ebb-and-flow system and obtained a higher number of bulblets (1025) compared with semisolid culture (82), applying 15 min of immersion 4 times per day.

While the TIS culture has shown promising results for many plant species, it does not always yield superior outcomes. For instance, Mirmasoumi and Bakhshaie [66] found that shoot organogenesis from the callus of *L. ledebourii* was less efficient when callus cultures were grown in the Rita® system compared to a semisolid medium. Similarly, Nesi et al. [67]

observed that *Lilium* hybrids achieved better propagation through a continuous immersion system than the TIS using an ebb and flood system.

Figure 3. TIS culture of saffron at the beginning (**a**) and after 90 days (**b**) of culture in Plantform™ bioreactor. Explant after 30 days (**c**), 60 days (**d**), and 90 days (**e**); Bars = 1.5 cm. (S: Shoot; MC: Microcorm; R: Root).

For the in vitro production of *Solanum tuberosum* L., Wongket and Pumisutapon [68] achieved 82–90% of microtubers using TIS (twin-flasks system) with an immersion every 12 h for 2 min than in the solid medium (20%). TIS conditions gave 4.1–4.5 times more microtubers respecting solid medium. The use of Rita® container in potato led to obtaining vigorous seedlings with a higher number of shoots [69], and Plantform™ was allowed to achieve a maximum number of tubers [70] rather than the conventional culture system. A study conducted by Andriani et al. [71] highlighted the significance of using another type of TIS for the micropropagation of *S. tuberosum*. It was found that explants grown in the Setis™ bioreactor produced a larger number, size, and weight of tubers than the semisolid medium. This improvement was attributed to the direct contact between the explants and the liquid medium during immersion periods, leading to enhanced nutrient uptake and consequently more efficient tuber induction.

The aforementioned findings suggest that the effectiveness of TIS may vary depending on the specific requirements of the plant species. The control and monitoring of different parameters, such as the optimization of the medium culture, immersion, and aeration time are fundamental for customized approaches in plant tissue culture. Particularly, an accurate fitting of the immersion time and dry periods can considerably reduce the hyperhydricity of the tissue, establishing optimal conditions for humidity. In our study, the application of immersion time 5 min every 4 h, which was in line with the reports for the bulbous species.

3.2. Machine Learning Analysis

A comparison of the performance metrics of four different machine learning models used to predict features in TIS is shown in Table 3. The number of microcorms, shoots, and roots was examined with different models: RF, SVM, GP, and MLP. The validity of the model was assessed using MAE, R^2, and RMSE metrics. The R^2 scores are between 0 and 1, where 1 represents the best possible prediction, and 0 indicates no explanation potential. The RMSE numbers, which typically vary from zero to positive infinity, show the precision of the model. Lower values indicate better performance. Similarly, the MAE scales from zero to positive infinity, with lower values indicating more accuracy; it represents projected accuracy.

Table 3. Performance matrices of machine learning models for the Temporary Immersion System (TIS).

Parameter	Performance Matrices	RF	SVM	GP	MLP
Microcorm number	R^2	0.81	0.56	0.67	0.70
	MAE	0.12	0.14	0.17	0.16
	RMSE	0.16	0.23	0.21	0.20
Shoot number	R^2	0.75	0.63	0.71	0.69
	MAE	0.13	0.14	0.17	0.15
	RMSE	0.17	0.20	0.19	0.20
Root number	R^2	0.64	0.51	0.65	0.78
	MAE	0.18	0.17	0.18	0.14
	RMSE	0.21	0.21	0.21	0.18

RF: Random Forest; SVM: Support Vector Machines; GP: Gaussian Process; MLP: Multilayer Perceptron; R^2: Coefficient of determination; MAE: Mean absolute error; RMSE: Root mean square.

RF showed the best predictive performance with an R^2 value of 0.81 for the microcorm number, and its value can explain about 81% of the variance in microcorms. Moreover, RF produced the lowest results at 0.12 and 0.16 for MAE and RMSE, respectively, indicating the highest predictive accuracy level and the least divergence between the expected and actual values. Compared to the other models, SVM demonstrated comparatively poorer performance metrics, as evidenced by its R-squared value of 0.56, indicating a lower level of predictive potential.

Concerning shoot number prediction, RF once more demonstrated the strongest performance, with an R^2 value of 0.75. Among the models, RF had the lowest MAE (0.13) and RMSE (0.17), indicating better overall accuracy than other models, while SVM and MLP models showed the highest RMSE of 0.20.

MLP proved to be the most effective method for estimating root number; its R^2 value of 0.78 meant that it could account for almost 78% of the variation in this parameter. Additionally, MLP had the lowest RMSE and MAE of all the models, indicating excellent prediction accuracy and low error in root number prediction. On the other hand, with an R^2 value of 0.51, SVM showed the least favorable performance metrics due to a rather poor predictive efficacy in this parameter prediction task. The findings revealed that the RF model outperformed other models in every parameter category, offering the TIS high prediction accuracy and low error rates in forecasting the number of microcorms, shoots, and roots.

However, each model's performance could vary based on the studied parameters; thus, choosing machine learning methods that are unique to the intended prediction task is important.

Table 4 also provides a detailed examination of RF, SVM, GP, and MLP in semisolid medium culture systems. Regarding microcorm predictions, both RF and SVM showed similar R^2 values of 0.75, which can explain almost 75% of the variance in this growth parameter. However, GP and MLP with 0.77 and 0.78, respectively, had higher R^2 values, indicating a better capacity to explain variance in microcorms. The MAE ranged from

0.11 to 0.12 for all models and was consistently low, showing a minor average divergence between the predicted and actual values. Similarly, the models were constant regarding the RMSE, which had values between 0.16 and 0.17, showing reliable accuracy in microcorm number prediction.

Table 4. Performance matrices of machine learning models for a semisolid medium.

Parameter	Performance Matrices	RF	SVM	GP	MLP
Microcorm number	R^2	0.75	0.75	0.77	0.78
	MAE	0.11	0.12	0.11	0.12
	RMSE	0.16	0.17	0.16	0.16
Shoot number	R^2	0.65	0.68	0.69	0.70
	MAE	0.14	0.13	0.14	0.12
	RMSE	0.19	0.18	0.19	0.17
Root number	R^2	0.50	0.60	0.59	0.54
	MAE	0.19	0.17	0.17	0.18
	RMSE	0.23	0.21	0.21	0.21

RF: Random Forest; SVM: Support Vector Machines; GP: Gaussian Process; MLP: Multilayer Perceptron; R^2: Coefficient of determination; MAE: Mean absolute error; RMSE: Root mean square.

GP and MLP led the field in shoot number prediction with the greatest R^2 values of 0.69 and 0.70, respectively, indicating their greater ability to explain variance. RF and SVM showed adequate R^2 values of 0.65 and 0.68, respectively, remaining slightly behind. Among the models, MLP had the lowest MAE of 0.12, which indicates that it predicts shoot numbers more accurately than the other models. Moreover, MLP had the least overall error with the lowest RMSE of 0.17.

Regarding root number, SVM has the greatest R^2 value (0.60), followed by GP (0.59). Both can explain the variance more effectively than other models. The R^2 values for RF and MLP were decreased (0.50 and 0.54); this suggests that their prediction power for root numbers is slightly lower. Notably, SVM and GP showed the lowest RMSE and MAE, indicating less error and higher accuracy in root number prediction.

In the case of the semisolid medium, MLP exhibited competitive performance in predicting microcorm and shoot numbers, even though each model displayed a unique performance across many attributes, while SVM seemed to be more successful in predicting the root number. These results highlight the need to choose suitable machine learning algorithms for predictive modeling in semisolid medium by considering particular attributes and corresponding model performances.

When evaluating the performance of the machine learning models, the RF algorithm consistently demonstrated superior accuracy in predicting parameters, particularly within the TIS culture system. Both the GP and the MLP models also performed effectively, showing good results with only slight adjustments in context.

SVM performed competitively in the semisolid medium, especially in root number prediction, while RF performed better in TIS; MLP continuously demonstrated higher accuracy in both system cultures, particularly regarding microcorm and shoot number prediction. Although RF was reliable, model selection should consider the features and culture conditions to maximize prediction efficacy.

Figures 4–6 show the actual and predicted values, making the comparison of observed data points with model forecasts easy. The model performance may be evaluated in the graphic representations, which highlight how well it predicts actual results.

In the general view of machine learning applications within the agricultural domain, particularly focusing on plant tissue culture, our study delves into the comparative effectiveness of the four models used in two distinct culture systems, namely the TIS and semisolid medium. This analysis is in line with other studies [35,37,72–74], which have explored various dimensions of machine learning applicability in predicting plant growth parameters under different experimental conditions. Our research underscores the superior predictive performance of RF in the TIS culture, particularly in forecasting the number of microcorms, shoots, and roots. This finding is consistent with the literature, such as Şimşek [35], which highlighted the robustness of RF in detecting water stress effects in strawberries, and Aasim et al. [37], which identified the efficacy of RF in predicting germination and morphological traits in hemp seedlings. The ability of RF to manage complex datasets with high accuracy highlights its value in agricultural predictive modeling, providing a basis for optimizing plant tissue culture conditions. The machine learning analysis section of our study reveals nuanced performances across different models and culture systems. For instance, in the semisolid medium, GP and MLP showed heightened effectiveness in predicting microcorms, aligning with observations of Aasim et al. [72] on the promising applications of quantum computing-enhanced machine learning models in plant tissue culture. This suggests that while RF generally offers a robust performance, the intricate nature of biological data might sometimes favor other models or necessitate a hybrid approach for enhanced prediction accuracy. Moreover, as explored by Aasim et al. [72], the integration of quantum computing techniques in machine learning introduces a groundbreaking perspective on future agricultural research. In our study, the quantum-enhanced algorithms were not applied, but the reported success in other research indicates a potential way for elevating the predictive modeling capabilities in plant science, thereby enhancing the efficiency of plant tissue culture protocols. The lesser predictive potential of SVM which is noted in TIS system is in agreement with Pepe et al. [74], who reported the limitations of SVM in complex plant growth modeling. This reinforces the necessity for a customized selection of models based on the specific dataset characteristics and predictive needs, highlighting the importance of exploring advanced machine learning techniques and potentially quantum computing methods to overcome the current modeling limitations. In essence, our study, alongside the referenced literature, emphasizes the dynamic nature of machine learning applications in agriculture and illustrates the need for constant methodological progress and adaptability to specific research contexts. The evolving complexity of biological data and the emergence of quantum computing herald a new era of possibilities that necessitate ongoing exploration and adaptation of machine learning methodologies in plant tissue culture and wider agricultural sciences.

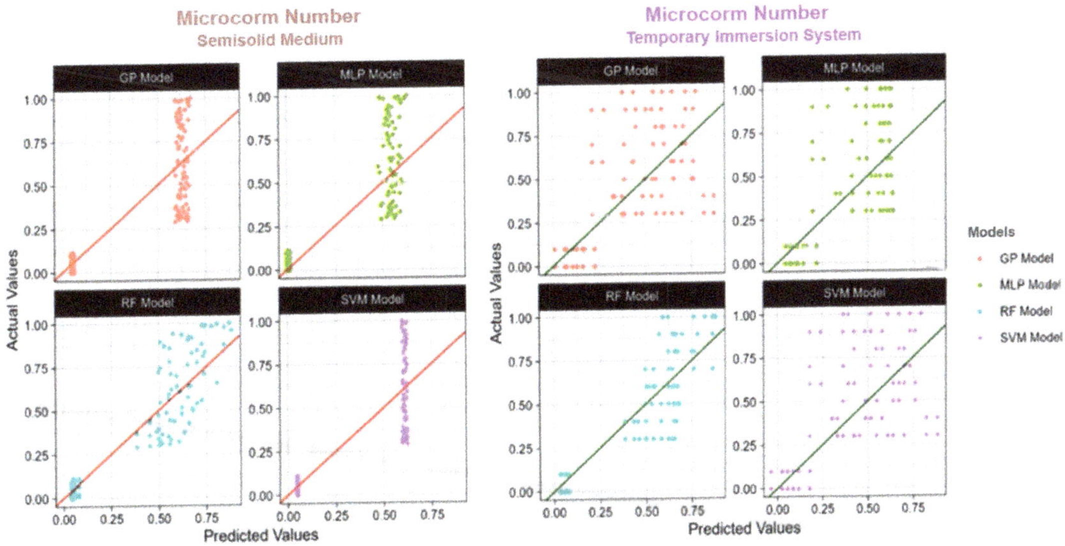

Figure 4. The actual and predicted value of TIS and semisolid medium microcorm number using four models: RF: Random Forest; SVM: Support Vector Machines; GP: Gaussian Process; MLP: Multilayer Perceptron; R^2: Coefficient of determination.

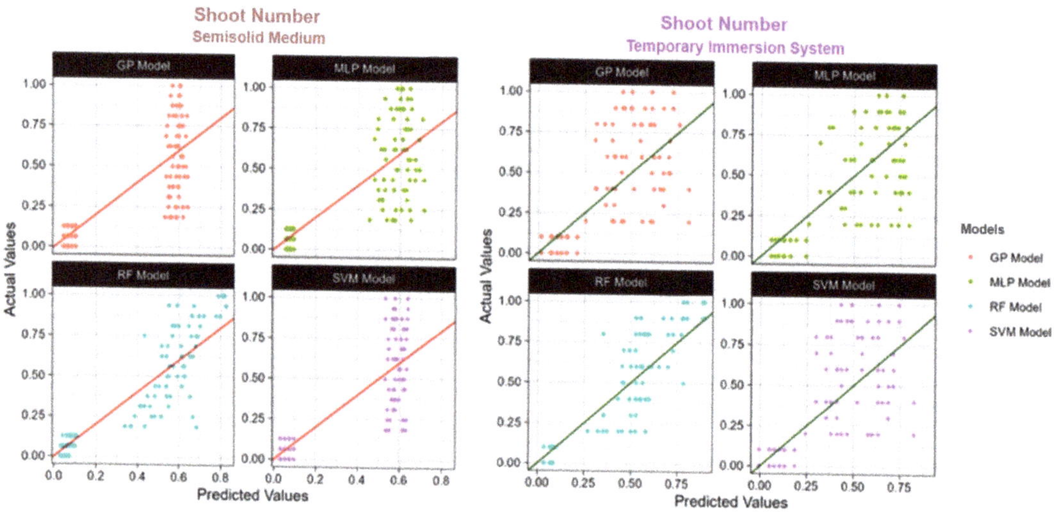

Figure 5. The actual and predicted value of TIS and semisolid medium shoot number using four models: RF: Random Forest; SVM: Support Vector Machines; GP: Gaussian Process; MLP: Multilayer Perceptron; R^2: Coefficient of determination.

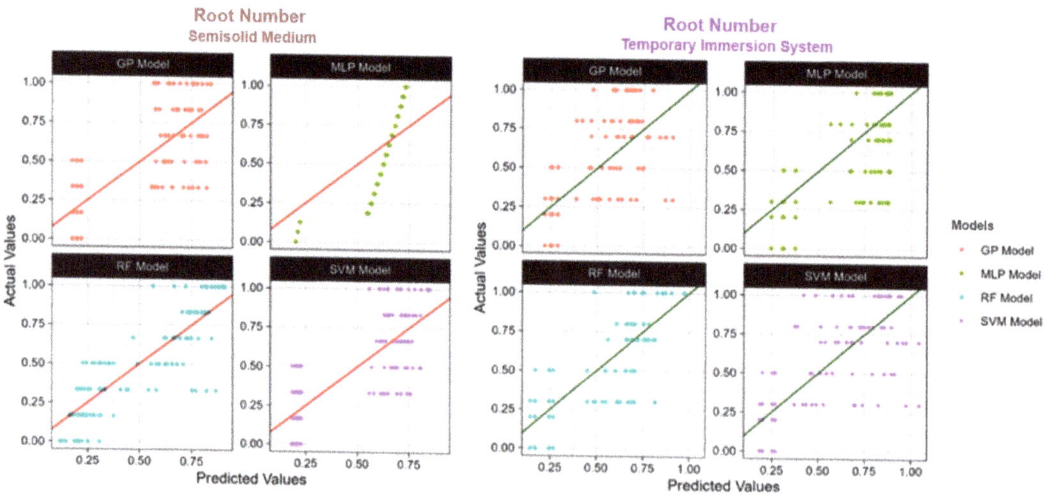

Figure 6. The actual and predicted value of TIS and semisolid medium root number using four models: RF: Random Forest; SVM: Support Vector Machines; GP: Gaussian Process; MLP: Multilayer Perceptron; R^2: Coefficient of determination.

4. Conclusions

Although in vitro propagation presents a significant opportunity to multiply disease-free saffron microcorms, achieving large-scale production remains a challenge. The Temporary Immersion System (TIS) offers a promising approach to enhance production and support the acclimatization phase for this species. This study represents the first effort to propagate saffron corms using the TIS technique with the Plantform^{TM} bioreactor. Indeed, the microcorm propagation was optimized in the TIS over the conventional in vitro culture on semisolids, with no abnormalities or hyperhydricity in corms production during the culture period. Therefore, the adoption of this innovative system could be a significant step toward producing large quantities of disease-free corms at reduced costs, enhancing the ease of operation and handling of plant material through a semiautomated system. Additionally, TIS might play a crucial role in acclimatization phase since the explants, during periods without immersion in the culture medium, begin to adapt to ex vitro conditions. In this sense, further attention and investigation are needed in future research to improve the acclimatization of saffron. Furthermore, integrating machine learning to predict outcomes in this field opens up novel approaches to solving complex biological questions. These findings provide important insights into the biological mechanisms and contextual factors that influence growth parameters across different experimental conditions.

Author Contributions: Conceptualization, C.B. and W.T.; methodology W.T., T.İ. and C.B.; software, Ö.Ş.; data curation, W.T., Ö.Ş. and T.İ.; writing—original draft preparation, W.T., T.İ. and C.B.; writing—review and editing, W.T., T.İ., N.C., Ö.Ş. and C.B.; visualization, T.İ.; supervision, C.B.; funding acquisition, C.B. All authors have read and agreed to the published version of the manuscript.

Funding: This research received no external funding.

Data Availability Statement: Data are contained within the article.

Acknowledgments: The authors acknowledge the National Council Research of Italy (Bilateral Project 2022–2023: 'Improving cultivation, quality, and conservation of saffron accessions from Morocco and Italy').

Conflicts of Interest: The authors declare no conflicts of interest.

References

1. Cardone, L.; Castronuovo, D.; Perniola, M.; Cicco, N.; Candido, V. Saffron (*Crocus sativus* L.), the king of spices: An overview. *Sci. Hortic.* **2020**, *272*, 109560. [CrossRef]
2. Khawar, M.K.; Yildirim, M.U.; Sarihan, E. Ex vitro macropropagation of saffron (*Crocus sativus* L.) corms. In *Saffron: The Age-Old Panacea in a New Light*; Sarwat, M., Sumaiya, S., Eds.; Academic Press: Cambridge, MA, USA, 2020; pp. 45–56.
3. Yildirim, M.U.; Sarihan, E.O.; Khawar, K.M. Ethnomedicinal and traditional usage of saffron (*Crocus sativus* L.) in Turkey. In *Saffron: The Age-Old Panacea in a New Light*; Sarwat, M., Sumaiya, S., Eds.; Academic Press: Cambridge, MA, USA, 2020; pp. 21–31.
4. Menia, M.; Sadaf Iqbal, I.; Zahida, I.R.; Tahir, I.S.; Kanth, I.R.; Aashq Hussian, I.; Iqbal, S.; Hussian, A. Production technology of saffron for enhancing productivity. *J. Pharmacogn. Phytochem.* **2018**, *7*, 1033–1039.
5. Taheri-Dehkordi, A.; Naderi, R.; Martinelli, F.; Salami, S.A. A robust workflow for indirect somatic embryogenesis and cormlet production in saffron (*Crocus sativus* L.) and its wild allies; *C. caspius* and *C. speciosus*. *Heliyon* **2020**, *6*, e05841. [CrossRef]
6. Chib, S.; Thangaraj, A.; Kaul, S.; Dhar, M.K.; Kaul, T. Development of a system for efficient callus production, somatic embryogenesis and gene editing using CRISPR/Cas9 in Saffron (*Crocus sativus* L.). *Plant Methods* **2020**, *16*, 47. [CrossRef]
7. Lagram, K.; Ben, M.; Caid, E.; El Aaouam, S.; Lachheb, M.; El, A.; Serghini, M.A. In vitro shoot regeneration and development of microcorms of moroccan saffron (*Crocus sativus* L.). *Atlas J. Plant Biol.* **2016**, 50–55. [CrossRef]
8. Sevindik, B.; Mendi, Y.Y. Somatic embryogenesis in *Crocus sativus* L. In *Methods in Molecular Biology*; Humana Press Inc.: Totowa, NJ, USA, 2016; Volume 1359, pp. 351–357.
9. Yang, B.M.; Huang, Y.L.; Xu, W.J.; Bao, L.X. Explant selection and cluster buds induction in vitro of saffron (*Crocus sativus* L.). *Agric. Sci. Technol. Commun* **2015**, *2*, 106–108.
10. Renau-Morata, B.; Moyá, L.; Nebauer, S.G.; Seguí-Simarro, J.M.; Parra-Vega, V.; Gómez, M.D.; Molina, R.V. The use of corms produced under storage at low temperatures as a source of explants for the in vitro propagation of saffron reduces contamination levels and increases multiplication rates. *Ind. Crops Prod.* **2013**, *46*, 97–104. [CrossRef]
11. Berthouly, M.; Etienne, H. Temporary immersion system: A new concept for use liquid medium in mass propagation. In *Liquid Culture Systems for In Vitro Plant Propagation*; Hvoslef-Eide, A.K., Preil, W., Eds.; Springer: Dordrecht, The Netherlands, 2005; pp. 165–195.
12. Yan, H.; Yang, L.; Li, Y. Improved growth and quality of *Dioscorea fordii* Prain et Burk and *Dioscorea alata* plantlets using a temporary immersion system. *Afr. J. Biotechnol.* **2011**, *10*, 19444–19448.
13. Yang, S.-H.; Yeh, D.-M. In vitro leaf anatomy, ex vitro photosynthetic behaviors and growth of *Calathea orbifolia* (Linden) Kennedy plants obtained from semi-solid medium and temporary immersion systems. *Plant Cell Tissue Organ Cult.* **2008**, *93*, 201–207. [CrossRef]
14. De Carlo, A.; Tarraf, W.; Lambardi, M.; Benelli, C. Temporary immersion system for production of biomass and bioactive compounds from medicinal plants. *Agronomy* **2021**, *11*, 2414. [CrossRef]
15. Pérez-Alonso, N.; Jiménez, E.; de Feria, M.; Capote, A.; Barbón, R.; Quiala, E.; Chávez, M. Effect of inoculum density and immersion time on the production of potato microtubers in temporary immersion systems and field studies. *Biotecnol. Veg.* **2007**, *7*, 149–154.

16. Businge, E.; Trifonova, A.; Schneider, C.; Rödel, P.; Egertsdotter, U. Evaluation of a new temporary immersion bioreactor system for micropropagation of cultivars of *Eucalyptus*, birch and fir. *Forests* **2017**, *8*, 196. [CrossRef]
17. Ruffoni, B.; Savona, M. The temporary immersion system (T.I.S.) for the improvement of micropropagation of ornamental plants. *Acta Hortic.* **2005**, *683*, 445–454. [CrossRef]
18. Elazab, D.; Capuana, M.; Ozudogru, E.A.; Anichini, M.; Lambardi, M. Use of liquid culture with the electis bioreactor for faster recovery of blackberry (*Rubus fruticosus* L.) shoots from conservation at 4 °C. *Horticulturae* **2023**, *9*, 680. [CrossRef]
19. Vives, K.; Andújar, I.; Lorenzo, J.C.; Concepción, O.; Hernández, M.; Escalona, M. Comparison of different in vitro micropropagation methods of *Stevia rebaudiana* B. including temporary immersion bioreactor (BIT®). *Plant Cell Tissue Organ Cult.* **2017**, *131*, 195–199. [CrossRef]
20. Ramírez-Mosqueda, M.A.; Iglesias-Andreu, L.G. Evaluation of different temporary immersion systems (BIT®, BIG, and RITA®) in the micropropagation of *Vanilla planifolia* Jacks. *Vitr. Cell. Dev. Biol. Plant* **2016**, *52*, 154–160. [CrossRef]
21. Nongdam, P.; Beleski, D.G.; Tikendra, L.; Dey, A.; Varte, V.; EL Merzougui, S.; Pereira, V.M.; Barros, P.R.; Vendrame, W.A. Orchid micropropagation using conventional semi-solid and temporary immersion systems: A review. *Plants* **2023**, *12*, 1136. [CrossRef] [PubMed]
22. Gatti, E.; Sgarbi, E.; Ozudogru, E.A.; Lambardi, M. The effect of Plantform™ bioreactor on micropropagation of *Quercus robur* in comparison to a conventional in vitro culture system on gelled medium, and assessment of the microenvironment influence on leaf structure. *Plant Biosyst.* **2017**, *151*, 1129–1136. [CrossRef]
23. Vidal, N.; Blanco, B.; Cuenca, B. A temporary immersion system for micropropagation of axillary shoots of hybrid chestnut. *Plant Cell Tissue Organ Cult.* **2015**, *123*, 229–243. [CrossRef]
24. Benelli, C.; De Carlo, A. In vitro multiplication and growth improvement of *Olea europaea* L. cv Canino with temporary immersion system (Plantform™). *3 Biotech* **2018**, *8*, 317. [CrossRef]
25. Georgiev, V.; Schumann, A.; Pavlov, A.; Bley, T. Temporary immersion systems in plant biotechnology. *Eng. Life Sci.* **2014**, *14*, 607–621. [CrossRef]
26. Şimşek, Ö.; Sekerci, A.D.; Isak, M.A.; Bulut, F.; Izgü, T.; Tütüncü, M.; Dönmez, D. Optimizing micropropagation and rooting protocols for diverse lavender genotypes: A synergistic approach integrating machine learning techniques. *Horticulturae* **2024**, *10*, 52. [CrossRef]
27. Jan, Z.; Ahamed, F.; Mayer, W.; Patel, N.; Grossmann, G.; Stumptner, M.; Kuusk, A. Artificial intelligence for industry 4.0: Systematic review of applications, challenges, and opportunities. *Expert Syst. Appl.* **2023**, *216*, 119456. [CrossRef]
28. Demirel, F.; Uğur, R.; Popescu, G.C.; Demirel, S.; Popescu, M. Usage of machine learning algorithms for establishing an effective protocol for the in vitro micropropagation ability of black chokeberry (*Aronia melanocarpa* (Michx.) Elliott). *Horticulturae* **2023**, *9*, 1112. [CrossRef]
29. Sadat-Hosseini, M.; Arab, M.M.; Soltani, M.; Eftekhari, M.; Soleimani, A.; Vahdati, K. Predictive modeling of Persian walnut (*Juglans regia* L.) in vitro proliferation media using machine learning approaches: A comparative study of ANN, KNN and GEP models. *Plant Methods* **2022**, *18*, 48. [CrossRef]
30. Lozano-Milo, E.; Landin, M.; Gallego, P.P.; García-Pérez, P. Machine learning deciphers genotype and ammonium as key factors for the micropropagation of *Bryophyllum* sp. medicinal plants. *Horticulturae* **2022**, *8*, 987. [CrossRef]
31. Özcan, E.; Atar, H.H.; Ali, S.A.; Aasim, M. Artificial neural network and decision tree–based models for prediction and validation of in vitro organogenesis of two hydrophytes—*Hemianthus callitrichoides* and *Riccia fluitans*. *Vitr. Cell. Dev. Biol. Plant* **2023**, *59*, 547–562. [CrossRef]
32. Hesami, M.; Jones, A.M.P. Application of artificial intelligence models and optimization algorithms in plant cell and tissue culture. *Appl. Microbiol. Biotechnol.* **2020**, *104*, 9449–9485. [CrossRef] [PubMed]
33. Rezaei, H.; Mirzaie-Asl, A.; Abdollahi, M.R.; Tohidfar, M. Enhancing petunia tissue culture efficiency with machine learning: A pathway to improved callogenesis. *PLoS ONE* **2023**, *18*, e0293754. [CrossRef]
34. Murashige, T.; Skoog, F. A revised medium for rapid growth and bioassays with tobacco tissue cultures. *Physiol. Plant* **1962**, *15*, 473–497. [CrossRef]
35. Şimşek, Ö. Machine learning offers insights into the impact of in vitro drought stress on strawberry cultivars. *Agriculture* **2024**, *14*, 294. [CrossRef]
36. Hu, J.; Sun, Y.; Li, G.; Jiang, G.; Tao, B. Probability analysis for grasp planning facing the field of medical robotics. *Meas. J. Int. Meas. Confed.* **2019**, *141*, 227–234. [CrossRef]
37. Aasim, M.; Katırcı, R.; Akgur, O.; Yildirim, B.; Mustafa, Z.; Nadeem, M.A.; Baloch, F.S.; Karakoy, T.; Yılmaz, G. Machine learning (ML) algorithms and artificial neural network for optimizing in vitro germination and growth indices of industrial hemp (*Cannabis sativa* L.). *Ind. Crops Prod.* **2022**, *181*, 114801. [CrossRef]
38. Sharma, K.D.; Rathour, R.; Sharma, R.; Goel, S.; Sharma, T.R.; Singh, B.M. In vitro cormlet development in *Crocus sativus*. *Biol. Plant.* **2008**, *52*, 709–712. [CrossRef]
39. Zeybek, E.; Önde, S.; Kaya, Z. Improved in vitro micropropagation method with adventitious corms and roots for endangered saffron. *Cent. Eur. J. Biol.* **2012**, *7*, 138–145. [CrossRef]
40. Mahmoud, K.B.; Rebaii, J.; Jemai, N.; Jedidi, E. In vitro micropropagation attempts of the high value spice saffron (*Crocus sativus* L.). *Rev. Des. Regions Arid.* **2020**, 727–734. Available online: https://www.researchgate.net/publication/352056079_In_vitro_micropropagation_attempts_of_the_high_value_spice_saffron_Crocus_sativus_L (accessed on 29 April 2024).

41. Mereu, A.; Dorsaf, K.; Scarpa, G. In vitro culture of saffron: Hormones influence on the development of new shoots and callus. *Plant Cell Biotechnol. Mol. Biol.* **2019**, *20*, 511–520.
42. Ahmad, M.; Zaffar, G.; Habib, M.; Arshid, A.; Dar, N.A.; Dar, Z.A. Saffron (*Crocus sativus* L.) in the light of biotechnological approaches: A review. *Sci. Res. Essays* **2014**, *9*, 13–18.
43. Gantait, S.; Vahedi, M. In vitro regeneration of high value spice *Crocus sativus* L.: A concise appraisal. *J. Appl. Res. Med. Aromat. Plants* **2015**, *2*, 124–133. [CrossRef]
44. Tahiri, A.; Mazri, M.A.; Karra, Y.; Ait Aabd, N.; Bouharroud, R.; Mimouni, A. Propagation of saffron (*Crocus sativus* L.) through tissue culture: A review. *J. Hortic. Sci. Biotechnol.* **2023**, *98*, 10–30. [CrossRef]
45. Salami, S.A. Tissue culture techniques for saffron improvement. In *The Saffron Genome. Compendium of Plant Genomes*; Vakhlu, J., Ambardar, S., Salami, S.A., Kole, C., Eds.; Springer: Cham, Switzerland, 2022; pp. 195–220.
46. Devi, K.; Sharma, M.; Singh, M.; Singh Ahuja, P. In vitro cormlet production and growth evaluation under greenhouse conditions in saffron (*Crocus sativus* L.)—commercially important crop. *Eng. Life Sci.* **2011**, *11*, 189–194. [CrossRef]
47. Yasmin, S.; Nehvi, F.A.; Wani, S.A. Tissue culture as an alternative for commercial corm production in saffron: A heritage crop of Kashmir. *Afr. J. Biotechnol.* **2013**, *12*, 3940–3946.
48. Teixeira da Silva, J.A. Disinfection of explants for saffron (*Crocus sativus*) tissue culture. *Environ. Exp. Biol.* **2016**, *14*, 183–198. [CrossRef]
49. Raja, W.; Zaffer, G.; Wani, S.A. In vitro microcorm formation in saffron (*Crocus sativus* L.). *Acta Hortic.* **2007**, *739*, 291–296. [CrossRef]
50. Soukrat, S.; Benlhabib, O.; Alfaiz, C.; Lage, M. Boosting saffron (*Crocus sativus* L.) micro-propagation through in vitro corm production. *Acta Hortic.* **2017**, *1184*, 87–95. [CrossRef]
51. Lagram, K.; El Merzougui, S.; Boudadi, I.; Ben El Caid, M.; El Boullani, R.; El Mousadik, A.; Serghini, M.A. In vitro shoot formation and enrooted mini-corm production by direct organogenesis in saffron (*Crocus sativus* L.). *Vegetos* **2023**. Available online: https://link.springer.com/article/10.1007/s42535-023-00639-9 (accessed on 29 April 2024). [CrossRef]
52. Datta, S.K.; Chakrabarty, D.; Janakiram, T. Low cost tissue culture: An overview. *J. Plant Sci. Res.* **2017**, *33*, 181–199.
53. Naik, R.; Bhushan, A.; Gupta, R.K.; Walia, A.; Gaur, A. Low cost tissue culture technologies in vegetables: A review. *Int. J. Biochem. Res. Rev.* **2020**, *29*, 66–78. [CrossRef]
54. Sota, V.; Benelli, C.; Çuko, B.; Papakosta, E.; Depaoli, C.; Lambardi, M.; Kongjika, E. Evaluation of ElecTIS bioreactor for the micropropagation of *Malus sylvestris* (L.) Mill., an important autochthonous species of Albania. *Hortic. Sci.* **2021**, *48*, 12–21. [CrossRef]
55. Lambardi, M.; Roncasaglia, R.; Previati, A.; De Carlo, A.; Dradi, G.; Da Re, F.; Calamai, L. In vitro slow growth storage of fruit rootstocks inside gas-tight or gas-permeable containers. *Acta Hortic.* **2006**, *725*, 483–488. [CrossRef]
56. Ozudogru, E.A.; Benelli, C.; Dradi, G.; Lambardi, M. Effect of culture container and carbohydrate content on in vitro slow growth storage of the cherry rootstock 'Gisela®5'. *Acta Physiol. Plant.* **2017**, *39*, 2–9. [CrossRef]
57. Gianguzzi, V.; Sottile, F. Temporary immersion system as an innovative approach for in vitro propagation of *Sorbus domestica* L. *Horticulturae* **2024**, *10*, 164. [CrossRef]
58. Blázquez, S.; Piqueras, A.; Sema, M.D.; Casas, J.L.; Fernández, J.A. Somatic embryogenesis in saffron: Optimisation through temporary immersion and polyamine metabolism. *Acta Hortic.* **2004**, *650*, 269–276. [CrossRef]
59. Ilczuk, A.; Winkelmann, T.; Richartz, S.; Witomska, M.; Serek, M. In vitro propagation of *Hippeastrum × chmielii* Chm. Influence of flurprimidol and the culture in solid or liquid medium and in temporary immersion systems. *Plant Cell Tissue Organ Cult.* **2005**, *83*, 339–346. [CrossRef]
60. Ptak, A. *Leucojum aestivum* L. in vitro bulbs induction and acclimatization. *Cent. Eur. J. Biol.* **2014**, *9*, 1011–1021. [CrossRef]
61. Ruffoni, B.; Savona, M.; Barberini, S. Biotechnological support for the development of new *Gladiolus* hybrids. *Floric. Ornam. Biotechnol.* **2012**, *6*, 45–52.
62. Lian, M.L.; Chakrabarty, D.; Paek, K.Y. Bulblet formation from bulbscale segments of *Lilium* using bioreactor system. *Biol. Plant.* **2003**, *46*, 199–203. [CrossRef]
63. Seon, J.H.; Kim, Y.S.; Son, S.H.; Paek, K.Y. The fed-batch culture system using bioreactor for the bulblets production of oriental lilies. *Acta Hortic.* **2000**, *520*, 53–59. [CrossRef]
64. Barberini, S.; Savona, M.; Ruffoni, B. Temporary immersion culture of *Lilium bulbiferum*. *Acta Hortic.* **2011**, *900*, 377–384. [CrossRef]
65. Lian, M.; Chakrabarty, D.; Paek, K.Y. Growth and uptake of sucrose and mineral ions by bulblets of *Lilium* oriental hybrid "Casablanca" during bioreactor culture. *J. Hortic. Sci. Biotechnol.* **2002**, *77*, 253–257. [CrossRef]
66. Mirmasoumi, M.; Bakhshaie, M. Effects of liquid, temporary immersion bioreactor and solid culture systems on micropropagation of *Lilium ledebourii* via bulblet microscales—An endangered valuable plant with ornamental potential. *Prog. Biol. Sci.* **2015**, *5*, 169–180.
67. Nesi, B.; Lazzereschi, S.; Pecchioli, S.; Grassotti, A.; Burchi, G.; Cardarelli, M.; Cardona Suarez, C.M.; Colla, G. Development, selection and propagation of interspecific hybrids of *Lilium*. *Acta Hortic.* **2014**, *1027*, 155–164. [CrossRef]
68. Wongket, A.; Pumisutapon, P. Effect of feeding frequency and period in temporary immersion system on microtuberization of potato. In Proceedings of the 24th Tri-University International Joint Seminar and Symposium 2017 Mie University, Tsu, Japan, 23–27 October 2017; pp. 123–125.

69. Peña-Rojas, G.; Sanchez, H.; Barahona, I.R.; Ayme, V.A.; Segura-Turkowsky, M.; Jimenez, R.E. Alternative inputs for micropropagation of *Solanum tuberosum, Ullucus tuberosus* and *Oxalis tuberosa* in semisolid and liquid medium and temporary immersion system. *Trop. Subtrop. Agroecosyst.* **2020**, *23*, 1–15. [CrossRef]
70. Al-Shareefi, M.J.H.; Abbass, J.A.; Abdulhussein, M.A.A. Effect of light sources and culture systems on microtubers production of potato (*Solanum Tuberosum* L.) in vitro. *Int. J. Agric. Stat. Sci.* **2020**, *16*, 679–686.
71. Andriani, S.; Siregar, L.A.M.; Safni, I. Microtubers production by using Temporary Immersion System (TIS) bioreactor to potato varieties. *IOP Conf. Ser. Earth Environ. Sci.* **2021**, *886*, 012005. [CrossRef]
72. Aasim, M.; Katırcı, R.; Acar, A.Ş.; Ali, S.A. A comparative and practical approach using quantum machine learning (QML) and support vector classifier (SVC) for light emitting diodes mediated in vitro micropropagation of black mulberry (*Morus nigra* L.). *Ind. Crops Prod.* **2024**, *213*, 118397. [CrossRef]
73. Kirtis, A.; Aasim, M.; Katırcı, R. Application of artificial neural network and machine learning algorithms for modeling the in vitro regeneration of chickpea (*Cicer arietinum* L.). *Plant Cell Tissue Organ Cult.* **2022**, *150*, 141–152. [CrossRef]
74. Pepe, M.; Hesami, M.; Small, F.; Jones, A.M.P. Comparative analysis of machine learning and evolutionary optimization algorithms for precision micropropagation of *Cannabis sativa*: Prediction and validation of in vitro shoot growth and development based on the optimization of light and carbohydrate sou. *Front. Plant Sci.* **2021**, *12*, 757869. [CrossRef]

horticulturae

MDPI

Article

Establishment of an In Vitro Micropropagation System for *Cannabis sativa* 'Cheungsam'

Sang-Cheol Baek [1,†], Sang-Yoon Jeon [1,2,†], Yoon-Jung Choi [1,†], Bo-Hyun Byun [1], Da-Hoon Kim [1], Ga-Ram Yu [1,3,4], Hyuck Kim [1,3,4] and Dong-Woo Lim [1,3,4,*]

1 TOPO Lab., Co., Ltd., Goyang 10326, Republic of Korea; 9494bbaek@gmail.com (S.-C.B.); panax1218@gmail.com (S.-Y.J.); choeyunjeong97@gmail.com (Y.-J.C.); macbo2001@gmail.com (B.-H.B.); lilwayne224@gmail.com (D.-H.K.); kalama2@dongguk.edu (G.-R.Y.); hyuckkim@dongguk.ac.kr (H.K.)
2 Department of Life Science, Dongguk University, Goyang 10326, Republic of Korea
3 Department of Diagnostics, College of Korean Medicine, Dongguk University, Goyang 10326, Republic of Korea
4 Institute of Korean Medicine, Dongguk University, Goyang 10326, Republic of Korea
* Correspondence: greatwoodong@dongguk.edu; Tel.: +82-31-961-5833; Fax: +82-31-961-5835
† These authors contributed equally to this work.

Abstract: Cannabis has been cultivated for thousands of years for a variety of purposes, including fiber, seeds, oil, and medicinal compounds. The cannabis industry is growing rapidly because several countries have recently legalized the use of cannabis. In these countries, the industry related to cannabinoid ingredients such as cannabidiol (CBD) and Δ^9-tetrahydrocannabinol (THC) is steadily increasing every year. High concentrations of cannabinoids are mainly produced in unfertilized female flowers. Maintaining plants with high cannabinoid content is essential for producing uniform substances in large quantities. This study established an in vitro micropropagation protocol that can maintain the mother plant characteristics of *Cannabis sativa* 'Cheungsam'. As a result of this experiment, the shoot tips and nodes of *Cannabis sativa* 'Cheungsam' at various concentrations (0, 0.25, 0.5, 1.0 mg/L) of 2iP, BA, and mT plant growth regulators (PGRs), and all concentrations of 2iP showed better results compared to two other hormones. However, the cut surfaces turned black, and excessive hyperhydricity occurred. Based on these symptoms, activated charcoal was added to the medium with the assumption that necrosis and hyperhydricity occur due to the accumulation of reactive oxygen species (ROS). When treated with 0.5 g/L charcoal, hyperhydricity was not overcome, and there was no difference compared to the control. As a new alternative, we divided the experiments into MS (Murashige and Skoog) and DKW (Driver and Kuniyuki Walnut) medium, which were commercially available. As a result, the rate of hyperhydricity was reduced, the cut surface did not turn black, and the growth conditions were also improved. Subsequently, ½ MS medium and ½ DKW medium were treated with various concentrations of IBA alone and with combinations of IBA and NAA for rooting. As a result, ½ DKW with IBA 0.5 mg/L showed the highest rooting rate and the best root condition for Cheungsam. After 4 weeks, when considering rooted plants with a height above 5 cm that were acclimatized, the acclimatization rate reached 100%. In conclusion, the *Cannabis sativa* 'Cheungsam' plants used in this study produced healthy shoots on DKW medium containing 1.0 mg/L 2iP and 0.5 mg/L of IBA in ½ DKW medium showed the best rooting rate.

Keywords: cannabis industry; plant tissue culture; micropropagation; plant growth regulators; Cheungsam

1. Introduction

Cannabis sativa, commonly known as hemp, is an annual herbaceous plant in the Cannabaceae family [1–3]. It has been cultivated for thousands of years for various purposes, including fiber, seed, oil, and medicinal compounds [4].

The global cannabis industry is growing rapidly and with the legalization of medical and industrial cannabis worldwide, the potential of cannabis is being recognized in various applications. As of 2024, Canada, Uruguay, and 18 U.S. states have legalized both medical and recreational cannabis. Additionally, several European countries including Germany and the UK, as well as New Zealand and Thailand, have legalized medical cannabis. In Korea, since the amendment of the Narcotics Control Act in 2019, research and industrial efforts related to medical cannabis have been increasing [5].

Cannabis can be broadly classified into marijuana and hemp. Marijuana, derived from *Cannabis sativa* subsp. *sativa* or *Cannabis sativa* subsp. *indica* plants, contains high levels of THC, a psychoactive compound classified as a narcotic in many countries [3]. Some countries use marijuana for both recreational and medical purposes. Hemp, a variant of *Cannabis sativa* subsp. *sativa*, contains very low levels of THC (typically less than 0.3% on a dry weight) and is primarily used for medical and industrial purposes [6]. Hemp is characterized by its high CBD content, another major cannabinoid found in cannabis, which has various medical benefits such as treating epilepsy syndromes in medications like Epidiolex®, alleviating anxiety and depression, as well as having anti-inflammatory and analgesic properties, without the psychoactive effects of THC [2,7–9]. High concentrations of cannabinoids are primarily produced in young leaves and unfertilized female flowers. The value of the crop is determined by cannabinoid content in the flowers.

The increasing demand for clean propagated cannabis material necessitates the development of scalable production systems, such as vertical farms, plant factories, and smart farms [10]. Micropropagation is a tissue culture technique that produces numerous identical clone plants from tissue sections of a mother plant. This allows for the mass production of genetically identical plants in a shorter time than traditional propagation methods [11]. Plants are cultured in bottles and grown in a culture room equipped with LED lighting, enabling the maintenance of a large number of plants in a small space. Additionally, cultivation in sterile media allows for the production and maintenance of disease-free plants.

Cannabis is a typical dioecious plant and anemophilous, relying on wind for pollination. Thus, it is difficult to maintain genetically identical plants to the mother plant using traditional propagation methods. High concentrations of cannabinoids are primarily produced in young leaves and unfertilized female flowers. However, cannabinoid production decreases when pollination occurs through wind [12]. Moreover, infections such as Hops Latent viroid (HLVd), *Fusarium oxysporum*, and Powdery Mildew, as well as pests like spider mites and aphids, can reduce quality and yield in traditional propagation methods [13,14].

'Cheungsam' is a F_1 hybrid cultivar developed by crossing IH3, a low-narcotic genetic resource among 44 introduced germplasms from CPRO in the Netherlands in 1997, with a Korean native strain. 'Cheungsam' is characterized by its low narcotic content, with a THC content of 0.34% and a CBD content of 1.34%. In Korea, 'Cheungsam' is the most widely cultivated outdoor cultivar due to strict regulations on the use of cannabis and its derived ingredients and products. It was primarily cultivated in Korea for fiber production, but the industrial use of its by-products, such as roots and stems, has recently been growing.

A previous study screened various conditions to optimize in vitro clonal propagation and ex vitro rooting and acclimatization of fiber hemp via shoot tips. However, low rooting rates of 74.6% were reported [15]. Another study investigated benefit of using DKW medium instead of MS. DKW basal salts improved micropropagation as demonstrated by the canopy area, multiplication, and callogenesis in five different cultivars (BA-21, BA-41, BA-49, BA-61, and BA-71) [16]. Another study tested the impacts of different compositions of basal media with PGRs on hemp cultivar TJ's CBD during initiation, multiplication, in vitro rooting, ex vitro rooting, and acclimatization [17]. Despite the studies above, there has been no studies on the development of optimized systems of micropropagation for 'Cheungsam', which is widely cultivated in the Republic of Korea for various industrial purposes.

Therefore, this study selected micropropagation as a method to maintain the traits of the mother cannabis plant. We addressed the issues at each stage and established a protocol tailored to *Cannabis sativa* 'Cheungsam' to obtain normal plantlets.

2. Materials and Methods

2.1. Seed Germination

The cannabis used in this study was provided by the cultivation farmland of the Korea Hemp Industry Association, an incorporated association located in Imha-myeon, Andong-si Gyeongsangbuk-do, Republic of Korea.

Seeds of *Cannabis sativa* 'Cheungsam' were surface sterilized in 75% ethanol for 2 min 30 s followed by rinsing with distilled water. They were then sterilized with 10% sodium hypochlorite (NaClO, Daejung, Siheung, Republic of Korea) solution for 10 min, before being rinsed with sterile distilled water eight times. The seeds were then subjected to sonication for 10 min. After sonication, the seeds were primed with 1% hydrogen peroxide (H_2O_2, Daejung, Siheung, Republic of Korea) solution cultivated in darkness (25 ± 2 °C) for 3 days. The seeds were germinated in Phytohealth plant culturing containers (lid size: 96.65 mm [diameter] × 41.00 mm [height], dish size: 91.02 mm [diameter] × 81.60 mm [height], SPL Life Sciences, Pocheon, Republic of Korea) containing 50 mL MS medium. The MS medium contained full MS, including a Gamborg B5 vitamin mixture, 3% sucrose (MB cell, Seoul, Republic of Korea), and 0.8% plant agar (Duchefa Biochemie, The Netherlands). The pH was adjusted to 5.8 with 1N NaOH or HCl. Medium and culture vessels were autoclaved at 121 °C for 15 min. The culture plate was cultured in a culture room set at a temperature at 25 ± 2 °C under a 16/8 h (light/dark) photoperiod with a light intensity of 120.25 μmol m^{-2} s^{-1} LED daylight (Figure 1).

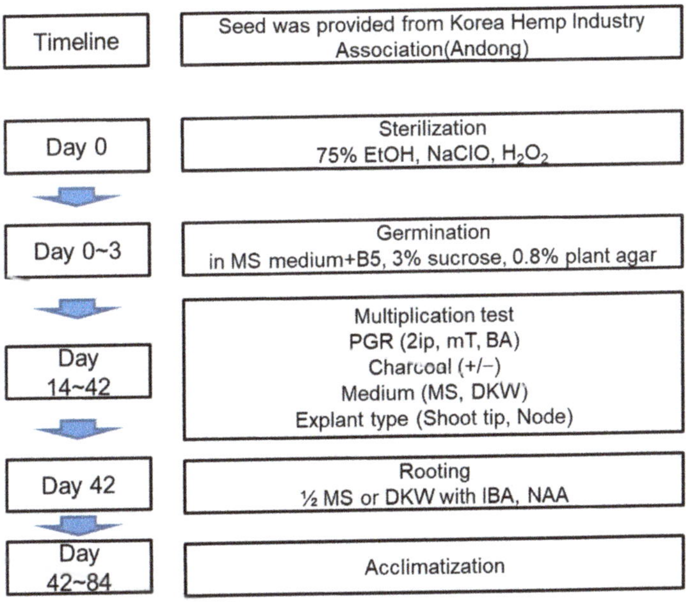

Figure 1. The overall process and timeline for micropropagation of *Cannabis sativa* 'Cheungsam' in this study.

2.2. Multiplication

Sections of 0.5–1 cm length of explants (shoot tip and node) were cut from two-week-old in vitro seedlings and used as the explants. The explants were cultured on MS or DKW medium supplemented with various concentration of 2iP, mT, and BA, with or without charcoal. After four weeks of culture, the plant growth parameters (plant height, number of leaves, leaf chlorosis rate, and number of shoots) were assessed in the resulting cannabis plants. All cultures were placed in a culture room set at a temperature at 25 ± 2 °C under a 16/8 h (light/dark) photoperiod with a light intensity of 120.25 μmol m^{-2} s^{-1} LED daylight. Each measurement included ten bottles with three replications.

2.3. Effect of Plant Growth Regulators (PGRs) and Activated Charcoal on In Vitro Plant Multiplication of the Cannabis sativa 'Cheungsam'

Two-week-old hemp plants were cultured in vitro on MS media supplemented with different concentrations of 6-(γ,γ-Dimethylallylamino)purine (2iP; PhytoTech Labs, Lenexa, KS, USA), meta-Topoline (mT; Duchefa Biochemie, The Netherlands), and 6-Benzylaminopourine (BA; Sigma-Aldrich, St. Louis, MO, USA). Furthermore, charcoal (Daejung, Siheung, Republic of Korea) was added for plant growth and suppression of hyperhydricity to some samples. Each bottle contained five plants, and there were ten bottles per treatment with three replications.

2.4. Effect of Medium Formulation on In Vitro Plant Multiplication of the Cannabis sativa 'Cheungsam'

Several commercial basal nutrient formulations were tested for their effect on in vitro cannabis explant growth. In this study, we used MS or DKW medium. Each medium was evaluated at their recommended concentrations of 4.4 and 5.22 g/L, respectively. Each bottle contained five plants, and there were ten bottles per treatment with three replications.

2.5. Effect of Explants Type on In Vitro Plant Multiplication of the Cannabis sativa 'Cheungsam'

Shoot tip and nodal segments were compared as explant types for their effect on in vitro propagation. Each bottle contained five plants, and there were ten bottles per treatment with three replications.

2.6. Rooting of Shoots

Shoots grown to over 5 cm were cultured for four weeks on a half-strength MS or DKW medium supplemented with various concentration of Indole-3-butyric acid (IBA; Sigma-Aldrich, St. Louis, MO, USA) and 1-Naphthaleneacetic acid (NAA; MB cell, Seoul, Republic of Korea) for root formation. Each bottle contained five plants, and there were five bottles per treatment with three replications.

2.7. Acclimatization

The plantlets with well-developed roots were removed from the culture vessels and washed several times with running tap water. The plantlets were then transplanted into pots containing mixed soil (cocopeat:peatmoss:perlite = 4:4:2, *w*/*w*/*w*) and acclimatized for two weeks in clean plastic trays. Then, the plastic trays of successfully adapted plants were removed. Following one month, the plant survival rate was assessed by outer appearance.

2.8. Statistical Analysis

The statistical analysis of data was performed using GraphPad Prism version 5.03 (La Jolla, CA, USA) and presented as mean ± standard errors of three replicates. The significance of differences between means was determined using Tukey's multiple range test and the least significant difference test, with a significance level of $p < 0.05$

3. Results

3.1. Effect of PGRs on In Vitro Shoot Induction

Shoot tip and node explants of *Cannabis sativa* 'Cheungsam' cultured on MS medium without PGRs struggled to maintain a healthy state for up to four weeks. When explants were cultured on media containing mT (0.25 or 0.5 mg/L) and BA (0.25 or 0.5 mg/L), their condition was better compared to those without PGRs, but the results were not as favorable as those observed with 2iP (0.25, 0.5 or 1.0 mg/L). In contrast, healthy shoot formation occurred in media containing 2iP regardless of the concentration. However, the maximum number of shoots per explant did not show significant differences among all treatments (Tables 1 and 2). Unfortunately, the shoots developed from some treatments (mT and BA), shoot tips, and all treatments nodes exhibited symptoms of hyperhydricity, such as a glassy appearance, thin stems, and translucent leaves (Table 2). When these plants were subcultured in the same composition media, all of them died. Therefore, among the three PGRs (2iP, mT, BA), 2iP was confirmed to be the most suitable hormone, and subsequent experiments were conducted using only the 2iP (0.25, 0.5, and 1.0 mg/L).

Table 1. Comparison of plant growth of various PGRs concentration on shoot induction from shoot tip explants of *Cannabis sativa* 'Cheungsam' after four weeks of culture.

Explant	PGRs (mg/L) 2IP	mT	BA	No. of Shoot/Explant	Plant Height (cm)	No. of Leaf	Leaf Chlorosis Rate (%)	Hyperhydricity
Shoot tip	-	-	-	1.12 ± 0.12 b	1.30 ± 0.30 d	4.94 ± 0.65 c	23.17 ± 7.13 bc	+
	0.25	-	-	1.05 ± 0.05 b	3.64 ± 0.48 a	7.81 ± 0.60 b	10.49 ± 3.00 d	-
	0.50	-	-	1.18 ± 0.13 b	3.31 ± 0.30 ab	8.94 ± 0.57 ab	26.55 ± 5.86 bc	-
	1.00	-	-	1.00 ± 0.00 b	3.46 ± 0.36 ab	8.18 ±0.57 b	6.36 ± 2.51 d	-
	-	0.25	-	1.26 ± 0.06 ab	2.73 ± 0.21 abc	5.54 ± 0.13 bc	36.11 ±13.60 b	+
	-	0.50	-	1.59 ± 0.23 ab	2.99 ± 0.20 abc	5.82 ± 0.23 bc	48.99 ± 4.18 a	+
	-	-	0.25	1.77 ± 0.10 ab	1.80 ± 0.16 d	8.60 ± 0.38 ab	32.52 ± 2.31 b	+
	-	-	0.50	2.13 ± 0.12 a	1.44 ± 0.21 d	5.47 ± 0.39 c	30.13 ± 2.46 b	+

These data represent the means of three replicates with standard error. Means with the same letters are not significantly different (LSDT, $p < 0.05$). -: normal plant; +: mild hyperhydricity symptoms.

Table 2. Comparison of plant growth of various PGRs concentration on shoot induction from node explants of *Cannabis sativa* 'Cheungsam' after four weeks of culture.

Explant	PGRs (mg/L) 2IP	mT	BA	No. of Shoot/Explant	Plant Height (cm)	No. of Leaf	Leaf Chlorosis Rate (%)	Hyperhydricity
Node	-	-	-	1.94 ± 0.06 a	1.60 ± 0.12 bc	8.22 ± 0.53 a	45.13 ± 5.58 ab	++
	0.25	-	-	1.90 ± 0.07 a	2.77 ± 0.24 a	9.05 ± 0.60 a	39.67 ± 4.53 b	++
	0.50	-	-	1.80 ± 0.20 a	2.92 ± 0.81 a	8.00 ± 1.41 a	24.83 ± 6.87 bc	++
	1.00	-	-	2.00 ± 0.00 a	3.10 ± 0.48 a	9.78 ± 1.28 a	36.81 ± 7.01 b	++
	-	0.25	-	1.78 ± 0.08 a	2.33 ± 0.12 ab	3.95 ± 0.06 c	51.27 ± 0.31 a	++
	-	0.50	-	1.86 ± 0.14 a	1.87 ± 0.18 b	3.80 ± 0.07 c	31.99 ± 6.57 a	++
	-	-	0.25	1.47 ± 0.09 ab	0.76 ± 0.07 d	4.47 ± 0.30 c	31.42 ± 3.30 bc	++
	-	-	0.50	1.55 ± 0.10 ab	0.73 ±0.06 d	5.32 ± 0.33 b	25.41 ± 2.87 bc	++

These data represent the means of three replicates with standard error. Means with the same letters are not significantly different (LSDT, $p < 0.05$). -: normal plant; ++: hard hyperhydricity symptoms.

3.2. Effect of Activated Charcoal on In Vitro Shoot Induction and Overcome Hyperhydricity

Activated charcoal is known to help plant growth and development. Thus, activated charcoal (0.5 g/L) was added to the medium to overcome hyperhydricity and prevent the cut surface from blackening. However, almost all characteristics were significantly better without charcoal across all treatments (Figures 2A–H and 3A–H). In the case of the leaf chlorosis rate, an increase was observed (Tables 3 and 4). Additionally, hyperhydricity was not attenuated and was remained at similar levels (Table 4). Therefore, it was concluded that charcoal is not a suitable component for overcoming hyperhydricity.

Table 3. Comparison of the impact of activated charcoal on plant growth and shoot induction from shoot tip explants of *Cannabis sativa* 'Cheungsam' after four weeks of culture.

Explant	PGRs (mg/L) 2IP	Charcoal (g/L)	No. of Shoot/Explant	Plant Height (cm)	No. of Leaf	Leaf Chlorosis Rate (%)	Hyperhydricity
Shoot tip	-	-	1.17 ± 0.17 a	1.53 ± 0.40 b	5.33 ± 0.87 c	13.38 ± 5.97 ab	+
	-	0.5	1.00 ± 0.00 a	2.38 ± 0.28 ab	6.50 ± 0.34 c	21.83 ± 7.04 ab	-
	0.25	-	1.06 ± 0.06 a	3.69 ± 0.54 a	7.94 ± 0.76 b	6.73 ± 2.77 b	-
	0.25	0.5	1.06± 0.20 a	1.82 ± 0.41 b	5.00 ± 0.45 cd	48.11 ± 6.83 a	-
	0.50	-	1.25 ± 0.18 a	3.41 ± 0.28 a	9.17 ± 0.58 a	22.61 ± 6.98 ab	-
	0.50	0.5	1.17 ± 0.20 a	2.32 ± 0.34 ab	7.60 ± 0.83 bc	26.25 ± 5.69 ab	-
	1.00	-	1.00 ± 0.00 a	3.54 ± 0.29 a	8.38 ± 0.84 ab	6.36 ± 2.51 b	-
	1.00	0.5	1.00 ± 0.00 a	1.14 ±0.25 b	4.60 ± 0.40 d	24.00 ± 7.97 ab	-

These data represent the means of three replicates with standard error. Means with the same letters are not significantly different (LSDT, $p < 0.05$). -: normal plant; +: mild hyperhydricity symptoms.

Figure 2. Illustration of growth status of *Cannabis sativa* 'Cheungsam' shoot tip explants cultured on with or without activated charcoal. These images were taken after four weeks of culture: (**A**) without PGRs, (**B**) 2iP 0.25 mg/L, (**C**) 2iP 0.5 mg/L, (**D**) 2iP 1.0 mg/L, (**E**) charcoal 0.5 g/L, (**F**) 2iP 0.25 mg/L + charcoal 0.5 g/L, (**G**) 2iP 0.5 mg/L + charcoal 0.5 g/L, (**H**) 2iP 1.0 mg/L + charcoal 0.5 g/L.

Figure 3. Illustration of growth status of *Cannabis sativa* 'Cheungsam' node explants cultured on with or without activated charcoal. These images were taken after four weeks of culture: (**A**) without PGRs, (**B**) 2iP 0.25 mg/L, (**C**) 2iP 0.5 mg/L, (**D**) 2iP 1.0 mg/L, (**E**) charcoal 0.5 g/L, (**F**) 2iP 0.25 mg/L + charcoal 0.5 g/L, (**G**) 2iP 0.5 mg/L + charcoal 0.5 g/L, (**H**) 2iP 1.0 mg/L + charcoal 0.5 g/L.

Table 4. Comparison of the impact of activated charcoal on plant growth and shoot induction from node explants of *Cannabis sativa* 'Cheungsam' after four weeks of culture.

Explant	PGRs (mg/L) 2IP	Charcoal (g/L)	No. of Shoot/Explant	Plant Height (cm)	No. of Leaf	Leaf Chlorosis Rate (%)	Hyperhydricity
Node	-	-	2.00 ± 0.00 a	1.64 ± 0.50 a	8.22 ± 0.53 ab	42.43 ± 6.69 a	++
	-	0.5	1.80 ± 0.13 a	2.70 ± 0.72 a	7.90 ± 0.57 ab	48.92 ± 4.94 a	++
	0.25	-	1.87 ± 0.09 a	2.77 ± 0.32 a	9.05 ± 0.60 ab	35.56 ± 5.61 a	++
	0.25	0.5	1.90 ± 0.10 a	1.39 ± 0.20 a	6.70 ± 0.63 b	47.58 ± 6.20 a	++
	0.5	-	1.80 ± 0.20 a	2.92 ± 0.81 a	8.00 ± 1.41 ab	24.83 ± 6.87 a	++
	0.5	0.5	1.90 ± 0.10 a	2.22 ± 0.28 a	10.80 ± 0.68 a	47.42 ± 4.17 a	++
	1.0	-	2.00 ± 0.00 a	2.08 ± 0.36 a	9.78 ± 1.28 ab	35.55 ± 7.01 a	++
	1.0	0.5	1.88 ± 0.13 a	1.99 ± 0.39 a	6.50 ± 0.93 b	55.54 ± 9.44 a	++

These data represent the means of three replicates with standard error. Means with the same letters are not significantly different (LSDT, $p < 0.05$). -: normal plant; ++: hard hyperhydricity symptoms.

3.3. Effect of Media Formulation on In Vitro Shoot Induction

To determine the optimal medium composition for multiplication, shoot tips and nodes were cultured on MS or DKW media containing various concentration of 2iP and observed after four weeks.

For shoot tips, there were no significant difference in the number of shoots per explant across all treatment. This is likely due to the presence of apical meristems in shoot tips, which generally produce only one shoot. The plant height was smaller in media without 2iP compared to media with 2iP, indicating that 2iP, a type of cytokinin, plays a role in promoting plant elongation. The number of leaves was highest in MS and DKW media containing 0.5 mg/L 2iP, while leaf chlorosis was least observed in MS and DKW media containing 1.0 mg/L of 2iP (Table 5). This suggests that higher concentrations of 2iP have a positive effect on shoot formation compared to lower concentrations. No significant differences were observed between MS and DKW media (Figure 4A–H). Therefore, for shoot tip multiplication, the optimal protocol is to culture shoot tips in MS or DKW media containing high concentration (1.0 mg/L) of 2iP.

Table 5. Comparison of the impact of two different media formulation on plant growth and shoot induction from shoot tip explants of *Cannabis sativa* 'Cheungsam' after four weeks of culture.

Explant	Media Type	2IP	No. of Shoot/Explant	Plant Height (cm)	No. of Leaf	Leaf Chlorosis Rate (%)	Hyperhydricity
Shoot tip	MS	-	1.12 ± 0.12 a	1.30 ± 0.30 a	4.94 ± 0.65 b	23.17 ± 7.13 ab	-
	MS	0.25	1.05 ± 0.05 a	3.64 ± 0.48 d	7.81 ± 0.60 a	10.49 ± 3.00 b	-
	MS	0.50	1.18 ± 0.13 a	3.31 ± 0.30 cd	8.94 ± 0.57 a	26.55 ± 5.86 a	-
	MS	1.00	1.00 ± 0.00 a	3.46 ± 0.36 cd	8.18 ±0.57 a	6.36 ± 2.51 bc	-
	DKW	-	1.00 ± 0.00 a	1.53 ± 0.11 a	5.06 ± 0.25 b	11.89 ± 2.60 b	-
	DKW	0.25	1.03 ± 0.03 a	2.39 ± 0.17 bc	7.14 ± 0.27 a	10.47 ± 1.47 b	-
	DKW	0.50	1.11 ± 0.06 a	2.82 ± 0.17 cd	8.29 ± 0.32 a	6.61 ± 1.21 bc	-
	DKW	1.00	1.09 ± 0.05 a	2.69 ± 0.19 cd	7.95 ± 0.35 a	7.65 ± 1.22 bc	-

These data represent the means of three replicates with standard error. Means with the same letters are not significantly different (LSDT, $p < 0.05$). -: normal plant.

Similarly for nodes, there were no significant differences in the number of shoots per explant across all treatments. However, compared to shoot tips, approximately twice as many adventitious shoots were formed. This is likely due to the axillary buds present in the nodes. In terms of plant height, as with shoot tips, both MS and DKW media containing 2iP showed greater growth compared to media without 2iP (Figure 5A–H). For the number of leaves, DKW media produced statistically significantly more leaves compared to MS media, regardless of 2iP concentration (0.25, 0.5 and 1.0 mg/L). In terms of leaf chlorosis rate, the MS control showed statistically significant higher chlorosis, while DKW with 2iP 1.0 mg/L treatment exhibited the lowest chlorosis. There were no statistically significant differences when comparing MS with 2iP 0.25 mg/L and DKW with 2iP (0, 0.25, 0.5 mg/L). Regarding hyperhydricity, the DKW media had lower occurrence rates compared to MS media, regardless of the 2iP concentration (Table 6). Therefore, the optimal medium composition for node multiplication was determined to be DKW with 2iP 1.0 mg/L.

Figure 4. Illustration of growth status of *Cannabis sativa* 'Cheungsam' shoot tip explants cultured on two different types (MS and DKW) of media and various 2iP concentration. These images were taken after four weeks of culture: (**A**) MS without PGRs, (**B**) MS with 2iP 0.25 mg/L, (**C**) MS with 2iP 0.5 mg/L, (**D**) MS with 2iP 1.0 mg/L, (**E**) DKW without PGRs, (**F**) DKW with 2iP 0.25 mg/L, (**G**) DKW with 2iP 0.5 mg/L, (**H**) DKW with 2iP 1.0 mg/L.

Figure 5. Illustration of growth status of *Cannabis sativa* 'Cheungsam' node explants cultured on two different types of media and various 2iP concentration. These images were taken after four weeks of culture: (**A**) MS hormone free, (**B**) MS + 2iP 0.25 mg/L, (**C**) MS + 2iP 0.5 mg/L, (**D**) MS + 2iP 1.0 mg/L, (**E**) DKW hormone free, (**F**) DKW + 2iP 0.25 mg/L, (**G**) DKW + 2iP 0.5 mg/L, (**H**) DKW + 2iP 1.0 mg/L.

Table 6. Comparison of the impact of two different media formulation on plant growth and shoot induction from node explants of *Cannabis sativa* 'Cheungsam' after four weeks of culture.

Explant	Media Type	2IP	No. of Shoot/Explant	Plant Height (cm)	No. of Leaf	Leaf Chlorosis Rate (%)	Hyperhydricity
Node	MS	-	1.94 ± 0.06 a	1.60 ± 0.12 c	8.22 ± 0.53 b	45.13 ± 5.58 a	++
	MS	0.25	1.90 ± 0.07 a	2.77 ± 0.24 ab	9.05 ± 0.60 b	39.67 ± 4.53 a	++
	MS	0.50	1.80 ± 0.20 a	2.92 ± 0.81 ab	8.00 ± 1.41 b	24.83 ± 6.87 ab	++
	MS	1.00	2.00 ± 0.00 a	3.10 ± 0.48 a	9.78 ± 1.28 b	36.81 ± 7.01 ab	++
	DKW	-	1.96 ± 0.03 a	1.27 ± 0.09 c	9.18 ± 0.52 b	19.68 ± 2.85 b	+
	DKW	0.25	1.92 ± 0.04 a	2.69 ± 0.16 ab	12.64 ± 0.47 a	22.79 ± 1.94 b	-
	DKW	0.50	1.90 ± 0.04 a	2.86 ± 0.14 ab	13.23 ± 0.41 a	21.23 ± 1.84 b	-
	DKW	1.00	1.92 ± 0.04 a	2.57 ± 0.15 b	13.22 ± 0.48 a	17.79 ± 2.03 b	-

These data represent the means of three replicates with standard error. Means with the same letters are not significantly different (LSDT, $p < 0.05$). -: normal plant; +: mild hyperhydricity symptoms; ++: hard hyperhydricity symptoms.

3.4. Effect of Explants Type on In Vitro Shoot Induction

Shoot tip and node explants were compared for their effect on propagation productivity. After four weeks of culture, comparing the shoot tip and node revealed that the shoot tip exhibited relatively longer shoot length and lower chlorosis rates, while the node produced a higher number of shoots per explant and number of leaves. The best results for the shoot tip were observed in MS medium with 1.0 mg/L 2iP, achieving the greatest plant height of 3.46 cm and the lowest chlorosis rate of 6.36% (Table 5). The highest number of shoots per explant was observed in nodes cultured in MS medium with 1.0 mg/L 2iP, producing an average of 2.0 shoots. The greatest number of leaves was formed in nodes cultured in DKW medium with 0.5 mg/L 2iP with an average of 13.23 leaves (Table 6).

However, all node treatments cultured in MS medium exhibited hyperhydricity, while no hyperhydricity occurred in the shoot tips (Figures 4 and 5). Therefore, while it is difficult to definitively determine the superior tissue, nodes, which produced more shoots per explant, are deemed more suitable for the purpose of mass propagation in micropropagation.

3.5. Rooting of Shoots

The derived shoots were cultured in MS or DKW media supplemented with 2iP and then transferred to media containing IBA (0.25 and 0.5 mg/L) for four weeks. As a result, rooting did not occur in most media, and the plants became necrotic. Consequently, a new solution was implemented by reducing the concentration of MS and DKW media to half. This adjustment resulted in a higher rooting rate compared to full-concentration media. When comparing MS and DKW media, DKW showed a higher rooting rate at all IBA concentrations (0.1, 0.25, and 0.5 mg/L). Specifically, compared to MS medium, DKW media with IBA 0.25 mg/L exhibited over four times higher rooting rate than MS with IBA 0.25 mg/L, and DKW with IBA 0.5 mg/L showed approximately twice the rooting rate of MS with IBA 0.5 mg/L (Table 7).

Table 7. Comparison of various PGRs concentrations on root induction from *Cannabis sativa* 'Cheungsam' after four weeks of culture.

Media Type	PGRs (mg/L)		Rooting Rate (%)
	IBA	NAA	
MS	-	-	18.52 ± 10.36 c
MS	0.1	-	76.92 ± 6.35 ab
MS	0.25	-	19.51 ± 6.00 c
MS	0.5	-	48.15 ± 6.29 bc
MS	0.25	0.05	4.92 ± 7.50 d
DKW	-	-	61.54 ± 4.60 b
DKW	0.1	-	78.95 ± 6.59 ab
DKW	0.25	-	85.37 ± 6.14 a
DKW	0.5	-	91.49 ± 5.26 a

These data represent the means of three replicates with standard error. Means with the same letters are not significantly different (LSDT, $p < 0.05$).

Next, the rooting rates were compared between IBA alone and combination of IBA and NAA. In all concentrations, IBA alone resulted in better rooting rates when compared to the combination of IBA and NAA (Figure 6A–E). In fact, the IBA and NAA combination had lower rooting rates than treatments without PGRs. Based on these results, the combination of NAA with IBA was not used in the DKW media with IBA treatments. Among the IBA-alone treatments, DKW with IBA 0.5 mg/L showed the best result, with no statistically significant difference when compared to DKW with IBA 0.25 mg/L (Figure 6A–I). However, upon visual inspection, root condition, including root length and root thickness, appeared to be better in DKW with IBA 0.5 mg/L (Figure 6H,I). Therefore, the optimal medium composition for rooting was determined to be DKW with IBA 0.5 mg/L.

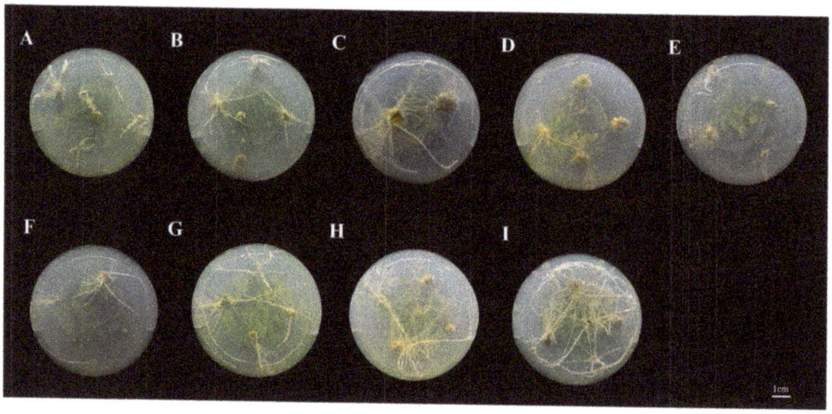

Figure 6. Illustration of root growth status of *Cannabis sativa* 'Cheungsam' explant cultured on various PGRs concentration. These images were taken after four weeks of culture: (**A**) ½ MS without PGRs, (**B**) ½ MS with IBA 0.1 mg/L, (**C**) ½ MS with IBA 0.25 mg/L, (**D**) ½ MS with IBA 0.5 mg/L, (**E**) ½ MS with IBA 0.25 mg/L and NAA 0.05 mg/L, (**F**) ½ DKW without PGRs, (**G**) ½ DKW with IBA 0.1 mg/L, (**H**) ½ DKW with IBA 0.25 mg/L, (**I**) ½ DKW with IBA 0.5 mg/L.

3.6. Acclimatization

After culture on rooting media for four weeks, in vitro grown plants with heights of over 5 cm were acclimatized. During acclimatization, the relative humidity was gradually decreased. As a result, all plants were successfully acclimatized (Figure 7A). Two weeks after acclimatization, the plants were transplanted into larger pots (Figure 7B). The survival rate was assessed again six weeks later, showing a 100% acclimatization success rate.

Figure 7. Illustration of successfully acclimatized *Cannabis sativa* 'Cheungsam' (**A**) after two weeks of acclimatization, (**B**) after six weeks of acclimatization.

4. Discussion

Previous studies have reported successful micropropagation of different cannabis cultivars [15–17]. In this study, we developed a protocol for micropropagation from shoot tip and node of cannabis 'Cheungsam' for industrial development purposes.

To determine the optimal multiplication medium of 'Cheungsam', we tested with various PGRs, such as 2iP, mT, and BA at concentration of 0.25, 0.5, and 1.0 mg/L. Shoot tips cultured on PGR-free media exhibited lower height and fewer leaves. Media containing 2iP and mT, regardless of concentration, showed statistically significant taller plant heights. All treatments with 2iP, mT, and BA resulted in a significant number of leaves. In the case of nodes, PGR-free treatments exhibited better growth than mT and BA, but had a higher leaf chlorosis rate. Among the cytokinin tested, media containing 2iP, irrespective of concentration, resulted in the tallest plant height, the highest number of leaves, and the lowest leaf chlorosis rate. BA, a common cytokinin, is used for multiplication of micropropagation in various crops (Cannabis, Quercus, Tea Tree, Ginger, Stevia) [18–22]. However, in our study, BA-treated plants showed similar plant heights and higher chlorosis rates compared to those without PGRs treatment. Although mT has been recommended as an excellent cytokinin for multiplication in other cannabis micropropagation studies [23–26], our results for mT-treated plants were not favorable in terms of plant height, number of leaves, and leaf chlorosis rate, similar to the results from BA-treated plants. Conversely, 2iP-treated shoot tip and node explants displayed the greatest plant height, greatest number of leaves, and lowest leaf chlorosis rate across all concentrations. While higher concentrations of 2iP produced better results, the differences were not statistically significant. This accords with the findings of Stephen et al. [17], who reported that 2iP at 5.0 μM yielded the best quality shoots of *Cannabis sativa* 'TJ's CBD'. Therefore, we concluded that 2iP is the most suitable PGRs for the multiplication phase of Cannabis.

However, all explants exhibited blackened cut surfaces regardless of hormone type and concentration, and node explants showed severe hyperhydricity. Hyperhydricity is a common issue in plant tissue culture where plants absorb excessive water, leading to softened tissues and physiological abnormalities [27]. Typical symptoms include translucent, fragile, and twisted leaves that are easily damaged, shortened internodes, and thickened stems [28,29]. Causes of hyperhydricity include high humidity in the culture vessel, excessive cytokinin use, poor vessel ventilation, influence of nitrogen ions, and accumulation of ethylene and ROS. We hypothesized that the hyperhydricity in our experiment was due to the accumulation of phenolic compounds in the culture vessel, leading to high ROS levels. ROS affects ethylene metabolism [30], and ethylene in turn increases ROS accumulation [31]. Thus, explants were likely due to phenolic compound accumulation. To address these issues, we added 0.5 g/L charcoal to the media.

Activated charcoal is composed of carbon arranged in a semi-graphitic form with small particle sizes. It removes non-carbon impurities and oxidizes the carbon surface. Activated charcoal is used in plant tissue culture primarily to adsorb growth-inhibiting substances (such as phenolic compounds) in the culture medium, thereby promoting plant growth [32]. According to a study by Lee and Chang [33], the addition of 200 mg/L of activated charcoal during the multiplication phase in micropropagation improved shoot length and prevented the regeneration of abnormal shoots. In other cannabis varieties, culturing shoot tip and node explants in a medium with 1 g/L of activated charcoal resulted in the formation of long shoots in some genotypes [34]. However, in this study, the addition of activated charcoal to all treatments resulted in inhibited plant growth compared to treatments without charcoal. Komalaballi and Rao [35], reported a decrease in the number of shoots when activated charcoal was added to the micropropagation medium of *Gymnema sylvestre* [35], and Tivarekar and Eapen [36], found that the addition of activated charcoal to the regeneration medium of mung beans inhibited shoot growth. These results suggest that activated charcoal not only adsorbs growth-inhibiting substances but also essential nutrients necessary for plant growth, leading to reduced growth. Therefore, it is highly suspected that the observed inhibition of the plant growth in our study also caused by the absorption of essential nutrients by charcoal.

When cultured on MS medium, plant cut surfaces turned black and necrotic, and node explants were hyperhydricity, causing growth problems. To solve this problem, experiments were conducted using different commercial media formulations. As a result, the shoot tip grew well in both MS and DKW media. In the case of nodes, DKW medium showed better results than MS medium in all growth indicators including height and number of leaves. In particular, severe hyperhydricity occurred when cultured on MS medium, but no hyperhydricity occurred when cultured on DKW medium. This suggests that the components in the medium have a negative effect on the plants when cultured on MS medium. MS medium is the most widely used medium in plant tissue culture and is effective for most plants, but there are some plants that do not grow well in MS medium. For this reason, a variety of commercial media formulations exists and was developed for specific plants. Compared to MS medium, DKW has a similar ammonium-to-nitrate ratio but lower total amount of nitrogen in the medium [37,38]. El-Dawayati and Zayed [39] reported that hyperhydricity was overcome during the culture process and redifferentiation was successful by reducing the nitrogen ratio of the culture medium. In Phillips and Garda [38], it was reported that excessive ammonium in MS medium can cause growth problems, and Liu et al. [40] showed that when garlic was cultured in MS medium, growth was slow and hyperhydricity occurred. DKW medium also contains significantly higher levels of sulfur (~7×), calcium (~3×) and copper (10×) [16]. However, no clear research has been conducted on which of these factors are important for micropropagation of *Cannabis sativa* 'Cheungsam'. Therefore, this is an area that requires further research.

In this study, auxin such as IBA and NAA were added to ½ MS and ½ DKW media to induce rooting. The rooting rate was better in half-concentration media than in full-concentration media. In the case of PGRs, treatment with IBA alone showed a higher rooting rate than treatment without PGRs or with a combination of NAA and IBA. In Stephen et al. [17], the result was contrary to the result that the control group and 0.05 μM IBA were the best. However, according to the study in Lata et al. [41], when 2.5 μM of IBA was applied to the rooting medium of the *Cannabis sativa* 'MX', a 95% rooting rate was achieved. This is the same IBA concentration as the best treatment in this study: ½ DKW with IBA 0.5 mg/L. It is believed that the concentration of hormones required varies depending on the cultivar. Through this study, it was confirmed that IBA induces high rooting in *Cannabis sativa*, although the cultivars are different. The fact that the DKW medium showed a better rooting rate than the MS medium is presumed to be due to the same factors affecting shoot induction. Therefore, further research is needed on the factors of DKW medium that affect rooting rate of *Cannabis sativa* 'Cheungsam'.

5. Conclusions

Cannabis exists as many genetically diverse cultivars. To establish the optimal in vitro micropropagation protocol, a series of experiments were conducted focusing on 'Cheungsam' cultivars. In conclusion, for proliferation, the best shoots were obtained in DKW medium containing 1.0 mg/L 2iP, and shoot elongation was achieved when subcultured in the same medium. Subsequently, when cultured in ½ DKW medium containing 0.5 mg/L IBA, the rooting rate was highest, and healthy roots were obtained. When the rooted plants were acclimatized, the acclimatization rate was 100%. Therefore, this protocol can increase the rate at which stable populations can be maintained through in vitro mass production of *Cannabis sativa* 'Cheungsam'.

Author Contributions: Designed the study, S.-C.B. and B.-H.B.; conducted the experiments, S.-C.B., S.-Y.J. and Y.-J.C.; wrote and revised the manuscript, S.-C.B.; project administration, D.-H.K., G.-R.Y., D.-W.L. and H.K.; funding acquisition, D.-H.K., G.-R.Y., D.-W.L. and H.K.; supervised the project, D.-W.L. All authors have read and agreed to the published version of the manuscript.

Funding: This research was supported by a grant of the Korea Health Technology R&D Project through the Korea Health Industry Development Institute (KHIDI), funded by the Ministry of Health & Welfare, Republic of Korea (grant number: HF23C0112).

Data Availability Statement: The original contributions presented in the study are included in the article, further inquiries can be directed to the corresponding author.

Conflicts of Interest: All authors were employed by the company TOPO Lab., Co., Ltd. The authors declare that the research was conducted in the absence of any commercial or financial relationships that could be construed as a potential conflict of interest.

References

1. Booth, J.K.; Bohlmann, J. Terpenes in *Cannabis sativa*—From plant genome to humans. *Plant Sci.* **2019**, *284*, 67–72. [CrossRef] [PubMed]
2. Kisková, T.; Mungenast, F.; Suváková, M.; Jäger, W.; Thalhammer, T. Future aspects for cannabinoids in breast cancer therapy. *Int. J. Mol. Sci.* **2019**, *20*, 1673. [CrossRef] [PubMed]
3. Andre, C.M.; Hausman, J.-F.; Guerriero, G. *Cannabis sativa*: The plant of the thousand and one molecules. *Front. Plant Sci.* **2016**, *7*, 19. [CrossRef] [PubMed]
4. Li, H.-L. An archaeological and historical account of cannabis in China. *Econ. Bot.* **1974**, *28*, 437–448. [CrossRef]
5. Kim, H.-S.; Shin, M.-J.; Pan, Y.-H. A study on the improvement of cannabis production history management in Korea-focused on Gyeongbuk hemp regulation free zone. *J. Korea Converg. Soc.* **2022**, *13*, 249–259. [CrossRef]
6. Coppess, J.; Schnitkey, G.; Zulauf, C.; Paulson, N.; Gramig, B.; Swanson, K. The agriculture improvement act of 2018: Initial Review. *Farmdoc Dly.* **2018**, *8*, 227.
7. Romero, P.; Peris, A.; Vergara, K.; Matus, J.T. Comprehending and improving cannabis specialized metabolism in the systems biology era. *Plant Sci.* **2020**, *298*, 110571. [CrossRef]
8. Bonini, S.A.; Premoli, M.; Tambaro, S.; Kumar, A.; Maccarinelli, G.; Memo, M.; Mastinu, A. Cannabis sativa: A comprehensive ethnopharmacological review of a medicinal plant with a long history. *J. Ethnopharmacol.* **2018**, *227*, 300–315. [CrossRef]
9. Ligresti, A.; Petrosino, S.; Di Marzo, V. From endocannabinoid profiling to 'endocannabinoid therapeutics'. *Curr. Opin. Chem. Biol.* **2009**, *13*, 321–331. [CrossRef]
10. Monthony, A.S.; Page, S.R.; Hesami, M.; Jones, A.M.P. The past, present and future of *Cannabis sativa* tissue culture. *Plants* **2021**, *10*, 185. [CrossRef]
11. Scoggins, H.; Bridgen, M. *Plants from Test Tubes: An Introduction to Micropropogation*, 4th ed.; Timber Press: Portland, OR, USA, 2013; p. 270.
12. Yang, R.; Berthold, E.C.; McCurdy, C.R.; da Silva Benevenute, S.; Brym, Z.T.; Freeman, J.H. Development of cannabinoids in flowers of industrial hemp (*Cannabis sativa* L.): A pilot study. *J. Agric. Food Chem.* **2020**, *68*, 6058–6064. [CrossRef] [PubMed]
13. Punja, Z.K.; Rodriguez, G.; Chen, S. Assessing genetic diversity in Cannabis sativa using molecular approaches. In *Cannabis sativa L. Botany and Biotechnology*; Springer: Cham, Switzerland, 2017; pp. 395–418.
14. Warren, J.; Mercado, J.; Grace, D. Occurrence of hop latent viroid causing disease in *Cannabis sativa* in California. *Plant Dis.* **2019**, *103*, 2699. [CrossRef]
15. Wróbel, T.; Dreger, M.; Wielgus, K.; Słomski, R. Modified nodal cuttings and shoot tips protocol for rapid regeneration of *Cannabis sativa* L. *J. Nat. Fibers* **2022**, *19*, 536–545. [CrossRef]
16. Page, S.R.; Monthony, A.S.; Jones, A.M.P. DKW basal salts improve micropropagation and callogenesis compared with MS basal salts in multiple commercial cultivars of *Cannabis sativa*. *Botany* **2021**, *99*, 269–279. [CrossRef]
17. Stephen, C.; Zayas, V.A.; Galic, A.; Bridgen, M.P. Micropropagation of hemp (*Cannabis sativa* L.). *HortScience* **2023**, *58*, 307–316. [CrossRef]
18. Iiyama, C.M.; Cardoso, J.C. MicropropagatioAvailable online: n of *Melaleuca alternifolia* by shoot proliferation from apical segments. *Trees* **2021**, *35*, 1497–1509. [CrossRef]
19. Ioannidis, K.; Dadiotis, E.; Mitsis, V.; Melliou, E.; Magiatis, P. Biotechnological approaches on two high CBD and CBG *Cannabis sativa* L.(Cannabaceae) varieties: In vitro regeneration and phytochemical consistency evaluation of micropropagated plants using quantitative 1H-NMR. *Molecules* **2020**, *25*, 5928. [CrossRef]
20. Iannaccone, M.; Di Santo, P.; Buhagiar, J.A.; Paura, B.; Cocozza, C. Enhancement of Sprouting and Rooting of Quercus Pubescens by Benzylaminopurine and Indole-Butyric Acid in Micropropagation. *Fresenius Environ. Bull.* **2020**, *29*, 10287–10293.
21. Miri, S.M. Micropropagation, callus induction and regeneration of ginger (*Zingiber officinale* Rosc.). *Open Agric.* **2020**, *5*, 75–84. [CrossRef]
22. Rokosa, M.T.; Kulpa, D. Micropropagation of *Stevia rebaudiana* plants. *Ciência Rural.* **2019**, *50*, e20181029. [CrossRef]
23. Mestinšek-Mubi, Š.; Svetik, S.; Flajšman, M.; Murovec, J. In vitro tissue culture and genetic analysis of two high-CBD medical cannabis (*Cannabis sativa* L.) breeding lines. *Genetika* **2020**, *52*, 925–941. [CrossRef]
24. Boonsnongcheep, P.; Pongkitwitoon, B. Factors affecting micropropagation of *Cannabis sativa* L.: A review. *Pharm. Sci. Asia* **2020**, *47*, 21–29. [CrossRef]

25. Codesido, V.; Meyer, S.; Casano, S. Influence of media composition and genotype for successful *Cannabis sativa* L. in vitro introduction. *Acta Hortic.* **2020**, *1285*, 75–80. [CrossRef]
26. Lata, H.; Chandra, S.; Techen, N.; Khan, I.A.; ElSohly, M.A. In vitro mass propagation of Cannabis sativa L.: A protocol refinement using novel aromatic cytokinin meta-topolin and the assessment of eco-physiological, biochemical and genetic fidelity of micropropagated plants. *J. Appl. Res. Med. Aromat. Plants* **2016**, *3*, 18–26. [CrossRef]
27. Polivanova, O.B.; Bedarev, V.A. Hyperhydricity in plant tissue culture. *Plants* **2022**, *11*, 3313. [CrossRef] [PubMed]
28. Kevers, C.; Franck, T.; Strasser, R.J.; Dommes, J.; Gaspar, T. Hyperhydricity of micropropagated shoots: A typically stress-induced change of physiological state. *Plant Cell Tissue Organ Cult.* **2004**, *77*, 181–191. [CrossRef]
29. Gaspar, T.; Kevers, C.; Franck, T.; Bisbis, B.; Billard, J.-P.; Huault, C.; Le Dily, F.; Petit-Paly, G.; Rideau, M.; Penel, C. Paradoxical results in the analysis of hyperhydric tissues considered as being under stress: Questions for a debate. *Bulg. J. Plant. Physiol.* **1995**, *21*, 80–97.
30. Li, T.; Yun, Z.; Zhang, D.; Yang, C.; Zhu, H.; Jiang, Y.; Duan, X. Proteomic analysis of differentially expressed proteins involved in ethylene-induced chilling tolerance in harvested banana fruit. *Front. Plant Sci.* **2015**, *6*, 845. [CrossRef]
31. Wi, S.J.; Jang, S.J.; Park, K.Y. Inhibition of biphasic ethylene production enhances tolerance to abiotic stress by reducing the accumulation of reactive oxygen species in *Nicotiana tabacum*. *Mol. Cells* **2010**, *30*, 37–49. [CrossRef]
32. Thomas, T.D. The role of activated charcoal in plant tissue culture. *Biotechnol. Adv.* **2008**, *26*, 618–631. [CrossRef]
33. Lee, Y.-C.; Chang, J.-C. Development of an improved micropropagation protocol for red-fleshed pitaya 'Da Hong'with and without activated charcoal and plant growth regulator combinations. *Horticulturae* **2022**, *8*, 104. [CrossRef]
34. Holmes, J.E.; Lung, S.; Collyer, D.; Punja, Z.K. Variables affecting shoot growth and plantlet recovery in tissue cultures of drug-type *Cannabis sativa* L. *Front. Plant Sci.* **2021**, *12*, 732344. [CrossRef] [PubMed]
35. Komalavalli, N.; Rao, M.V. In vitro micropropagation of Gymnema sylvestre—A multipurpose medicinal plant. *Plant Cell Tissue Organ Cult.* **2000**, *61*, 97–105. [CrossRef]
36. Tivarekar, S.; Eapen, S. High frequency plant regeneration from immature cotyledons of mungbean. *Plant Cell Tissue Organ Cult.* **2001**, *66*, 227–230. [CrossRef]
37. Driver, J.A.; Kuniyuki, A.H. In vitro propagation of Paradox walnut rootstock. *HortScience* **1984**, *19*, 507–509. [CrossRef]
38. Phillips, G.C.; Garda, M. Plant tissue culture media and practices: An overview. *In Vitro Cell. Dev. Biol. Plant* **2019**, *55*, 242–257. [CrossRef]
39. El-Dawayati, M.M.; Zayed, Z.E. Controlling hyperhydricity in date palm in vitro culture by reduced concentration of nitrate nutrients. In *Date Palm Biotechnology Protocols Volume I: Tissue Culture Applications*; Springer: Berlin/Heidelberg, Germany, 2017; Volume 1637, pp. 175–183.
40. Liu, M.; Jiang, F.; Kong, X.; Tian, J.; Wu, Z.; Wu, Z. Effects of multiple factors on hyperhydricity of *Allium sativum* L. *Sci. Hortic.* **2017**, *217*, 285–296. [CrossRef]
41. Lata, H.; Chandra, S.; Khan, I.A.; El Sohly, M.A. High frequency plant regeneration from leaf derived callus of high Δ9-tetrahydrocannabinol yielding *Cannabis sativa* L. *Planta Med.* **2010**, *76*, 1629–1633. [CrossRef]

horticulturae

MDPI

Article

An Efficient In Vitro Shoot Organogenesis and Comparative GC-MS Metabolite Profiling of *Gaillardia pulchella* Foug

Mahima Bansal [1], A. Mujib [1,*], Yashika Bansal [1], Yaser Hassan Dewir [2] and Nóra Mendler-Drienyovszki [3]

[1] Cellular Differentiation and Molecular Genetics Section, Department of Botany, Jamia Hamdard, New Delhi 110062, India; bansalmahima617@gmail.com (M.B.); yashikab333@gmail.com (Y.B.)
[2] Plant Production Department, College of Food and Agriculture Sciences, King Saud University, Riyadh 11451, Saudi Arabia; ydewir@ksu.edu.sa
[3] Research Institute of Nyíregyháza, Institutes for Agricultural Research and Educational Farm (IAREF), University of Debrecen, P.O. Box 12, 4400 Nyíregyháza, Hungary; mendlerne@agr.unideb.hu
* Correspondence: amujib3@yahoo.co.in

Abstract: *Gaillardia pulchella* Foug. is a widely studied plant because of its high pharmacological and ornamental value. The leaves of *G. pulchella* were used for inducing callus and subsequent plant regeneration as it is the primary source of phytocompounds. The purpose of the present investigation was to formulate an in vitro propagation method for *Gaillardia* by using leaf explants in MS (Murashige and Skoog) medium. The best callus induction was observed on high (2.0 mg/L) α-naphthalene acetic acid (NAA) and a low (0.5 mg/L) 6-benzylaminopurine (BAP) with callus induction frequency of 91.66%. The leaf callus also demonstrated high caulogenesis ability (95.83%), with an average 5.2 shoots/callus mass at 0.5 mg/L BAP and 2.0 mg/L NAA. Indole Acetic acid (IAA) at 1.0 mg/L had the maximum rooting percentage (79.17%) with 12.4 roots per shoot. Rooted plantlets were later transferred to greenhouse conditions, showing a survivability rate of 75–80%. The physiological parameters, i.e., phenolic compounds and the flavonoids' level, in the DPPH assay were higher in leaves obtained in vitro compared to callus formed from leaves and field-obtained (mother) leaves. Gas chromatography–mass spectrometry (GC–MS) analysis of methanol extracts of leaves (in vivo and in vitro) and leaf callus presented a wide array of compounds. In callus extract, some 34 phytocompounds were identified. Some of them were 3-hydroxy-2,3-dihydromaltol (25.39%), isoamyl acetate (11.63%), palmitic acid (11.55%), 4-methyloxazole (7.54%), and 5-methoxypyrrolidin-2-one (7.49%). Leaves derived in vivo and in vitro had 45 and 28 phytocompounds, respectively, belonging to different classes like lignans, phenols, terpenoids, alkaloids and fatty acids, etc. Those findings demonstrated that the leaf derived callus and the leaves are the potential stable source of several compounds with medicinal importance. The developed protocol may provide an alternative source of compounds without affecting wild flora.

Keywords: indirect organogenesis; biochemical; antioxidant; phytochemical fingerprinting; chromatography

1. Introduction

Gaillardia pulchella Foug is an ornamental, short-lived, annual, flowering plant. It is commonly known as Indian blanket flower, sundance, or fire wheel [1]. *G. pulchella* is an indigenous species to North and South America and belongs to the Asteraceae family [2]. It is cultivated in several parts of India as one of the major commercial flower crops [3]. This plant thrives in dry, sunny regions and can spread to disrupted habitat [4]. Its easy cultivation, high tolerance to a variety of soil types and climates, extended flowering period, and vibrant colors (red, yellow, scarlet) make it one of the most popular flowers in India [5].

Due to the presence of a wide array of phytocompounds such as tannins, flavonoids, terpenoids, sterols, amino acids, polysaccharides, etc., with potential pharmacological uses, it is regarded as a therapeutically important plant [6]. *G. pulchella* has antitumoral, antiparasitic, chemotaxonomic indicator and cytotoxic properties [7]. Among several reported phytocompounds, pulchelloid A (sesquiterpene lactone) is a characteristic compound found in *Gaillardia* spp., which possess anti-mitotic activity [8]. Because of the immense phytocompounds present, classical breeding methods of *Gaillardia* spp. have recently been replaced by advanced biotechnological approaches [9]. Plant tissue culture is used for fast propagation of plants under a sterile environment, without harming natural vegetation [10]. It offers a number of advantages over conventional techniques like fast mass propagation, genetic transformation, procurement of microbe free healthy plants, important phyto-compounds production, preservation of germplasm. [11,12]. Organogenesis (via direct or indirect pathway) is one of the methods by which plant parts like roots, shoots, and even flowers can be generated from cultured explants [13,14].

The regular sub-culturing, genetic makeup, extended exposure to plant growth regulators (PGRs), adverse physical environments like humidity and photoperiod are some of the factors that alter physio-biochemical attributes of cultured plantlets [15]. The biochemical changes may affect phenolic compounds' content, sugar level, flavonoid content, and excess accumulation of reactive oxygen species (ROS) [16]. Thus, such biochemical characteristics and other related phytochemical composition of in vitro raised tissues need regular monitoring. Gas chromatography–mass spectrometry (GC-MS) requires a small quantity of plant extracts and is one of the best, fastest, and most accurate methods for detecting a wide range of phytochemicals, including alcohols, alkaloids, nitro compounds, long chain hydrocarbons, organic acids, steroids, esters, and amino acids [17]. GC-MS investigation has recently been conducted in several medicinally important plants like *Tecoma stans*, *Pluchea lanceolate*, and others [18,19].

To date, there is only one report available regarding in vitro regeneration of *G. pulchella*, in which authors optimized PGRs level in bud breaking response [20]. This study, therefore, describes for the first time an efficient in vitro plant regeneration method via indirect organogenesis from leaf derivedcallus in *G. pulchella*. The current study describes and compares phytochemicals and biochemical attributes present in different tissues (leaf callus, in vivo and in vitro leaves) of *G. pulchella*. It enables the pharmaceutical industry to use in vitro cultivation technologies to obtain a steady supply of therapeutically significant bioactive compounds.

2. Materials and Methods

2.1. Plant Material and Surface Sterilization

Leaves of *Gaillardia pulchella* Foug were procured from the Jamia Hamdard herbal garden (identification number: JH/BOT/DAC/2023/01) and were used as a primary source of explants. Leaves were cut into pieces 1.5–2.0 cm in length and their surface disinfected by following the protocol described earlier [21] with some modifications. The excised leaves were washed with running tap water, later rinsed in Teepol B-300 (Reckitt Benkiser, Gurugram, India) (a liquid detergent) for 30 min, and stirred in running tap water for 4 to 5 min to physically eradicate microorganisms. The explants were dipped in 70% ethanol for 2–3 min, and finally washed with sterilized double-distilled water to reduce the adverse effect of ethanol. The cut leaves were later treated with 0.1% mercuric chloride ($HgCl_2$), for 2–3 min, and washed with sterilized double-distilled water 3–4 times to remove any traces of sterilizing solution.

2.2. Induction of Callus and Culture Conditions

Callus was induced from excised leaves (1.5–2.0 cm) when inoculated on MS (Murashige and Skoog) medium [22] comprising 3% sucrose as a carbon source and 0.7% agar (*w*/*v*) as a gelling agent. The medium was additionally supplemented with 0.5–2.0 mg/L of auxins [α-naphthalene acetic acid (NAA), 2,4-dichlorophenoxyactic acid (2,4-D)] and cytokinin [6-benzylamino purine (BAP)]. The medium pH was adjusted to 5.6 to 5.7 with 0.1 N sodium hydroxide (NaOH) or1 N hydrochloric acid (HCl), prior to autoclaving. The test tubes were kept in an incubation room with the temperature at 25 ± 2 °C, 55% relative humidity (RH), 60 μmolm^{-2} s^{-1} light intensity and 16/8 h photoperiod provided by producer bulbs. In control, no PGR was added to the medium. After four weeks, the calli underwent regular subculturing on the same PGR containing media. The callus formation frequency (%) and callus biomass (g) were defined after 4 weeks of culture.

The following formula was used:

$$\text{Callus induction frequency (\%)} = \text{number of explants producing callus/total number of explants cultured} \times 100.$$

2.3. *Indirect Shoot Organogenesis and Proliferation*

After 3–4 weeks, the leaf-derived calli were placed into 0.5–1.0 mg/L BAP and 0.5–2.0 mg/L NAA containing MS for shoot induction. The shoots started to appear, and the shoot forming ability (%) and the average shoot number/callus mass were estimated after four weeks of incubation. The regenerated shoots were placed in a medium containing either 0.5–2.0 mg/L BAP or kinetin (Kn) for more proliferation of shoots. After five weeks, the average length of the regenerated shoots and shoot proliferation percentage (total number of shoots emerged/total explants × 100) per treatment were counted.

2.4. *Induction of Roots and Acclimatization*

The in vitro regenerated shoots measuring about 3–5 cm in length were removed and placed on a root-inducing MS medium containing different indole-3-acetic-acid (IAA),indole-3-butyric-acid (IBA)orNAA concentrations. After three weeks of culture, the average number of roots per shoot and the root induction frequency (%) were recorded. After washing with autoclaved double-distilled water to remove the adherent culture media, the rooted plantlets were transferred to plastic pots filled with an autoclaved 1:1:1 mixture of sand, soil, and soilrite. The potted plants were covered with clear plastic bags and kept in an incubation room at 25 ± 2 °C temperature, 70 ± 10% RH and 60 μmol m^{-2} s^{-1} light intensity, and were finally transferred to outdoor conditions.

2.5. *Preparation of Extracts*

The in vivo and in vitro grown leaves of *G. pulchella* and the leaf-derived callus were collected, and dried in the shade for three days at room temperature, and then 1.0 g samples (dry weight) were crushed into a fine powder using a mortar and pestle. Each sample was then separately extracted with 10 mL of methanol solvent in a rotary shaker for 48 h. Then, Whatman No. 1 filter paper was used to filter the methanolic extracts. The filtered samples were centrifuged at 10,000 rpm for five min, and the supernatant was kept at 4 °C before use.

2.6. *Gas Chromatography–Mass Spectrometry Analysis*

The GC–MS analysis of tissues (callus, in vivo and in vitro leaves) was performed with the use of a GC-MS-QP-2010 apparatus (Shimadzu, Tokyo, Japan) with the following program specifications: an oven temperature 100 °C for 3 min and then gradually increased to 300 °C, for 17 min. Helium was given as carrier gas at a continuous flow of 1.21 mL/min; the injection temperature was kept at 260 °C. For separation of compounds, Rxi-5Sil MS GC Capillary Column, 30 m, 0.25 mm ID, 0.25 μm df was used. Operating temperatures for the ion source and interface were 220 °C and 270 °C, respectively, with a 2.5 min solvent cut time and a GC–MS running time for every sample was 35 min. The GCMS solution software (Version 4.45 SP 1) was used to identify the bioactive compounds present in each sample by analyzing retention index, peak area, and area % with earlier identified phytocompounds and consulting mass spectral database with the National Institute of Standards and Technology (NIST) library.

2.7. Biochemical Parameters

2.7.1. Determination of Total Phenol Content

The Folin–Ciocalteu method [23] was used to determine the total phenolic content (TPC) of the extract. The tissue extract (0.5 mL) was thoroughly mixed with 2.5 mL (10%, v/v) of the Folin–Ciocalteu reagent (Sigma-Aldrich, New York, NY, USA). The mixture was kept at room temperature for five min; then, 2.0 mL sodium carbonate (7%) was added and the mixture was kept for 90 min. The absorbance was read at 765 nm against the extract-free blank using a spectrophotometer (BL-295 Biolinkk, Delhi, India). The calibration curve equation for the standard gallic acid solution was used to calculate the total phenolic content, which was measured in triplicate. The outcome information was presented as mg GAE/g Dw.

2.7.2. Determination of Flavonoid Level

The procedure described by Aryal et al. [24] was followed in order to measure the total flavonoid content (TFC). Firstly, 1.0 mL of extract solutions was mixed with 10% aluminum chloride (0.2 mL) and 1 M potassium acetate solution (0.2 mL). After an incubation time of 30 min at room temperature, 3.6 mL of distilled water was added to the final reaction volume, which was 5.0 mL. After mixing the solution, each sample's absorbance at 415 nm was measured in relation to a blank. There were three repetitions of the measurement made. Different quantities of quercetin were plotted against corresponding absorbances on a standard graph. The TFC of the samples was presented in mg quercetin per g dry weight (mg QE/g DW).

2.7.3. DPPH Assay for Scavenging Activity

With the help of stable radical 2,2-diphenyl-1-picrylhydrazyl (DPPH), the free radical scavenging activity (FRSA) of *G. pulchella* sample extract was determined following the method described by Baliyan et al. [23]. DPPH (0.024%, w/v) 3.0 mL and 0.1 mL methanol was mixed and was used as standards, and 0.1 mL of extract solutions were immediately added to each test tube. Following that, the samples were incubated for 90 min at room temperature in complete darkness. The absorbance was measured at 517 nm. The following formula [10] was used to assess the antioxidant capacity of each sample: $AC - AS/AC \times 100$. AC is control absorbance and AS is sample absorbance, giving the scavenging activity percentage.

2.8. Statistical Analysis

A completely randomized design (CRD) was used for in vitro trials with eight explants per PGR treatment, conducted in triplicates. The biochemical analyses were also performed in triplicates. The mean $\pm$ standard error was used to express the data. The differences in the samples' means were estimated or calculated with Duncan's Multiple Range Test (DMRT) at $p \leq 0.05$.

3. Results

3.1. Callus Induction

The leaves of *G. pulchella* were used as explants for inducing callus (Figure 1A). White friable callus began to emerge from the leaf after two weeks of culturing (Figure 1B). The control treatment lacking PGRs failed to produce any callus; the formation of callus was observed in all the media supplemented with PGRs. The NAA and BAP combinations (2.0 mg/L + 0.5 mg/L and 0.5 mg/L + 0.5 mg/L) were found to be more effective in inducing callus (91.66–79.17%) compared to treatments containing 2,4-D, which exhibited a callus induction range of 62.50%-8.33% (Table 1 and Figure 1C). The highest number of leaves produced callus (91.66%) on BAP (0.5 mg/L) and NAA (2.0 mg/L) fortified medium. The maximum callus biomass (3.8 g/explant) was achieved on this very medium. In contrast, very little callusing (8.33%) and biomass (0.9 g/explant) were observed on 2.0 mg/L 2,4-D.

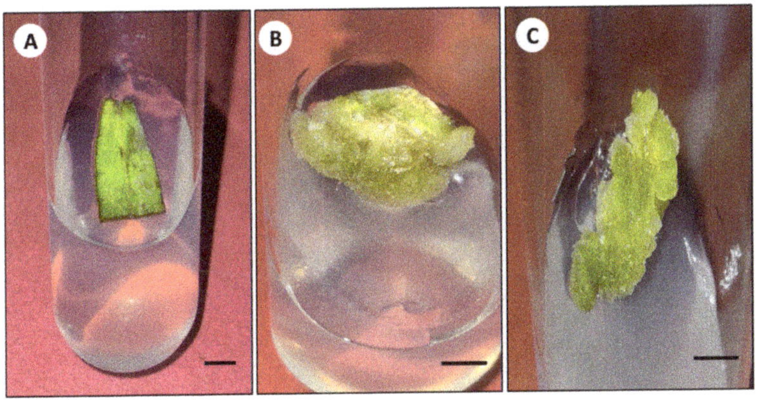

Figure 1. Callus induction and proliferation from leaf explant of *G.pulchella*. (**A**) Leaf explant inoculated on PGR (2.0 mg/L NAA and 0.5 mg/L BAP) containing MS medium. (**B**) Callus initiation after 2-week period. (**C**) Callus proliferation after 4-week period (bars (**A**) = 1.0 cm, (**B**,**C**) = 0.5 cm).

Table 1. Effects of various PGR's concentrations on callus induction from leaves in *G. pulchella*.

PGRs	Concentration (mg/L)	Callus Induction Percentage (%)	Callus Biomass (g)
Control	0	0^{e}	0^{d}
NAA + BAP	0.5 + 0.5	79.17 ± 4.17^{ab}	3.3 ± 0.5^{a}
	0.5 + 1.0	58.33 ± 8.33^{c}	2.6 ± 0.4^{ab}
	1.0 + 2.0	54.17 ± 4.17^{c}	2.7 ± 0.3^{ab}
	2.0 + 0.5	91.66 ± 8.33^{a}	3.8 ± 0.8^{a}
2,4-D	0.5	62.50 ± 7.22^{bc}	2.9 ± 0.4^{ab}
	1.0	45.83 ± 4.16^{cd}	1.8 ± 0.2^{bc}
	1.5	33.33 ± 11.02^{d}	1.1 ± 0.2^{c}
	2.0	8.33 ± 4.16^{e}	0.9 ± 0.1^{c}

Values represent means ± S.E. of eight replicates of three experiments. Within each column, means followed by different letters are significantly different at $p \leq 0.05$ according to DMRT. The superscripts represent ranking, i.e., 'a' indicates the best response and 'e/d' indicates the worst response.

3.2. Indirect Shoot Organogenesis and Proliferation

The MS medium without PGRs was ineffective for shoot induction, but the inclusion of NAA and BAP in media successfully transformed the callus into an organogenic one (Figure 2A), which induced shoots from the calli. Friable callus on the MS medium containing NAA (0.5–2.0 mg/L) and BAP (0.5–2.0 mg/L) induced a shoot (indirect) within 3–5 weeks' time (Figure 2B,C). Among various tested BAP and NAA levels, the 0.5 mg/L BAP and 2.0 mg/L NAA was the most efficient treatment for producing maximum numbers of shoots. The callus showed a shoot regeneration ability of 95.83%, with an average 5.2 shoot numbers/callus mass at 0.5 mg/L BAP and 2.0 mg/L NAA concentration (Table 2). At higher BAP level, a significant reduction in shoot number was noted. The lowest shoot formation rate (33.33%) with an average 1.8 shoot number/callus mass was recorded on BAP (1.0 mg/L) and NAA (2.0 mg/L) containing an MS medium.

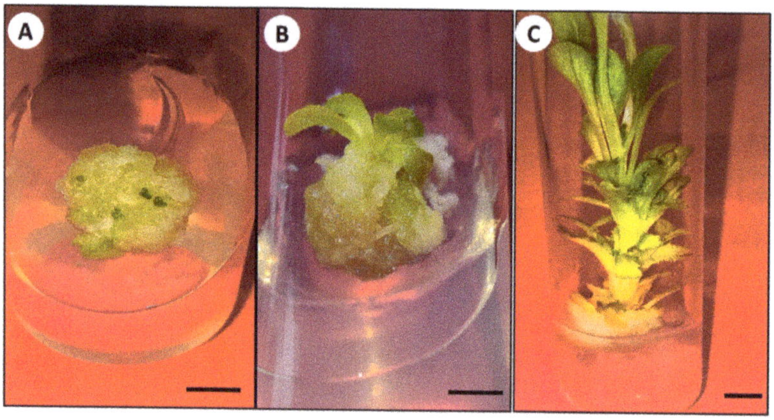

Figure 2. Indirect shoot organogenesis from callus of *G. pulchella* on MS media supplemented with 2.0 mg/L NAA and 0.5 mg/L BAP. (**A**) Organogenic callus. (**B**) Indirect shoot formation. (**C**) Elongated shoot after 6 weeks of culture (bars (**A**,**B**) = 0.5 cm, (**C**) = 1.0 cm).

Table 2. Influence of BAP and NAA on callus shoot formation in *G. pulchella*.

PGRs	Concentration (mg/L)	Shoot Regeneration Percentage (%)	Shoot Numbers/Callus Mass
Control	0	0^{d}	0^{c}
BAP + NAA	0.5 + 1.0	79.17 ± 11.02^{ab}	4.9 ± 0.8^{a}
	0.5 + 2.0	95.83 ± 4.16^{a}	5.2 ± 0.5^{a}
	1.0 + 0.5	54.17 ± 11.02^{bc}	4.1 ± 0.2^{a}
	1.0 + 1.5	37.50 ± 7.21^{c}	2.6 ± 0.3^{b}
	1.0 + 2.0	33.33 ± 11.02^{c}	1.8 ± 0.4^{b}

Values represent mean ± S.E. of eight replicates of three experiments. Within each column, means followed by different letters are significantly different at $p \leq 0.05$ according to DMRT. The superscripts represent ranking, i.e., 'a' indicates the best response and 'c/d' indicates worst response.

3.3. *Induction of Root and Acclimatization*

The regenerated shoots were subsequently transferred to a root induction medium. Two auxins, namely IBA and IAA, were added to the MS medium at three different concentrations (0.5, 0.75, and 1.0 mg/L) in order to achieve rooting of regenerants. With the exception of control, in every rooting media, the roots were formed at the ends of the shoots. In terms of average root numbers per shoot and root induction percentage, IAA treatments outperformed IBA. Furthermore, 1.0 mg/L IAA caused the maximum rooting percentage (79.17%) with 12.4 roots per shoot (Figure 3A), whereas 0.75 mg/L and 1.0 mg/L IBA allowed us to obtain the lowest rooting percentage with mean number of roots per shoot (Table 3). Plants treated with IAA formed longer roots, while shoots produced in IBA formed thin and lower numbers of roots. Healthy plantlets were later transferred to a greenhouse for acclimatization and had a survivability rate of 75–80% (Figure 3B).

Figure 3. Root induction and transfer of in vitro derived shootlets of *G. pulchella*, (**A**) development of roots in 1.0 mg/L IAA fortified MS medium, (**B**) potted plants in outdoor (bars (**A**) = 1.0 cm, (**B**) = 2.5 cm).

Table 3. Effect of IBA and IAA on rooting of in vitro developed *G. pulchella* shoots. The data were scored after 3 weeks of cultivation.

PGRs	Concentration (mg/L)	Root Induction Frequency (%)	Root Numbers/Shoot
Control	0	0^{d}	0^{e}
IAA	0.5	54.16 ± 11.02^{abc}	9.7 ± 0.4^{bc}
	0.75	70.83 ± 15.02^{ab}	10.6 ± 0.6^{b}
	1	79.17 ± 8.33^{a}	12.4 ± 0.2^{a}
IBA	0.5	41.66 ± 4.16^{bc}	8.4 ± 0.4^{c}
	0.75	29.16 ± 11.02^{c}	6.1 ± 0.5^{d}
	1	29.17 ± 4.17^{c}	5.8 ± 0.5^{d}

Values represent mean ± S.E. of eight replicates of three experiments. Within each column, means followed by different letters are significantly different at $p < 0.05$ according to DMRT. The superscripts represent ranking, i.e., 'a' indicates the best response and 'e/d' indicates worst response.

3.4. GC-MS Analysis

The phytocompounds present in methanolic extracts of in vivo and in vitro grown leaves, and callus of *G. pulchella* were identified by using GC–MS. The phytocompounds with their retention time (RT), retention area (%),molecular formula and molecular weight are presented in Tables 4–6. The NIST library consulted and the GC–MS chromatograms are presented in Figures 4–6. The phytochemical composition of *G. pulchella's* leaf-derived callus was identified and examined. Upon analyzing the methanolic callus extract, 34 phytocompounds were identified (Table 4 and Figure 4), several of which were found in trace amounts. Compounds like 3-hydroxy-2,3-dihydromaltol (25.39%), isoamyl acetate (11.63%), palmitic acid (11.55%), 4-methyloxazole (7.54%), 5-methoxypyrrolidin-2-one (7.49%) were observed as the major phytocompounds present in higher quantities.

Table 4. Phytocompounds identified in leaf-derived callus extract of *G. pulchella* through GC-MS.

Peak#	R. Time (min)	Area%	Name	Molecular Formula	Molecular Weight
1	4.160	3.38	pyranone	$C_6H_8O_4$	144
2	4.552	7.54	4-methyloxazole	C_4H_5NO	83
3	5.103	1.93	levulinic acid	$C_5H_8O_3$	116
4	5.311	0.05	2,5-dimethyl-3(2h) furanone	$C_6H_8O_3$	128
5	5.675	3.28	cyanuramide	$C_3H_6N_6$	126
6	5.910	4.03	1-aminocyclopropane-1-carboxylic acid	$C_4H_7NO_2$	101
7	6.344	0.52	3-methyloctane	C_9H_{20}	128
8	6.645	25.39	3-hydroxy-2,3-dihydromaltol	$C_6H_8O_4$	144
9	6.837	0.69	di-2-ethylhexyl amine	$C_{16}H_{35}N$	241
10	6.957	0.69	neopentyl acetate	$C_7H_{14}O_2$	130
11	7.172	7.49	5-methoxypyrrolidin-2-one	$C_5H_9NO_2$	115
12	7.843	11.63	isoamyl acetate	$C_7H_{14}O_2$	130
13	8.159	1.56	glycerol 1-acetate	$C_5H_{10}O_4$	134
14	8.749	1.04	2-methylpentyl propyl ether	$C_9H_{20}O$	144
15	8.975	0.38	1-methylpiperidine	$C_6H_{13}N$	99
16	10.183	1.03	bis(2-ethylhexyl) amine	$C_{16}H_{35}N$	241
17	11.032	3.39	di-sec-butyl oxalate	$C_{10}H_{18}O_4$	202
18	11.703	0.77	1,2,3,5-cyclohexanetetrol	$C_6H_{12}O_4$	148
19	12.696	0.73	2-aminobenzoyl azide	$C_7H_6N_4O$	162
20	13.165	1.65	methyl hexopyranoside	$C_7H_{14}O_6$	194
21	13.640	0.47	santalol	$C_{15}H_{24}O$	220
22	14.510	0.94	caprinic acid	$C_{10}H_{20}O_2$	172
23	15.238	0.34	(−)-cis-pinan	$C_{10}H_{18}$	138
24	15.729	0.33	(s)-(+)-3-methyl-1-pentanol	$C_6H_{14}O$	102
25	16.590	11.55	palmitic acid	$C_{16}H_{32}O_2$	256
26	18.215	0.79	cyclooctene, 3-methyl-	C_9H_{16}	124
27	18.467	0.56	cyclohexyl laurate	$C_{18}H_{34}O_2$	282
28	19.473	0.25	bis(2-(dimethylamino)ethyl) ether	$C_8H_{20}N_2O$	160
29	20.493	1.11	neophytadiene	$C_{20}H_{38}$	278
30	21.205	1.18	cis-z-.alpha.-bisabolene epoxide	$C_{15}H_{24}O$	220
31	21.531	0.76	1,2-benzenedicarboxylic acid, diisooctyl e	$C_{24}H_{38}O_4$	390
32	23.909	2.16	squalene	C30H50	410
33	31.099	1.97	1,16-dichlorohexadecane	$C_{16}H_{32}Cl_2$	294
34	32.623	0.40	6(e),9(z),13(e)-pendectriene	$C_{15}H_{26}$	206

Table 5. GC–MS analysis of in vivo grown plant's leaves of *G. pulchella* showing phytocompounds.

Peak#	Retention Time (min)	Retention Area (%)	Phytocompounds	Formula	Molecular Weight
1	7.811	6.28	Isoamyl Acetate	$C_7H_{14}O_2$	130
2	11.061	0.38	(−)-Globulol	$C_{15}H_{26}O$	222
3	11.307	0.07	4iodobis(Bicyclo [2.2.1])Hexane	$C_{12}H_{17}I$	288
4	11.349	0.25	2-Tridecynyl2,6-Difluorobenzoate	$C_{20}H_{26}F_2O_2$	336
5	11.543	0.11	(3e)-6-Methyl-3,5-Heptadien-2-One	$C_8H_{12}O$	124
6	15.170	0.12	(5e)-3-Methyl-5-Undecene	$C_{12}H_{24}$	168
7	15.239	3.63	3-Methylene-7,11,15-Trimethylhexadec-1-Ene	$C_{20}H_{38}$	278
8	15.494	0.72	Neophytadiene	$C_{20}H_{38}$	278
9	15.688	1.15	3,7,11,15 Tetramethyl-2-Hexadecen-1-Ol	$C_{20}H_{40}O$	296
10	15.800	0.32	1-Hexadecyloctahydro-1h-Indene	$C_{25}H_{48}$	348
11	16.575	0.77	1,4-Cyclohexanediol, (Z)-, Tms Derivative	$C_{16}H_{32}O_2$	256
12	16.798	0.15	Pentadecanal	$C_{15}H_{30}O$	226
13	17.954	3.64	Phytol	$C_{20}H_{40}O$	296
14	19.000	0.12	14-Heptadecenal	$C_{17}H_{32}O$	252

Table 5. *Cont.*

Peak#	Retention Time (min)	Retention Area (%)	Phytocompounds	Formula	Molecular Weight
15	20.299	1.14	Vitamin A Acetate	$C_{22}H_{32}O_2$	328
16	20.554	1.23	2-Methyl-1-Phenyl-1-Butanol	$C_{11}H_{16}O$	164
17	20.613	0.44	6-Epi-Shyobunol	$C_{15}H_{26}O$	222
18	20.808	0.37	Xanthinin	$C_{17}H_{22}O_5$	306
19	20.896	0.45	Limonene Dioxide	$C_{10}H_{16}O_2$	168
20	20.942	1.25	Isoborneol, Allyldimethylsilyl Ether	$C_{15}H_{28}OSi$	252
21	21.187	13.23	Stanolone Acetate	$C_{21}H_{32}O_3$	332
22	21.260	34.53	1-Heptatriacotanol	$C_{37}H_{76}O$	536
23	21.467	0.35	5,8,11-Eicosatrienoic Acid, (Z)-, Tms Derivative	$C_{23}H_{42}O_2Si$	378
24	21.539	0.40	5-Hydroxy-4-Nitro-1-Decalinone	$C_{10}H_{15}NO_4$	213
25	21.719	0.20	Diazoprogesterone	$C_{21}H_{30}N_4$	338
26	22.453	0.48	Cis-Z-.Alpha.-Bisabolene Epoxide	$C_{15}H_{24}O$	220
27	22.657	0.12	Cyclopropanecarboxylic Acid, 1-Amino-	$C_4H_7NO_2$	101
28	22.740	0.95	Periplogenin	$C_{23}H_{34}O_5$	390
29	22.824	0.42	2-Bromo Dodecane	$C_{12}H_{25}Br$	248
30	23.162	12.26	Chrysanin	$C_{20}H_{24}O_5$	344
31	23.503	0.27	1,2-Dibromoethane	$C_2H_4Br_2$	186
32	23.702	0.34	N,N′-Di(2-Propenyl)-1,2,4,5-Tetrazin-3,6-Diamine	$C_8H_{12}N_6$	192
33	23.774	0.77	2-Hydroxy-1,1,10-Trimethyl-6,9-Epidioxy-7-Octalin	$C_{13}H_{20}O_3$	224
34	23.912	2.08	Squalene	$C_{30}H_{50}$	410
35	24.386	0.27	1-Chlorooctadecane	$C_{18}H_{37}Cl$	288
36	24.699	0.38	1-Heptatriacotanol	$C_{37}H_{76}O$	536
37	24.834	1.20	1-Tridecyn-4-Ol	$C_{13}H_{24}O$	196
38	25.287	1.06	Bromocyclohexane	$C_6H_{11}Br$	162
39	27.054	0.29	.Gamma.-Tocopherol	$C_{28}H_{48}O_2$	416
40	28.402	1.87	Vitamin E	$C_{29}H_{50}O_2$	430
41	30.647	0.64	2,2,4, 4-Tetramethyl-3-Pentanone	$C_9H_{18}O$	142
42	30.998	1.00	2,6-Dimethyl-6-Nitro-2-Hepten-4-One	$C_9H_{15}NO_3$	185
43	31.598	2.08	Isocitronellol	$C_{10}H_{20}O$	156
44	32.616	1.28	Sitosterol	$C_{29}H_{50}O$	414
45	32.864	0.97	(6e)-2,4-Dimethyl-2,6-Octadiene	$C_{10}H_{18}$	138

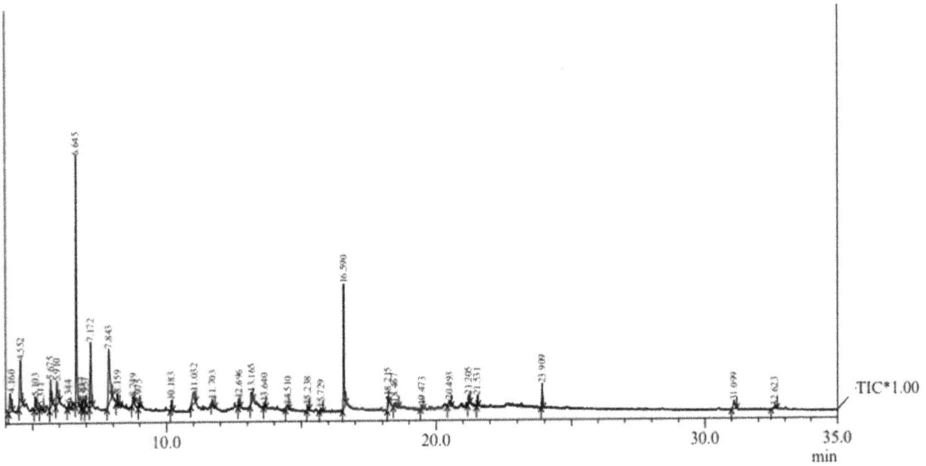

Figure 4. GC-MS chromatogram of methanolic extract of leaf-derived callus extract of *G. pulchella* revealing phytoconstituents with retention time.

Table 6. GC–MS analysis of in vitro grown plant's leaves showing phytocompounds.

Peak#	Retention Time (min)	Retention Area (%)	Phytocompound	Formula	Molecular Weight
1	7.573	1.70	Pyranone	$C_6H_8O_4$	144
2	11.856	15.72	Nitroisobutylglycerol	$C_4H_9NO_5$	151
3	13.636	0.41	1,4-Cyclohexanediol, (Z)-, TMS Derivative	$C_9H_{20}O_2Si$	188
4	16.225	3.20	Phytol, Acetate	$C_{22}H_{42}O_2$	338
5	16.477	0.28	3-Eicosyne	$C_{20}H_{38}$	278
6	16.675	1.39	Neophytadiene	$C_{20}H_{38}$	278
7	17.129	0.28	Octadecanoic Acid, Methyl Ester	$C_{19}H_{38}O_2$	298
8	18.284	0.93	Cholestan-3,26-Diol-22-One	$C_{27}H_{46}O_3$	418
9	18.637	1.74	2-Methylpropionic Acid, 3,4-Dichlorophenyl Ester	$C_{10}H_{10}Cl_2O_2$	232
10	18.831	0.37	Methyl Petroselinate	$C_{19}H_{36}O_2$	296
11	19.447	0.43	Hexane, 3,3-Dimethyl	C_8H_{18}	114
12	20.467	0.50	Octanoic Acid, 2-Dimethylaminoethyl Ester	$C_{12}H_{25}NO_2$	215
13	21.880	0.46	2-Decanone	$C_{10}H_{20}O$	156
14	21.953	1.09	Propylhexedrine	$C_{10}H_{21}N$	155
15	22.088	6.43	Longifolenaldehyde	$C_{15}H_{24}O$	220
16	22.165	13.16	Isocembrol	$C_{20}H_{34}O$	290
17	23.837	4.27	Tetracontane	$C_{40}H_{82}$	562
18	24.030	2.31	Chrysanin	$C_{20}H_{24}O_5$	344
19	24.598	1.31	Squalene	$C_{30}H_{50}$	410
20	25.415	14.56	17,18,19,20-Tetrahydro-16,21-Di-T-Butyl-1,6-Methano-32-Annulene	$C_{41}H_{44}$	536
21	26.447	6.48	Hexatriacontane	$C_{36}H_{74}$	506
22	26.865	0.90	Carbonic Acid, 2-Ethylhexyl Pentadecyl Ester	$C_{24}H_{48}O_3$	384
23	27.175	2.17	(+)-.Alpha.-Tocopherol	$C_{29}H_{50}O_2$	430
24	28.395	1.10	11-Hydroxypregn-4-Ene-3,20-Dione	$C_{21}H_{30}O_3$	330
25	28.695	2.90	6-Nitrocholesteryl Acetate	$C_{29}H_{47}NO_4$	473
26	29.222	3.98	2-Methylhexacosane	$C_{27}H_{56}$	380
27	29.465	4.98	Ergost-5-En-3-Ol, (3.Beta.)-	$C_{28}H_{48}O$	400
28	32.471	3.02	Eicosane	$C_{20}H_{42}$	282

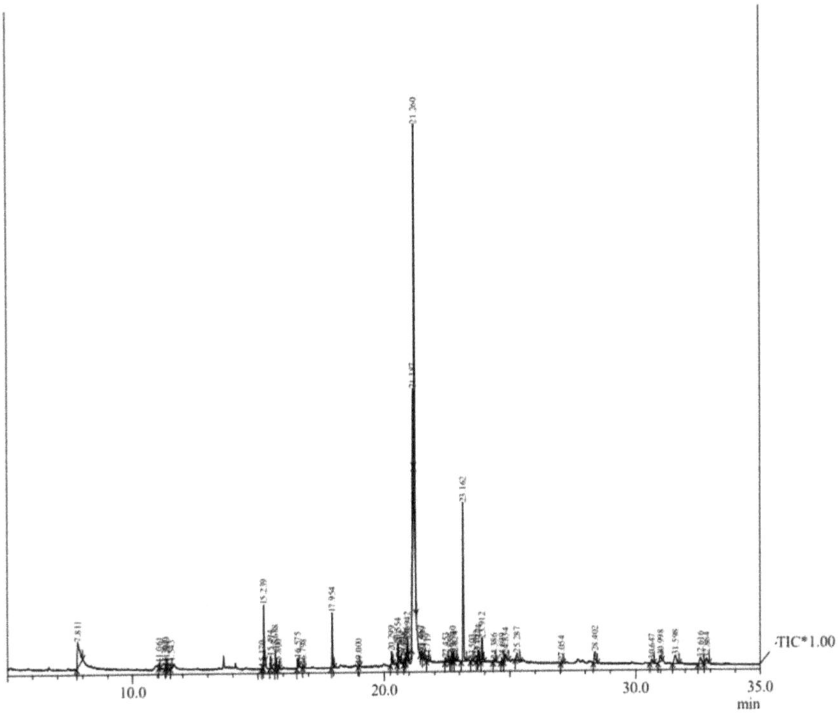

Figure 5. GC-MS chromatogramic view of methanolic extract of in vivo grown leaves of *G. pulchella* revealing phytoconstituents with retention time.

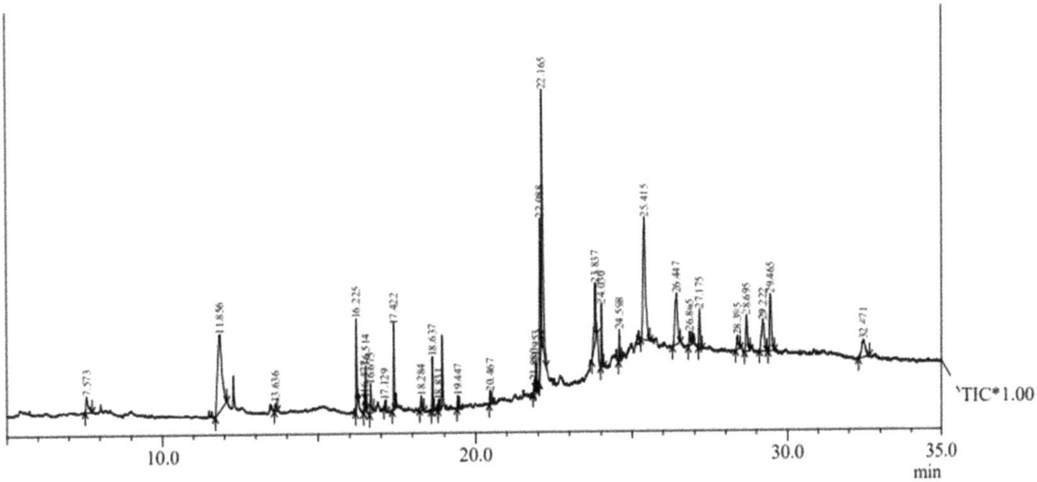

Figure 6. GC-MS chromatogram of methanolic extract of in vitro derived leaf tissue of *G. pulchella* revealing the phytoconstituents with their retention time.

Leaves derived in vivo and in vitro were exposed to chromatographic examination, in which 45 and 28 phytocompounds were identified, respectively, that belong to different classes like lignans, phenols, terpenoids, alkaloids and fatty acids, etc. Various medicinally important phytocompounds were found in methanolic extract of in vivo leaves samples such as stanolone acetate (13.23%), isoamyl acetate (6.28%), phytol (3.64%), isocitronellol (2.08%), vitamin E (1.87%), sitosterol (1.28%) vitamin A acetate (1.14%), (−)-globulol (0.38%), pentadecanal (0.15%) (Table 5 and Figure 5). Similarly, several bioactive compounds were exclusively found in in vitro regenerated leaf tissues in varied quantities. They were like nitroisobutylglycerol (15.72%), isocembrol (13.16%), hexatriacontane (6.48%), longifolenaldehyde (6.43%), ergost-5-en-3-ol (4.98%), tetracontane (4.27%), 2-methylhexacosane (3.98%), phytol acetate (3.20%), eicosane (3.02%), (+)-.alpha.-tocopherol (2.17%), and pyranone (1.70%) (Table 6 and Figure 6). In both extracts, the most common bioactives found were squalene, chrysanin, neophytadiene and 1,4-Cyclohexanediol, (Z)- and TMS derivative. The amount of chrysanin was found to be more than five times higher (12.26%) in in vivo grown leaves compared to in vitro grown ones (2.31%). Similarly, squalene content in in vivo grown leaves was nearly twice higher (2.08%) than in in vitro grown leaves (1.31%). Neophytadiene also showed a comparable variation, with field-grown leaves' amount of 0.72% and in vitro derived leaves' of 1.39%. A similar difference was noted for 1,4-Cyclohexanediol, (Z)-, TMS derivative (0.77% in field-grown and 0.41% in tissue-culture-obtained leaves).

3.5. Determination of Total Phenolic and Flavonoid Content and DPPH Activity

The Folin–Ciocalteu method was used to assess the total phenolic content of tissues (the callus and leaves extracts), with gallic acid serving as the standard. The in vitro leaves had the highest phenolic content, followed by in vivo leaves and the callus (Table 7). The in vitro developed leaves had a TPC (total phenolic content) value of 5.32 ± 0.1 mg GAE/g DW, while the field-grown leaves had a TPC value of 3.49 ± 0.1 mg GAE/g DW. The callus had the lowest TPC, which was 0.95 ± 0.1 mg GAE/g DW. The aluminum chloride method was used to quantify the total flavonoid content (TFC) of the tested samples. The values were represented as QE/gDW. The extracts contained flavonoids and the yield varied from 6.56 to 13.19 mg QE/g DW. Among all the samples tested, the in vitro leaves had the highest flavonoid content, followed by in vivo grown leaves and leaf-derived callus, respectively (Table 7). The in vitro developed leaves had a TFC value of 13.19 ± 0.1 mg QE/g DW, while the field ones grown had a TFC value of 11.20 ± 0.4 mg QE/g DW. The callus extract had the lowest TFC value 6.56 ± 0.2 mg QE/g DW. Using a DPPH free radical scavenging experiment, the antioxidant activity of each extract was measured, and the results are shown in Table 7. Just like TPC and TFC, DPPH demonstrated a similar trend in antioxidant potential of the tested tissues. The in vitro grown leaves' extract was found to have a greater level of scavenging activity (50.64%) in comparison to that of the in vivo leaves (36.22%). The callus extract had the lowest level of scavenging activity (30.28%).

Table 7. Total phenolic content (mg GAE/g DW), total flavonoid content (mg QE/g DW) and DPPH scavenging activity (%) of different tissues in *G. pulchella*.

Tissue Type	TPC	TFC	DPPH Scavenging Activity
Callus	0.95 ± 0.1 [c]	6.56 ± 0.2 [c]	30.28 ± 1.0 [c]
Leaves (in vivo)	3.49 ± 0.1 [b]	11.20 ± 0.4 [b]	36.22 ± 0.9 [b]
Leaves (in vitro)	5.32 ± 0.1 [a]	13.19 ± 0.1 [a]	50.64 ± 0.6 [a]

Values represent means ± S.E. of three replicates. Within each column, means followed by different letters are significantly different at $p \leq 0.05$ according to DMRT. Abbreviations used: TPC: total phenolic content, TFC: total flavonoid content, GAE: gallic acid equivalent, QE: quercetin equivalent, DW: dry weight.

4. Discussion

The current investigation described the influence of PGRs on in vitro morphogenic processes like the induction of callus, the formation of shoots and the rooting of *Gaillardia pulchella* shoots. In vitro generated tissue's photochemical and biochemical profiles were also examined. The leaves were first inoculated on MS with varying concentrations of PGRs for induction of callus. The response obtained suggests that NAA(2.0 mg/L) and

BAP (0.5 mg/L) synergistically induced prolific callus with maximum fresh biomass as compared to the 2, 4-D treatments. In recent years, Manokari et al. [20] noted in vitro shoot regeneration on MS medium supplemented with IAA and BAP in *G. pulchella*. The beneficial effects of NAA and BAP were observed on several other members of Asteraceae, like *Saussurea obvallata* [25] and *Artemesia annua* [26]. After further subculture of callus in the presence of the same PGRS, the shoots were formed and proliferated. Similar effects of BAP and NAA on callus-mediated shoot induction were observed in various plant species, such as *E. urophylla* × *E. grandis* and *Dendrocalamus latiflorus* [27,28], which is in accordance with the present study. Auxins, along with cytokinins, are frequently used in triggering shoots since these signaling components are known to influence cytokinins' overbearing effects [29]. Earlier reports on plant species like *Curcuma zedoaria* [30], *Basella rubra* [31], and *Santalum album* [32] demonstrated a similar promotive action of auxin with cytokinin in shoot production. Afterwards, in our study, the shoots were moved to a rooting medium with varying IAA and IBA concentrations. The rate of rooting was greater in shoots raised on IAA than IBA. Sholikhah et al. [33] reported the promotive effect of IAA compared to other auxin treatments in producing roots of in vitro derived Cavendish banana shoots. On the contrary, IBA was equally effective in promoting roots, observed in plants like *Dracaena sanderiana* [34], *Adenostyles alpina* [35] and many others.

Regenerated tissues and cell lines are under stress in in vitro culture environments, which affect growth, development and survival rate [36]. It is thus essential to evaluate cellular physiology periodically. In *G. pulchella*, the biochemical attributes of tissue-culture-derived plant tissues were studied and compared with the mother plant. There are several elements that affect up- and down-regulation of phenolic and flavonoid production, such as different PGRs used during cultivation [37]. In our investigation, it was found that in vitro leaf tissues had higher levels of phenolic and flavonoid contents than the mother plants. This is inconsistent with previously published biochemical research carried out on several plant species [38,39]. The data from the present investigations demonstrated that in vitro leaf tissues had enriched level of antioxidant activities than the other two tested samples. The antioxidant activities of in vitro raised tissues were assessed through DPPH assays. When DPPH, an organic free radical with a dark color, absorbs an electron or another free radical, it changes to a light-yellow color, signifying the scavenging action [40]. According to this study, the higher level of antioxidant potential found in laboratory-grown leaves is due to phenolics and flavonoids, which positively correlate with antioxidant activity. These molecules deactivate free radicals by donating hydrogen atoms to them [24]. Numerous plants like *Salvia hispanica* [41], *Tylophora indica* [42], etc., showed similar antioxidant ability.

Many phytoelements, like volatile oils, hydrocarbons, sugar alcohols, esters, alkaloids, flavonoids, and saponins, can be easily identified by GC–MS, a commonly used analytical approach [18]. By identifying variations in peak area percentage in the metabolite profiles of in vitro regenerated plants and their wild (mother) counterparts, this method offers significant insights into the impact of plants' growth and development. Different plants are frequently used for medicinal purposes like *Catharanthus roseus* [43], and *Amomum nilgiricums* [17] and were recently subject to the GC–MS technique. The chromatographic data obtained from the present research analysis demonstrated the presence of over 25 significant bioactive chemicals in all examined samples. Different levels of phytocompounds were found in methanolic leaf extracts produced in vivo and in vitro. The in vitro derived leaves tissues produced a nearly similar amount of phytocompounds when compared to the in vivo grown leaves. The in vitro derived leaves also produced pharmaceutically important phytocompounds that were not present in in vivo leaves and leaf-derived callus. This phenomenon could be attributed to a number of factors like genotype, temperature, photoperiod, media composition and PGRS concentration [44]. A range of phytochemicals, such as n-hexadecanoic acid, phytol, cyclopropaneoctanoic acid, limonene, platambin, etc., were previously reported to be present in field-grown leaves of *G. pulchella* [45]. Thus, it represents a more potent and dependable source of phytocompounds for pharmacological applications. *G. pulchella* contains a variety of phytoconstituents with therapeutic application. In this study, several phytocompounds were exclusively found in methanolic leaves extracts of in vivo grown plants. These were vitamin E, beta-sitosterol, vitamin A acetate, andisocitronellol, etc. Vitamin E acts as a liposoluble antioxidant and is used in the treatment of cardiovascular, neurological and aging-related diseases [46]. Beta-sitosterol is a phytosterol showing many beneficial effects against different diseases; it lowers urinary tract infection, hypercholesteremia, immunosuppression and inflammation, rheumatoid arthritis and androgenetic alopecia [47]. Some phytocompounds are exclusively found in methanolic leaf extract of in vivo grown plants, viz., eicosane, phytol acetate, 2-Methyl hexacosane, etc. Eicosane is a very strong anti-inflammatory, analgesic, and antipyretic agent [48]. Several phytocompounds such as squalene, neophytadiene, etc., were found in all methanolic samples. The terpenoid, squalene has several biological properties; it presents antioxidant, anticancer, detoxifying and moisturizing properties [49]. Neophytadiene is a diterpene, and it plays a role as an anti-inflammatory, antimicrobial, antipyretic agent, and is frequently exploited against headaches, rheumatism and in skin issues [50]. This observation shows that in vitro grown tissues like callus and leaves have enormous pharmacological potential, which is in agreement with previous observation [51]. Finding these compounds may make it possible to develop new drugs more quickly and in a cheaper way. Future research could be conducted to understand protein–ligand action of important bioactives/drugs by employing techniques like molecular docking and bio-prospecting.

5. Conclusions

The micropropagation method was successfully used to induce callus, shoots and roots in *G. pulchella*, an important medicinal plant. A combination of NAA and BAP was proven to be efficient in inducing callus from leaf explants with subsequent shoot formation (indirect organogenesis) on the same medium. The in vitro regenerated tissues were then analyzed through GC-MS for identifying phytocompounds. The present study identified a large number of phytocompounds including alkaloids, flavonoids, and terpenoids of medicinal uses. When tissues of different sources were compared for biochemical and antioxidant activities, the in vitro leaves demonstrated a higher level of phenolics and flavonoids and antioxidant capacity than the other studied tissues. A variety of therapeutically important bioactive compounds were found in in vitro grown plant tissues. The pharmaceutical industry can use these compounds on a large-scale by collaborating in vitro culture techniques as the cultures can be proven to be a continuous source of such important phytocompounds. Future studies like structural characterization and modification, investigations on ligand–protein interaction, and toxicological analyses under in vivo models can be performed for better development of novel drugs.

Author Contributions: Concept: M.B. and A.M.; Method: M.B. and Y.B.; Analysis: M.B and Y.B.; Data Curation: M.B. and Y.B.; Writing—Original Draft: M.B., Y.B. and A.M.; Writing—Review and Editing: M.B., A.M., Y.H.D. and N.M.-D.; Visualization: A.M., Y.H.D. and N.M.-D.; Project Implementation: A.M. All authors have read and agreed to the published version of the manuscript.

Funding: The authors are grateful to Department of Botany, Jamia Hamdard, New Delhi for receiving research facilities and Researchers Supporting Project number (RSP-2024R375), King Saud University, Riyadh, Saudi Arabia.

Data Availability Statement: The original contributions presented in the study are included in the article, further inquiries can be directed to the corresponding author.

Acknowledgments: The authors acknowledge Researchers Supporting Project number (RSP-2024R375), King Saud University, Riyadh, Saudi Arabia.

Conflicts of Interest: The authors declare that there are no conflicts of interest.

References

1. Gawade, N.; Bhalekar, S.G.; Bhosale, P.; Katwate, S.M.; Wadekar, V. Studies on different genotypes of Gaillardia (*Gaillardia pulchella* L.) for quantitative and qualitative performance. *Int. J. Curr. Microbiol. Appl. Sci* **2018**, *7*, 1030–1039. [CrossRef]
2. Marak, J.R.; Singh, S.; Saxena, A.K. Effect of various organic growing media on growth and yield of gaillardia (*Gaillardia pulchella* foug) at Dehradun valley of Uttarakhand. *Pharma Innov. J.* **2023**, *12*, 1437–1447.
3. Sowmyamala, B.V.; Kumar, R.V.; Nagaraj, H.T.; Kulkarni, B.S.; Hiremath, J.B.; Turamari, R.M. Effect of organic and inorganic fertilizers on plant height, spread and flower diameter of gaillardia (*Gaillardia pulchella*). *Int. J. Appl. Biol. Pharmaceut. Technol.* **2012**, *3*, 53–57.
4. Nagy, K.N.; Kardos, L.V.; Orbán, Z.; Bakacsy, L. The allelochemical potential of an invasive ornamental plant, the Indian blanket flower (*Gaillardia pulchella* Foug.). *Plant Species Biol.* **2023**, *39*, 102–108. [CrossRef]
5. Kadam, M.; Malshe, K.; Salvi, B.; Chavan, S. Effect of plant growth regulators on flowering and flower yield in Gaillardia (*Gaiilardia pulchella*) cv. Local Double. *Int. J. Chem. Studies* **2020**, *8*, 927–930. [CrossRef]
6. El-Khateeb, M.; Ashour, H.; Eid, R.; Mahfouze, H.; Abd Elaziz, N.; Radwan Ragab, M.S. Induction of genetic variability with gamma radiation and detection of DNA polymorphisms among radio mutants using sequence-related amplified polymorphism markers in *Gaillardia pulchella* Foug. plants. *Egypt. Pharmaceut. J.* **2023**, *22*, 272. [CrossRef]
7. Belal, R.; Eldesouky, M.; Elmasry, A.; Weheda, B.; Aggag, S. Gene expression profile of some antioxidant-related genes in five species of the family Asteraceae in Egypt. *J. Agric. Chem. Biotechnol.* **2023**, *14*, 57–64. [CrossRef]
8. Bosco, A.; Molina, L.; Kernéis, S.M.; Hatzopoulos, G.; Favez, T.; Gönczy, P.; Tantapakul, C.; Maneerat, W.; Yeremy, B.; Williams, D.E.; et al. Pulchelloid A, a sesquiterpene lactone from the Canadian prairie plant *Gaillardia aristata* inhibits mitosis in human cells. *Mol. Biol. Rep.* **2021**, *48*, 5459–5471. [CrossRef] [PubMed]
9. Kaňuková, Š.; Lenkavská, K.; Gubišová, M.; Kraic, J. Suspension culture of stem cells established of *Calendula officinalis* L. *Sci. Rep.* **2024**, *14*, 441. [CrossRef]
10. Bansal, Y.; Mujib, A.; Siddiqui, Z.H.; Mamgain, J.; Syeed, R.; Ejaz, B. Ploidy status, nuclear DNA content and start codon targeted (SCOT) genetic homogeneity assessment in *Digitalis purpurea* L., regenerated in vitro. *Genes* **2022**, *13*, 2335. [CrossRef]
11. Kim, D.H.; Gopal, J.; Sivanesan, I. Nanomaterials in plant tissue culture: The disclosed and undisclosed. *RSC Adv.* **2017**, *7*, 36492–36505. [CrossRef]
12. Espinosa-Leal, C.A.; Puente-Garza, C.A.; García-Lara, S. In vitro plant tissue culture: Means for production of biological active compounds. *Planta* **2018**, *248*, 1–18. [CrossRef] [PubMed]
13. Zhang, J.; Gai, M.; Li, X.; Li, T.; Sun, H. Somatic embryogenesis and direct as well as indirect organogenesis in *Lilium pumilum* dc. fisch., an endangered ornamental and medicinal plant. *Biosci. Biotechnol. Biochem.* **2016**, *80*, 1898–1906. [CrossRef] [PubMed]
14. Neves, M.; Correia, S.; Cavaleiro, C.; Canhoto, J. modulation of organogenesis and somatic embryogenesis by ethylene: An overview. *Plants* **2021**, *10*, 1208. [CrossRef] [PubMed]
15. Escobedo-GraciaMedrano, R.M.; Maldonado-Borges, J.I.; Burgos-Tan, M.J.; Valadez-González, N.; Ku-Cauich, J.R. Using flow cytometry and cytological analyses to assess the genetic stability of somatic embryo-derived plantlets from embryogenic *Musa acuminatacolla* (aa) ssp. malaccensis cell suspension cultures. *Plant Cell Tissue Organ Cult.* **2013**, *116*, 175–185. [CrossRef]
16. Khan, H.; Khan, T.; Ahmad, N.; Zaman, G.; Khan, T.; Ahmad, W.; Batool, S.; Hussain, Z.; Drouet, S.; Hano, C.; et al. Chemical elicitors-induced variation in cellular biomass, biosynthesis of secondary cell products, and antioxidant system in callus cultures of *Fagoniaindica*. *Molecules* **2021**, *26*, 6340. [CrossRef] [PubMed]
17. Konappa, N.; Udayashankar, A.C.; Krishnamurthy, S.; Pradeep, C.K.; Chowdappa, S.; Jogaiah, S. GC–MS analysis of phytoconstituents from *Amomum nilgiricum* and molecular docking interactions of bioactive serverogenin acetate with target proteins. *Sci. Rep.* **2020**, *10*, 16438. [CrossRef] [PubMed]
18. Hussain, S.A.; Ahmad, N.; Anis, M.; Alatar, A.A. Influence of meta-topolin on in vitro organogenesis in *Tecoma stans* l., assessment of genetic fidelity and phytochemical profiling of wild and regenerated plants. *Plant Cell Tissue Organ Cult.* **2019**, *138*, 339–351. [CrossRef]
19. Mamgain, J.; Mujib, A.; Syeed, R.; Ejaz, B.; Malik, M.Q.; Bansal, Y. Genome size and gas chromatography-mass spectrometry (GC–MS) analysis of field-grown and in vitro regenerated *Pluchealanceolata* plants. *J. Appl. Genet.* **2023**, *64*, 1–21. [CrossRef]
20. Manokari, M.; Raj, M.C.; Dey, A.; Faisal, M.; Alatar, A.A.; Joshee, N.; Shekhawat, M.S. Silver nanoparticles improved morphogenesis, biochemical profile and micro-morphology of *Gaillardia pulchella* Foug cv. 'Torch Yellow'. *Plant Cell Tissue Organ Cult.* **2023**, *155*, 433–445. [CrossRef]
21. Eliwa, G.I.; El-Dengawy, E.-R.F.; Gawish, M.S.; Yamany, M.M. Comprehensive study on in vitro propagation of some imported peach rootstocks: In vitro explant surface sterilization and bud proliferation. *Sci. Rep.* **2024**, *14*, 5586. [CrossRef] [PubMed]
22. Murashige, T.; Skoog, F. A Revised medium for rapid growth and bio assays with tobacco tissue cultures. *Physiol. Plant.* **1962**, *15*, 473–497. [CrossRef]
23. Baliyan, S.; Mukherjee, R.; Priyadarshini, A.; Vibhuti, A.; Gupta, A.; Pandey, R.P.; Chang, C.-M. Determination of antioxidants by DPPH radical scavenging activity and quantitative phytochemical analysis of *Ficusreligiosa*. *Molecules* **2022**, *27*, 1326. [CrossRef] [PubMed]

24. Aryal, S.; Baniya, M.K.; Danekhu, K.; Kunwar, P.; Gurung, R.; Koirala, N. Total phenolic content, flavonoid content and antioxidant potential of wild vegetables from western Nepal. *Plants* **2019**, *8*, 96. [CrossRef] [PubMed]
25. Dhar, U.; Joshi, M. Efficient plant regeneration protocol through callus for *Saussureaobvallata* (dc.) edgew. (asteraceae): Effect of explant type, age and plant growth regulators. *Plant Cell Rep.* **2005**, *24*, 195–200. [CrossRef] [PubMed]
26. Zayova, E.; Nedev, T.; Petrova, D.; Zhiponova, M.; Kapchina, V.; Chaneva, G. Tissue culture applications of *Artemisia annua* L. callus for indirect organogenesis and production phytochemical. *Plant Tissue Cult. Biotechnol.* **2020**, *30*, 97–106. [CrossRef]
27. Nazirah, A.; Nor-Hasnida, H.; Mohd-Saifuldullah, A.W.; Muhammad-Fuad, Y.; Ahmad-Zuhaidi, Y.; Rozidah, K. Development of an efficient micropropagation protocol for eucalyptus hybrid (*E. urophylla* × *E. grandis*) through axillary shoot proliferation. *J. Trop. For. Sci.* **2021**, *33*, 391–397. [CrossRef]
28. Tikendra, L.; Dey, A.; Jamir, I.; Sahoo, M.R.; Nongdam, P. Cytokinin influence on in vitro shoot induction and genetic stability assessment of *Dendrocalamus latiflorus munro*: A commercially important bamboo in Manipur, North-East India. *Vegetos* **2022**, *35*, 1085–1095. [CrossRef]
29. Novikova, T.I.; Asbaganov, S.V.; Ambros, E.V.; Zaytseva, Y.G. TDZ-induced axillary shoot proliferation of *Rhododendron mucronulatumturcz* and assessment of clonal fidelity using dna-based markers and flow cytometry. *In Vitro Cell. Dev. Biol.-Plant* **2020**, *56*, 307–317. [CrossRef]
30. Jena, S.; Ray, A.; Sahoo, A.; Sahoo, S.; Dash, B.; Kar, B.; Nayak, S. Rapid plant regeneration in industrially important *Curcuma zedoaria* revealing genetic and biochemical fidelity of the regenerants. *3 Biotech* **2020**, *10*, 17. [CrossRef]
31. Kumar, S.S.; Giridhar, P. In vitro micropropagation of *Basella rubra* L. through proliferation of axillary shoots. *Plant Cell Tissue Organ Cult.* **2021**, *144*, 477–483. [CrossRef]
32. Wardani, I.B. Effect of BAP (6-benzyl amino purine) and NAA (naphtalen acetic acid) on the induction of axillary shoots in sandalwood (*Santalum album* L.). *J. Sci. Technol. Educ.* **2022**, *1*, 23–30.
33. Sholikhah, R.I.; Makhziah, M.; Widiwurjani, W. Effect of IAA addition and some organic supplements on growth and rooting of cavendish banana (*Musa acuminata*, aaa) In-Vitro. *J. Tek. Pertan. Lampung (J. Agric. Eng.)* **2022**, *11*, 266. [CrossRef]
34. Mujib, A.; Aslam, J.; Bansal, Y. Low colchicine doses improved callus induction, biomass growth, and shoot regeneration in in vitro culture of *Dracaena sanderiana* sander ex mast. *Prop. Ornam. Plants* **2023**, *23*, 81–87.
35. Gianguzzi, V.; Barone, G.; Di Gristina, E.; Sottile, F.; Domina, G. Micropropagation of Endemic Endangered Taxa of the Italian Flora: *Adenostyles alpina* subsp. *macrocephala* (Asteraceae), as a Case Study. *Plants* **2023**, *12*, 1530. [CrossRef] [PubMed]
36. Malik, M.; Wachol, M.; Pawlowska, B. Liquid culture systems affect morphological and biochemical parameters during rosacanina plantlets in vitro production. *Not. Bot. HortiAgrobot. Cluj-Napoca* **2018**, *46*, 58–64. [CrossRef]
37. Ghosh, A.; Igamberdiev, A.U.; Debnath, S.C. Thidiazuron-induced somatic embryogenesis and changes of antioxidant properties in tissue cultures of half-high blueberry plants. *Sci. Rep.* **2018**, *8*, 16978. [CrossRef] [PubMed]
38. Jeong, B.R.; Sivanesan, I. Direct adventitious shoot regeneration, in vitro flowering, fruiting, secondary metabolite content and antioxidant activity of *Scrophulariatakesimensis* Nakai. *Plant Cell Tissue Organ Cult.* **2015**, *123*, 607–618. [CrossRef]
39. Bhattacharyya, P.; Kumaria, S.; Tandon, P. High frequency regeneration protocol for Dendrobium nobile: A model tissue culture approach for propagation of medicinally important orchid species. *S. Afr. J. Bot.* **2016**, *104*, 232–243. [CrossRef]
40. Khorasani Esmaeili, A.; Mat Taha, R.; Mohajer, S.; Banisalam, B. Antioxidant activity and total phenolic and flavonoid content of various solvent extracts from in vivo and in vitro grown *Trifolium pratense* L. (Red Clover). *BioMed Res. Int.* **2015**, *15*, 643285.
41. Zayova, E.; Nikolova, M.; Dimitrova, L.; Petrova, M. Comparative study of in vitro, ex vitro and in vivo propagated *Salvia hispanica* (chia) plants: Morphometric analysis and antioxidant activity. *AgroLife Sci. J.* **2016**, *5*, 166–173.
42. Mamgain, J.; Mujib, A.; Bansal, Y.; Gulzar, B.; Zafar, N.; Syeed, R.; Alsughayyir, A.; Dewir, Y.H. Elicitation induced α-amyrin synthesis in *Tylophora indica* in vitro cultures and comparative phytochemical analyses of in vivo and micropropagated plants. *Plants* **2024**, *13*, 122. [CrossRef]
43. Bansal, Y.; Mujib, A.; Mamgain, J.; Dewir, Y.H.; Rihan, H.Z. Phytochemical composition and detection of novel bioactives in anther callus of *Catharanthus roseus* L. *Plants* **2023**, *12*, 2186. [CrossRef] [PubMed]
44. Bansal, Y.; Mujib, A.; Mamgain, J.; Kumar, S.; Dewir, Y.H.; Magyar-Tábori, K. synthesis and accumulation of phytocompounds in field-, tissue-culture grown (stress) root tissues and simultaneous defense response activity in *Glycyrrhiza glabra* L. *Sustainability* **2024**, *16*, 1613. [CrossRef]
45. Yao, X.T.; Ling, P.X.; Jiang, S.; Lai, P.X.; Zhu, C.G. Analysis of the essential oil from *Gaillardia pulchella* Foug. and its antioxidant activity. *J. Oleo Sci.* **2013**, *62*, 329–333. [CrossRef]
46. Shahidi, F.; Pinaffi-Langley, A.C.C.; Fuentes, J.; Speisky, H.; de Camargo, A.C. Vitamin E as an essential micronutrient for human health: Common, novel, and unexplored dietary sources. *Free Radic. Biol. Med.* **2021**, *176*, 312–321. [CrossRef]
47. Loizou, S.; Lekakis, I.; Chrousos, G.P.; Moutsatsou, P. β-Sitosterol exhibits anti-inflammatory activity in human aortic endothelial cells. *Mol. Nutr. Food Res.* **2010**, *54*, 551–558. [CrossRef] [PubMed]
48. Okechukwu, P.N. Evaluation of anti-inflammatory, analgesic, antipyretic effect of eicosane, pentadecane, octacosane, and heneicosane. *Asian J. Pharmaceut. Clin. Res.* **2020**, *13*, 29–35. [CrossRef]
49. Bhat, M.P.; Rudrappa, M.; Hugar, A.; Gunagambhire, P.V.; Suresh Kumar, R.; Nayaka, S.; Almansour, A.I.; Perumal, K. In-vitro investigation on the biological activities of squalene derived from the soil fungus *Talaromyces pinophilus*. *Heliyon* **2023**, *9*, e21461. [CrossRef]

50. Gonzalez-Rivera, M.L.; Barragan-Galvez, J.C.; Gasca-Martínez, D.; Hidalgo-Figueroa, S.; Isiordia-Espinoza, M.; Alonso-Castro, A.J. In vivo neuropharmacological effects of neophytadiene. *Molecules* **2023**, *28*, 3457. [CrossRef]
51. Gulzar, B.; Mujib, A.; Malik, M.Q.; Mamgain, J.; Syeed, R.; Zafar, N. Plant tissue culture: Agriculture and industrial applications. In *Transgenic Technology Based Value Addition in Plant Biotechnology*; Academic Press: Cambridge, MA, USA, 2020; pp. 25–49.

horticulturae

MDPI

Article

Polyethylene Glycol (PEG)-Induced Dehydration Alters Enzymatic and Non-Enzymatic Components of the Antioxidant Defense System in *Nepeta nervosa* Royle ex Bentham

Jasmina Nestorović Živković *, Neda Aničić, Dragana Matekalo, Marijana Skorić, Biljana Filipović, Tijana Marković and Slavica Dmitrović *

Department of Plant Physiology, Institute for Biological Research "Siniša Stanković"—National Institute of the Republic of Serbia, University of Belgrade, Bulevar Despota Stefana 142, 11108 Belgrade, Serbia; neda.anicic@ibiss.bg.ac.rs (N.A.); dragana.bozic@ibiss.bg.ac.rs (D.M.); mdevic@ibiss.bg.ac.rs (M.S.); biljana.nikolic@ibiss.bg.ac.rs (B.F.); m3011_2022@stud.bio.bg.ac.rs (T.M.)

* Correspondence: jasmina.nestorovic@ibiss.bg.ac.rs (J.N.Ž.); slavica.dmitrovic@ibiss.bg.ac.rs (S.D.)

Abstract: Plants have developed a variety of defense mechanisms that allow them to recognize and respond to specific stressors by triggering complex signaling networks that cause appropriate biochemical changes to overcome the stress. In this study, we subjected in vitro grown *N. nervosa* plants to PEG-induced dehydration stress for 1 day (1DPT), 3 days (3DPT) and 6 days (6DPT). Our study investigated antioxidant enzyme activities, including catalase (CAT), peroxidase (POX) and superoxide dismutase (SOD), unveiling dynamic responses to PEG-induced water stress. CAT levels increased initially (1DPT) but declined with prolonged treatment; while POX activity significantly increased at 3DPT and 6DPT; and SOD, particularly the Mn-SOD3 isoform, demonstrated a substantial increase, emphasizing its role in the enzymatic free-radical scavenging activity. Furthermore, examination of the phenolic acid content revealed that rosmarinic acid (RA) was the predominant phenolic compound, followed by chlorogenic acid (CHLA), while ferulic acid (FA) and caffeic acid (CAFFA) were present in lower concentrations. Notably, PEG-induced dehydration significantly boosted RA content in *N. nervosa* plants at 3DPT. This increase highlights the plant's response to oxidative stress conditions and its role in non-enzymatic antioxidant defense mechanisms. These findings significantly contribute to our comprehension of *N. nervosa*'s adaptive strategies under PEG-induced dehydration stress, offering valuable insights into plant stress physiology within industrial and agricultural contexts.

Keywords: dehydration; polyethylene glycol; rosmarinic acid; antioxidant defense systems; *Nepeta nervosa*; in vitro culture; abiotic stress; antioxidant enzymes

1. Introduction

All stages of a plant's life cycle (germination, growth, development and reproduction) are closely linked to and dependent on a combination of environmental factors. If one or more factors endanger the survival of the plant, these conditions are defined as stressful [1]. As sessile organisms, they have evolved a plethora of defense mechanisms that enable them to recognize and respond to specific stressors by triggering complex signaling networks that induce appropriate biochemical changes to overcome the stress [2,3]. As a result, various physiological changes occur in plants [4].

The first visible change due to water dehydration is the reduction in leaf area, which is accompanied by a decrease in cell expansion, thereby reducing the transpiration area [5]. After stress perception by membrane proteins, the most rapid physiological process is abscisic acid (ABA) mediated stomata closure [6]. The ABA signal transduction system consists of protein phosphatases and kinases that induce the production of reactive oxygen species (ROS) [7]. Water stress also leads to reduced CO_2 fixation by the Calvin cycle, resulting in an accumulation of reduction equivalents ($NADPH^+$ and H^+) that generate various ROS [8]. Incompletely-reduced oxygen species, including the well-studied singlet oxygen ($^1O_2$), superoxide anion ($O_2^{\bullet -}$), hydrogen peroxide (H_2O_2) and hydroxyl radical ($^{\bullet}OH$), primarily relate to ROS [9,10], which on the one hand play an important signaling role in plants and control processes such as growth and development, while on the other hand triggering defense mechanisms against biotic and abiotic environmental stress.

The most important enzymes for scavenging ROS, including catalases (CAT), peroxidases (POX) and superoxide dismutases (SOD), form a highly efficient system for maintaining ROS homeostasis at various compartments of cell [11]. The first line of defense against ROS are the SOD, which eliminate $O_2^{\bullet -}$, forming H_2O_2 and O_2 [11]. Catalases are a group of metalloenzymes that catalyze the dismutation reaction of H_2O_2 to water and oxygen. Peroxidases are glycoproteins that catalyze the oxidation reaction of the substrate, resulting in water and an oxidized radical product [12].

Various polyphenolic compounds, including phenolic acids, are postulated to contribute to the non-enzymatic antioxidants [13], whose potential is mainly reflected in their highly effective ability to scavenge most oxidizing molecules involved in the production of free radicals in plants exposed to various environmental stresses [14–16]. The literature shows that water stress affects the production of phenolic compounds in plants and that this response can vary between and within species and plant parts [17,18]. Phenolic acids, including cinnamic acid derivatives, have a remarkably high antioxidant potential that is closely related to their molecular structure, which enables them to form stable phenoxyl radicals when they react with radical molecules [19,20]. The cinnamic acid derivatives, such as rosmarinic acid (RA), chlorogenic acid (CHLA), ferulic acid (FA) and caffeic acid (CAFFA), are particularly important due to their numerous health-promoting properties, such as anticancer, antimicrobial, antidiabetic, neuroprotective, hepatoprotective and anti-inflammatory effects [21–26]. The ability of cinnamic acid derivatives to scavenge free radicals, chelate prooxidant ions, prevent lipid peroxidation and bind to biomacromolecules has made them a valuable antioxidant used in the food industry [27–29]. Among these compounds, RA is found in many plant species, most notably in the Lamiaceae family which comprises specialized metabolite-rich aromatic plants. Currently, the most exploited species for RA isolation are *Salvia rosmarinus*, *S. officinalis*, *Mellisa officinalis*, *Ocimum basilicum* and *Mentha piperita* [30,31]. However, the production of this compound encounters several challenges, one of which is the extraction of highly pure RA, which is essential for various applications in industry. Obtaining RA from cultivated aromatic plants carries the risk of containing either externally-introduced impurities (i.e., pesticide or herbicide) or internal impurities originating from plant tissues (i.e., other specialized metabolites) [30–32]. Plant in vitro systems offer a valuable substitute for conventional field harvesting in overcoming the problem of external contaminants, in addition to other significant advantages such as controlled and enhanced biosynthesis, year-round production, genetic stability and scalability [33]. Selection of the most appropriate plant source with high contents of RA and significantly lower amounts of other groups of specialized metabolites could facilitate gaining high yields of highly pure RA. Previous studies have demonstrated a favorable compound ratio for RA derivation in *Nepeta nervosa* Royle ex Benth [34]. Plant species of the genus *Nepeta* (fam. Lamiaceae) are widely cultivated as horticultural and medicinal crops due to their high productivity of terpenes and phenolic compounds. Most species of this genus are characterized by the accumulation of the atypical monoterpenes, iridoids, both as aglycones (nepetalactones) and glycosides in their tissues, and only a few iridoid-lacking species are identified, among which is *N. nervosa* Royle ex Benth [34]. Moreover, *N. nervosa* has been shown to produce limited quantities and a poor diversity of other terpenoids [34]. However, this species has until now remained an unrecognized and underexploited candidate for RA isolation.

The present study uncovers the role of antioxidant molecules in the acclimation of *N. nervosa* plants cultivated in vitro to water deficit. This research aims to: (1) Explore antioxidative defense mechanisms during PEG-induced water stress to enhance understanding of the physiology of plant adaptability to stress conditions; (2) Evaluate the potential of short-term PEG treatment in scaling up the production of phenolic acids, particularly high purity RA in *N. nervosa* plants. This is the first attempt to analyze the antioxidative mechanisms of *N. nervosa* under PEG-induced dehydration, which is a promising way to obtain RA extracts with a favorable compound ratio for industrial application.

2. Materials and Methods

2.1. Plant Material

Nepeta nervosa Royle ex Benth. (Lamiaceae) seeds were surface sterilized in 50% (*v*/*v*) commercial bleach in sterile deionized water (2% *w*/*v* active chlorine) for 2 min and thoroughly washed 5 times with sterile deionized water. Seeds were transferred into plastic Petri dishes (9 cm diameter) containing 25 mL of ½ MS culture medium, named basal medium (BM) [35], made with modifications as in [36]. The pH of the BM was adjusted to 5.8. The seeds germinated ten days after placing on BM medium. One-node stem segments of selected 4-week-old seedlings were used as explants for the multiplication of different *N. nervosa* genotypes. Sub-culturing was performed every 4 weeks on BM medium in 370 mL glass jars. Three months of multiplication of selected genotypes provided enough plant material for experiments. All the cultures were kept in a growth chamber at 25 ± 2 °C, under a 16/8 h light/dark regime with a photon flux density of 70 μmol/m^{-2} s^{-1}, provided by white fluorescent tubes.

2.2. Experimental Design

The PEG experimental setup was performed as reported by [36]. Briefly, five rooted *N. nervosa* plants, four weeks old, were placed on a metal stand in 6.5 × 6.5 × 10 cm Magenta vessels (Sigma-Aldrich, Darmstadt, Germany) containing 30 mL of the ½ MS modified liquid medium without sucrose ± water-soluble PEG 8000. The osmotic pressure of PEG solutions was −3 MPa. This was achieved by adding 500 g of water-soluble PEG 8000 into 1 L of ½ MS liquid media without sucrose (50% *w*/*v*) after the sterilization of the medium by autoclaving at 114 °C for 25 min and subsequent cooling to room temperature. Michel's [37] equation: $\Psi = 1.29\,[PEG]^2T - 140\,[PEG]^2 - 4[PEG]$; [PEG] = g PEG/g H_2O at T = +25 °C, was used to calculate the final water potential of the culture media with the addition of PEG.

After 1, 3 and 6 days of PEG treatment, leaves of the non-treated and PEG-treated plants were cut, collected, ground in liquid nitrogen and stored at −80 °C until further use. The results were obtained using three biological replicates. Each biological replicate represented a pool of leaf tissues collected from five plants grown in the single Magenta vessels.

The following formula was used to calculate the percentage of decrease or increase in some values:

$$\% = \pm(1 - T/C) \times 100$$

T is the parameter of treated plants, while C is a parameter of control ones, and a positive value indicates an increase, while negative indicates a decrease in parameters.

2.3. Determination of Relative Water Content (RWC)

The relative water content (RWC) of plants was performed as described in [36] according to the formula for calculating the RWC of plants: RWC (%) = [(FW − DW)/(TW − DW)] × 100. The fresh weight (FW) of both non-treated and PEG-treated plants was measured on the day of sampling. To obtain the turgid weight (TW), plants were further immersed in distilled water for 24 h, at room temperature, and measured after that. Dry weight (DW) was measured after plants' drying at 70 °C for 72 h. Each treatment was conducted in triplicate.

2.4. Preparation of Methanol Extracts

Powdered leaf tissue (about 50 mg) was transferred into Eppendorf tubes (2 mL) and soaked in 1 mL of 96% methanol. Samples were strongly mixed at vortex for 10 s, sonicated in an ultrasonic bath (RK100, Bandelin, Berlin, Germany) for 15 min and further centrifuged at 10,000 g for 15 min at 4 °C. Supernatants were filtered through 0.2 μm syringe filters (Agilent Technologies, Santa Clara, CA, USA) and kept at 4 °C before analyses. Rosmarinic acid (RA), chlorogenic acid (CHLA), ferulic acid (FA) and caffeic acid (CAFFA) standards were obtained from Sigma-Aldrich, Darmstadt, Germany.

2.5. Metabolic Profiling of Targeted Phenolic Compounds

Chromatographic separations of phenolics in methanol extracts of *N. nervosa* plants were performed on a Dionex Ultimate 3000 UHPLC system (Thermo Fisher Scientific, Bremen, Germany) equipped with a diode array detector (DAD) and connected to a TSQ Quantum Access Max triple-quadrupole mass spectrometer (Thermo Fisher Scientific, Basel, Switzerland). Elution was performed at 40 °C on Syncronis C18 column (100 × 2.1 mm) with 1.7 μm particle size (Thermo Fisher Scientific, Bremen, Germany). Gradient elution of the mobile phase, consisting of 0.1% formic acid in H_2O (A), and acetonitrile (B), was as previously reported [13], and the injection volume was set to 10 μL. The vaporizer temperature of the mass spectrometer equipped with a heated electrospray ionization (HESI) source was kept at 300 °C, and the ion source settings were as follows: spray voltage 4000 V, sheath gas (N_2) pressure 28 AU, ion sweep gas pressure 1.0 AU, auxiliary gas (N_2) pressure 10 AU, capillary temperature 275 °C and skimmer offset 0 V. Mass spectrometry data were acquired in negative ionization mode, and collision-induced fragmentations were performed using argon, with the collision energy (cE) set to 30 eV. Quantification of chlorogenic acid (CHLA), caffeic acid (CAFFA), ferulic acid (FA) and rosmarinic acid (RA) in samples was performed in a selected reaction monitoring (SRM) experiment of the mass spectrometer, tracking the two characteristic MS2 fragments for each compound. The stock standard solution of phenolic acids was prepared by dissolving 1 mg of each of the compounds in 1 mL methanol. The working solution, prepared by mixing the stock solutions to obtain the concertation of 20 μg/mL, was further diluted with methanol down to 5 ng/mL, to obtain other calibration levels. Regression was calculated for each of the calibration curves, which all showed a good linearity ($r = 0.999$, $p < 0.001$).

2.6. Determination of CAT, POX and SOD Activities

Detailed procedures for antioxidant enzyme extraction and activity determination were carried out as described by [38], with some modifications in measuring the concentrations of isolated soluble proteins. Instead of the Bradford [39] method, soluble protein concentrations were accurately measured using a Qubit® 3.0 Fluorometer (Invitrogen, Thermo Fisher Scientific, Bremen, Germany) by highly sensitive Invitrogen Qubit quantitation assays.

Native polyacrylamide gel electrophoresis (Native-PAGE) separation and activity staining of CAT (EC 1.11.1.6), POX (EC 1.11. 1.7) and SOD (EC 1.15.1.1) were described by [38], with some modifications. POX activity was visualized after the incubation of gels in staining solution containing 0.17% 4-chloro-naphthol and 0.11% H_2O_2 in 50 mM K-phosphate buffer pH 6.5. Multiscan FC (Thermo Fisher Scientific, Bremen, Germany) microplate reader was used for recording absorbance at 540 nm for quantification of SOD activity.

Imaging System iBRIGHT CL1500 (Invitrogen, Thermo Fisher Scientific, Bremen, Germany) was used to record gels with developed enzyme activities and measure the intensity of POX bands, while ImageJ 1.32 software (W. Rasband, National Institute of Health, Bethesda, Maryland, USA) was used to determine the intensity of CAT and SOD bands, as described by [38].

2.7. Statistical Analysis

The StatGraphics software, version 4.2 (STSC Inc. and Statistical Graphics Corporation, 1985–1989, Rockville, Maryland, USA) was used for the calculation of means and statistical errors in PEG-treated and non-treated plants, at 1, 3 and 6 days of treatment, while an unpaired Students' *t*-test (Microsoft Excel for Microsoft 365 MSO (Version 2310 Build 16.0.16924.20054) 64-bit) was used to compare data, and a difference of $p \leq 0.05$ was considered significant.

3. Results and Discussion

Being an oxidative stress, water deficiency triggers the hyper-activation of non-enzymatic (phenolic compounds) and/or enzymatic ROS scavenging systems to prevent cellular damage and restore redox homeostasis. Consequently, levels of proteins responsible for ROS scavenging are found to be associated with the basal water stress tolerance [40]. A large amount of literature data has reported that abiotic stress leads to the increased production of ROS, which could increase production of plant defense secondary metabolites [41,42]. A substantial obstacle in the large-scale extraction of useful secondary metabolites arises from the low productivity of plants. This limitation stems from the fact that their content rarely exceeds 1% of dry weight [43]. Additionally, the yield of phenolic acids is subject to a multitude of influencing factors, including plant physiology, growth and development phases and pre- and post-harvest processes, as well as geographical and environmental factors. Much attention has been given lately to improving biosynthesis in Lamiaceae through the application of elicitors, chemical or physical tools (i.e., signal compounds and/or abiotic stress), and primarily on water stress [44–46], which is a common environmental phenomenon in the natural habitats of mints.

Herein, we investigated the impact of PEG-induced dehydration stress on enzymatic and non-enzymatic components of the antioxidant defense system of in vitro cultured *N. nervosa* plants and evaluated the potential of PEG treatment in scaling up the production of high-purity rosmarinic acid.

3.1. PEG-Induced Physiological Changes

Four-week-old in vitro grown *N. nervosa* plants, were exposed to PEG-induced dehydration stress for 1 (1 day PEG treatment—1DPT), 3 (3DPT) and 6 days (6DPT) (Figure 1B).

Tissue dehydration, turgor loss and reduction in leaf area (Figure 1B) and FW reduction (Figure 1C) in *N. nervosa* shoots were observed at all experimental points, which confirmed that the plants were experiencing water stress. When compared with respective non-treated *N. nervosa* (Figure 1A), the FW values of treated plants were significantly lowered by 39%, 67% and 68% at 1DPT, 3DPT and 6DPT, respectively (Figure 1C). These results are in accordance with the PEG-induced dehydration in *N. rtanjensis* and *N. argolica* ssp. *argolica* [36]. Likewise, Zulfiqar et al. 2021 [47] showed that shoot and root length, as well as the fresh and dry weight of sweet basil, were reduced under water stress.

The intensity of water stress in non-treated and PEG-treated plants was assessed by calculating the RWC (Figure 1D). This parameter is a useful indicator of the water status, as well as of the level of tolerance of plants to water stress. The results show that in *N. nervosa*, RWC values decrease significantly under the influence of PEG at all time-points of the experiment, which indicates that the PEG-treated plants experienced high-intensity water stress. The RWC values were significantly reduced (from 35% to 48% in comparison to the appropriate control) during the water-deficiency stress (Figure 1D). The water potential of the substrate was −3 MPa, which is a value that causes severe water stress in a number of plant species [48]. However, the RWC reduction was lower in *N. argolica* subsp. *argolica*, and especially in *N. rtanjensis* [36] than in *N. nervosa* plants, all exposed to PEG treatment under similar experimental conditions and water stress of the same intensity. Furthermore, the RWC values of non-treated *N. nervosa* plants (an average of 84%) were shown to be higher when compared to those in non-treated *N. rtanjensis* and *N. argolica* subsp. *argolica* plants (an average of 65% and 72%, respectively), which indicates structural and eco-physiological differences among related species. Moreover, it can be assumed that *N. nervosa* was less efficient in overcoming the dehydration-induced water stress under the current experimental conditions.

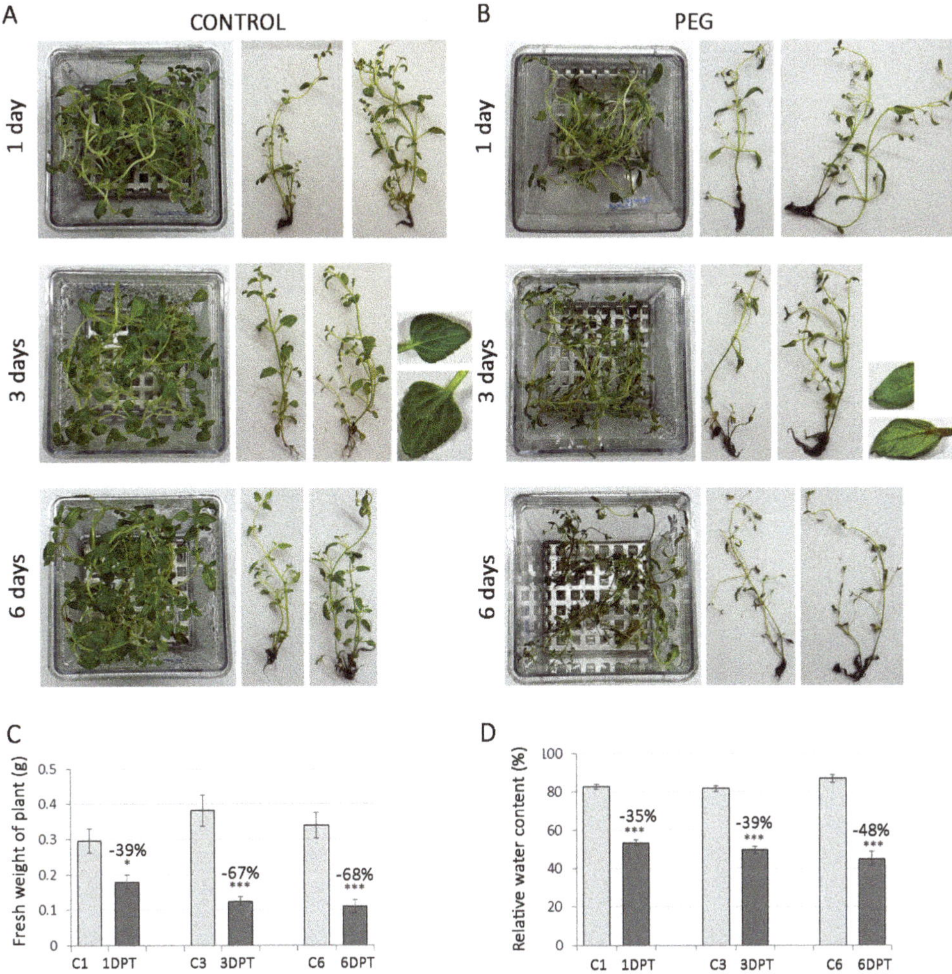

Figure 1. (**A**,**B**) Physiological changes in non-treated (control) and PEG-treated *N. nervosa* plants for 1, 3 and 6 days under in vitro conditions. (**C**,**D**) Parameters of dehydration stress: fresh weight (FW) of *N. nervosa* plants and their relative water content (RWC) were measured throughout 6 days of exposure to PEG-induced dehydration stress. *N. nervosa* plant immersed in distilled water for 24 h at room temperature to obtain turgid weight. PEG-treated *N. nervosa* plants (1DPT, 3DPT and 6DPT) are compared with their respective controls (C1, C3 and C6) at 1, 3 or 6 days of treatment. The results represent the mean ± SE of three biological replicates. Asterisks (*) indicate significant differences between PEG-treated and non-treated plants according to an unpaired *t*-test (* $p \leq 0.05$; *** $p \leq 0.001$). The numbers above the bars indicate the percent of reduction (−) relative to the control.

3.2. Content of Phenolic Acids

The chromatographic separation and quantification of phenolic compounds in samples was performed adopting an SRM (single reaction monitoring) experiment on the UHPLC/DAD/(±)HESI−MS^2 instrument operating in a negative mode. The analysis was targeted towards four phenolic acids belonging to the group of hydroxycinnamic acids (RA, CHLA, CAFFA, and FA), which were previously described as the most abundant phenolic compounds in *N. nervosa* [34]. *N. nervosa* is enriched with derivatives of CAFFA, including RA, which is structurally a dimer of CAFFA, as well as CHLA and nepetoidin

A or B (unpublished data). These compounds have previously been identified as abundant in other *Nepeta* species [34,49–52]. Peaks of CHLA, CAFFA, RA and FA were visible in UHPLC-DAD chromatograms at retention times of 5.18 min, 5.58 min, 6.98 min and 7.62 min, respectively (Figure 2A). The most abundant phenolic acid in *N. nervosa* leaves was RA, followed by CAFFA, while the observed concentrations of CHLA and FA were significantly lower. In parallel with phenolics, samples were also analyzed for the presence of iridoids (nepetalactone, nepetalactol 1,5,9-epideoxyloganic acid), and their concentrations were below the detection limits of the used analytical method. The obtained results are consistent with prior investigations on the constituents of *N. nervosa* extracts, which have consistently shown the absence of iridoids, while trace amounts of other terpenoid compounds were recorded [34,53,54]. Our findings further reinforce the hypothesis that *N. nervosa* represents a highly promising plant source for the extraction of RA, offering both a high yield and purity, due to its favorable quantitative and qualitative compound ratio.

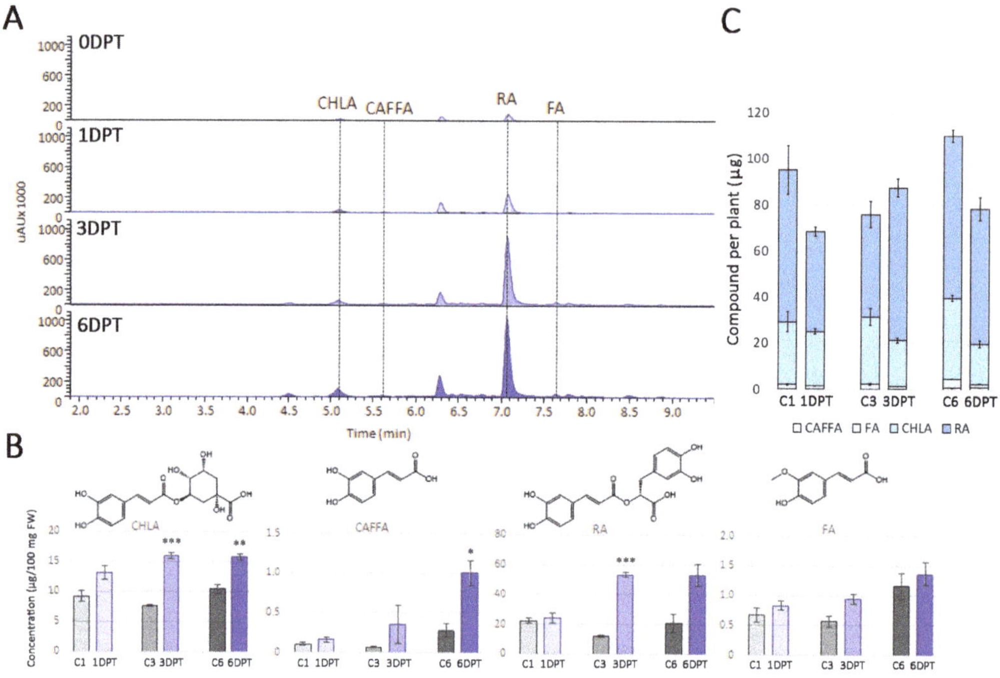

Figure 2. (**A**) Representative UHPLC/DAD chromatograms of the methanol extract of *N. nervosa* plants at the beginning of the experiment (0DPT), and one (1DPT), three (3DPT) and six (6DPT) days after the beginning of PEG treatment. (**B**) Content of phenolic acids: chlorogenic (CHLA), caffeic acid (CA), rosmarinic (RA) and ferulic (FA) in non-treated (grey bars) and PEG-treated (blue bars) *N. nervosa* plants for 1, 3 and 6 days of exposure. The results are expressed as µg/100 mg FW, and presented as the mean ± SE of three biological replicates. Asterisks (*) indicate significant differences between PEG-treated and non-treated plants according to an unpaired *t*-test (* $p \leq 0.05$; ** $p \leq 0.01$; *** $p \leq 0.001$). The numbers above the bars indicate the percent of enhancement (+) relative to the control. (**C**) Overall productivity of *N. nervosa* PEG-treated (1DPT, 3DPT and 6DPT) and non-treated plants (C1, C3, and C6) grown in vitro. A total compound yield is expressed as the µg of compound per plant.

In non-treated plants, the concentration of RA was determined to be on average 18.27 µg/100 mg FW (~0.02% FW), whereas in treated plants at 3DPT, the content of RA significantly increased to 52 µg/100 mg FW (~0.05% FW). Thus, PEG-induced dehydration led to an enhancement of RA amounts in *N. nervosa* plants from 9% (at 1DPT) to 355% (in 3DPT), in comparison to the respective controls (Figure 2B). Similarly to our findings, short-term PEG-induced water deficiency triggered a significant increase in RA content in *Salvia leriifolia* plantlets and calli grown in vitro [55]. Contrary to our results, severe drought stress did not affect the accumulation of rosmarinic acid in *Salvia milltiorrhiza* [44]. Furthermore, PEG showed negative effects on RA accumulation in *Thymus lotocephalus* shoots [56]. In summary, dehydration stress appears to modulate RA biosynthesis differently in related Lamiaceae species contingent upon the severity and duration of experimental conditions. A significant increase in the content of CHLA in *N. nervosa* was observed at 3DPT and 6DPT, while CAFFA content was significantly increased at 6DPT. No significant change in the amount of FA in response to PEG treatment was observed in *N. nervosa* plants (Figure 2B). These findings are consistent with expectations, as phenolic compounds play a vital role in regulating essential physiological functions in plants, providing resistance against a range of stress conditions. The pronounced antioxidant activity of phenolic compounds is attributed to their capacity to act as hydrogen atom donors and effectively scavenge free radicals, leading to the formation of less reactive phenoxyl radicals [57,58]. The formed radicals are more stable due to the delocalization of electrons and the existence of more resonant forms. Consequently, the stimulated accumulation of phenolic acids in response to oxidative stress conditions is also recorded in *N. pannonica* grown in the paraquat-enriched medium in vitro [51]. Likewise, in accordance with our results, PEG-induced dehydration elevated phenolic compound concentrations in several Lamiaceae species grown in vitro [59,60].

The quantitative assessment of 3DPT *N. nervosa* extracts suggests their potential value for industrial applications, as rosemary extracts containing RA within the range of 5–30% are widely recognized and utilized in the market as natural preservatives and antioxidants (Naturalin Bio-Resources Co., Ltd., Changsha City, Hunan Province, China). Furthermore, 3DPT induces no significant changes in the amounts of CAFFA and FA. Although CHLA is significantly elevated in the same PEG treatment, the chromatographic separation method adopted within the present study clearly distinguishes RA from CHLA, and can be isolated independently. Future research efforts should prioritize the identification of solvent systems that are both highly efficient and safety-conscious for the extraction of RA from *N. nervosa* samples, with potential applications across various industrial sectors.

Chlorogenic acid (9.04 μg/100 mg FW, an average value in non-treated plants) increased from 43% to 109% at 1DPT and 3DPT, respectively. Although the other two phenolic acids (FA and CAFFA) were detected at much lower values (about 0.80 μg/100 mg FW and 0.15 μg/100 mg FW, an average value in non-treated plants, respectively), PEG treatment also led to an increase in their content (66% FA at 3DPT, and 430% CAFFA at 3DPT). The largest increase in all detected phenolic acids occurred at 3DPT in comparison to the respective controls (Figure 2B). The highest concentrations of FA and CAFFA were observed at 6DPT, while the amounts of RA and CHLA accumulated at 3DPT appear to be maintained at 6DPT in treated plants. In the aerial parts of *Salvia officinalis*, a comparable halt in CAFFA to RA conversion was observed when contrasting mild and severe water-deficit conditions [61]. Ultimately, based on the growth parameters and phenolic compound production, it can be concluded that PEG-induced drought stress increased the overall productivity of *N. nervosa* plants after 3 days of exposure compared to the control, especially the RA yield (Figure 2C). Interestingly, the yields of FA and CHLA decreased. However, although greater phenolic compound accumulation was observed in PEG-treated plants (Figure 2B), their biomass decreased (Figure 1C), resulting in lower productivity at 1DPT and 6DPT compared to the controls (Figure 2C).

3.3. Antioxidant Enzymes Activity in *N. nervosa* Plants Influenced by PEG

Water-deficiency stress leads to the excessive production of ROS and oxidative damage, which induce the accumulation of osmolytes and the activation of antioxidant defense mechanisms [49,62]. Here, we analyzed the activities of three major antioxidative enzymes:

CAT, POX and SOD in the non-treated (C1, C3 and C6) and PEG-treated *N. nervosa* plants at 1DPT, 3DPT and 6DPT.

The in-gel assay unveiled the presence of four, two and one isoforms of CAT in non-treated *N. nervosa* plants at 1DPT, 3DPT and 6DPT, respectively. Conversely, in all PEG-treated *N. nervosa* plants, only one CAT isoform (CAT4) was detected at all experimental time points (Figure 3A,B). The spectrophotometric assay revealed a significant increase in CAT activity (29%) at 1DPT, whereas prolonged treatment for 3 and 6 days led to notable reductions in CAT activity (53% and 47% respectively) when compared to non-treated plants. The lowest CAT activity value (6.50 U/mg of total proteins) was recorded in *N. nervosa* plants at 6DPT (Figure 3C).

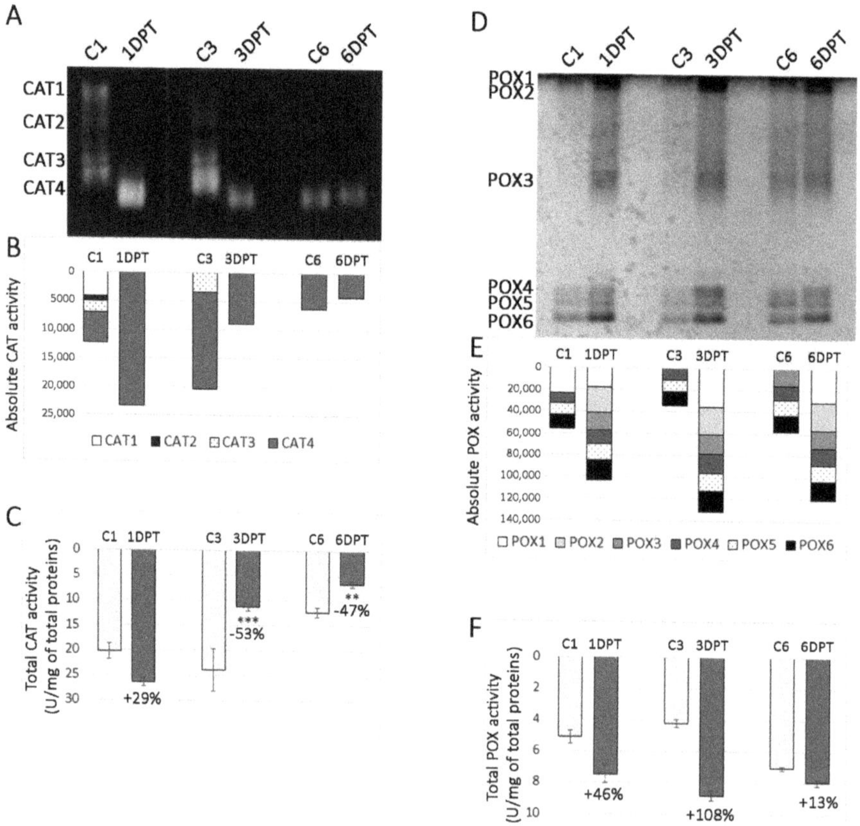

Figure 3. Catalases (CAT) and peroxidases (POX) activity in non-treated (C1, C3 and C6) and PEG-treated (1DPT, 3DPT and 6DPT) *N. nervosa* plants for 1, 3 and 6 days. (**A**) For the CAT assay, 5 µg of total soluble proteins extracted from the whole plant was loaded per lane on a 7% gel and separated by native-PAGE. (**B**) The detected activities were measured densitometrically by the Image J program and presented as absolute values. (**C**) Total CAT activity was measured spectrometrically. (**D**) For the POX assay, 20 µg of total soluble proteins extracted from the whole plant was loaded per lane on a 10% gel and separated by native-PAGE. (**E**) The detected activities were measured densitometrically by the Imaging System iBRIGHT CL1500 and presented as absolute values. (**F**) Total POX activity was measured spectrometrically. The results represent the mean ± SE of three biological replicates. Asterisks (*) indicate significant differences between PEG-treated and non-treated plants according to an unpaired *t*-test (** $p \leq 0.01$; *** $p \leq 0.001$). The numbers below the bars indicate the percent of enhancement (+)/reduction (−) relative to the control.

The staining of native-PAGE gels for POX activity revealed the presence of six isoforms in all PEG-treated *N. nervosa* plants, while non-treated plants exhibited three isoforms at 3DPT and four isoforms at the 1DPT and 6DPT (Figure 3D,E). Spectrophotometric measurement of POX activity demonstrated an increase ranging from 13% to 108% depending on the duration of PEG treatment. Specifically, the activity increased by 108% at 3DPT and by 13% at 6DPT compared to non-treated plants. The measured values of POX activity ranged from 7.47 to 8.82 U/mg of total proteins in *N. nervosa* plants subjected to PEG treatment for 1, 3 and 6 days (Figure 3F).

Staining native gels for SOD activity revealed the presence of four isoforms in both PEG-treated and non-treated *N. nervosa* plants at 1, 3 and 6 days (Figure 4A,B). Through the use of specific inhibitors (KCN and H_2O_2), Mn-SOD and Fe-SOD isoforms were distinguished (Figure 4D,E). Among the samples, one Fe-SOD and three Mn-SOD isoforms (Mn-SOD1, Mn-SOD2 and Mn-SOD3) were identified, with Mn-SOD3 exhibiting the highest activity in PEG-treated *N. nervosa* plants across all treatment durations (Figure 4A,B,D,E). The total SOD activities in non-treated plants remained similar in all experimental points, measuring approximately 5 U/mg of total proteins. However, in PEG-treated plants, the total SOD activities significantly increased, ranging from 78% to as high as 853% with the extension of the treatment duration from 1 to 6 days, compared to the non-treated control (Figure 4C).

Previous studies have extensively reported the efficient role of CAT and POX enzymes in effectively removing intracellular H_2O_2 by converting it into H_2O and O_2 [63]. In our study, we observed a significant increase in POX activity in the leaves of *N. nervosa* plants in all experimental points compared to the control. The observed pattern of POX activity correlates with the concentration of RA throughout the entire experiment. Interestingly, except for at 1DPT, CAT activity was significantly decreased upon PEG treatment, suggesting that H_2O_2 elimination is predominantly carried out by POX. This finding aligns with similar trends observed in wheat, where CAT activities increased during the early phase of drought stress but decreased with prolonged water-stress duration [64]. Moreover, [65] demonstrated an increase in CAT activity under PEG-mediated drought stress in the shoots and roots of two sunflower cultivars (Musala and Aydın), which is consistent with our results for *N. nervosa* PEG-treated plants at 1DPT. In *Nepeta pannonica* shoot cultures, the activities of the oxygen scavenging enzymes CAT and POX were progressively enhanced in response to oxidative stress induced by increasing concentrations of paraquat [51].

Superoxide dismutase is a well-known enzyme that plays a vital role in neutralizing harmful oxygen molecules ($O_2^{\cdot}$) by converting them into less reactive species such as H_2O_2 and H_2O. In our study, the activity of SOD, particularly the MnSOD3 isoform, was significantly elevated in the leaves of *N. nervosa* plants treated with PEG when compared to the control (up to an 8.5-fold increase at 6DPT). This substantial increase in SOD activity highlights the robust effort of the enzyme, particularly MnSOD3, in scavenging cellular free radicals. This finding aligns with previous research where PEG-induced water stress also resulted in increased SOD activities in sugarcane leaves [66].

The findings of our study reveal an interesting phenomenon in *N. nervosa* leaves, where PEG treatment leads to an increased number of POX and SOD isoforms compared to the corresponding controls. On the other hand, the number of CAT isoforms remains either the same or shows a lesser increase. Additionally, we observed that the most mobile isoforms of CAT, POX and SOD exhibited the highest activities in PEG-treated *N. nervosa* plants for 1, 3 and 6 days. This finding is significant because it indicates a dynamic response of antioxidative enzymes to PEG-induced water stress. The increased number of POX and SOD isoforms suggests an enhanced capacity to scavenge ROS and protect the plant cells from oxidative damage. Furthermore, the high activities of these isoforms in PEG-treated plants indicate their crucial role in combating ROS and maintaining cellular homeostasis under water-stress conditions.

These findings are consistent with previous research by [47], who demonstrated increased activities of CAT, SOD and POX in sweet basil leaves under drought stress.

Similarly, Sarker et al. 2018 [67] confirmed that CAT and SOD play important roles in detoxifying ROS in both sensitive and tolerant genotypes of *Amaranthus tricolor* under drought stress. Collectively, these studies emphasize the importance of antioxidative enzymes in mitigating the detrimental effects of drought stress and highlight their potential as targets for enhancing plant stress tolerance in agricultural practice.

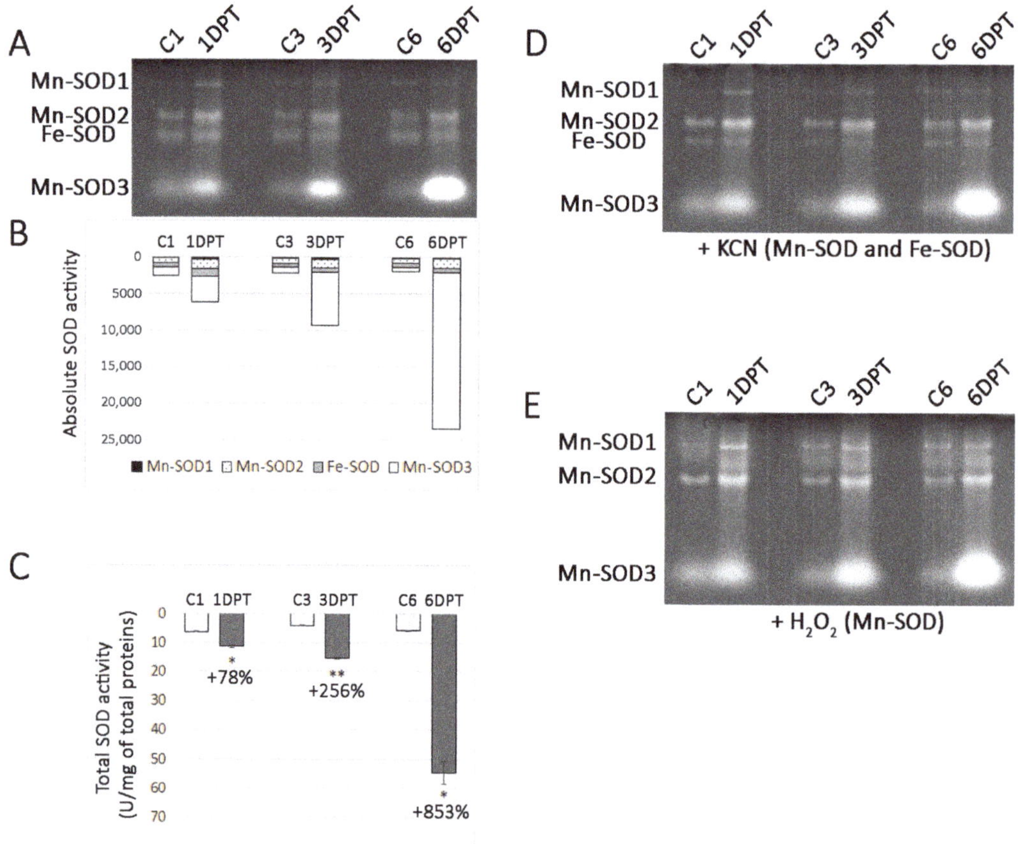

Figure 4. Superoxide dismutase (SOD) activity in non-treated (C1, C3 and C6) and PEG-treated (1DPT, 3DPT and 6DPT) *N. nervosa* plants for 1, 3 and 6 days. (**A**) For the SOD assay, 20 µg of total soluble proteins extracted from the whole plant was loaded per lane on 10% gel and separated by native-PAGE. (**B**) The detected activities were measured densitometrically by the Image J program and presented as absolute values. (**C**) Total SOD activity was measured spectrometrically. (**D**,**E**) Identification of SOD isoforms (Mn-SOD and Fe-SOD) based on differential sensitivity to inhibitors. The results represent the mean ± SE of three biological replicates. Asterisks (*) indicate significant differences between PEG-treated and non-treated plants according to an unpaired *t*-test (* $p \leq 0.05$; ** $p \leq 0.01$). The numbers below the bars indicate the percent of enhancement (+) relative to the control. The results represent the mean ± SE of three biological replicates.

4. Conclusions

In conclusion, our study on PEG-induced dehydration in *Nepeta nervosa* plants grown in vitro has unveiled a comprehensive understanding of both enzymatic and non-enzymatic antioxidant protective mechanisms. The enhanced enzymatic antioxidative defense mechanisms, particularly POX and SOD, provide insights into stress tolerance and plant adaptability. While CAT activity increased initially (1DPT) but decreased with prolonged treatment,

POX and SOD (especially the Mn-SOD3 isoform) activities exhibited a substantial increase at 3DPT and 6DPT, underscoring their roles in radical scavenging activity. Furthermore, the significant increases in RA at 3DPT and CHLA at 3DPT and 6DPT, along with CAFFA at 6DPT, have revealed the predominant non-enzymatic antioxidant components during water-stress conditions in *N. nervosa*. The increase in RA, a valuable phenolic compound, at 3DPT suggests the potential for scaling up its production for applications in pharmaceuticals, cosmetics and nutraceuticals. Overall, our findings contribute to the understanding of *N. nervosa* responses to dehydration and offer promising avenues for industrial applications related to phenolic compounds, antioxidative defense and plant stress physiology.

Author Contributions: Conceptualization, J.N.Ž., S.D. and N.A.; methodology, J.N.Ž., S.D., N.A., M.S., D.M., B.F. and T.M.; validation, J.N.Ž., S.D. and N.A.; formal analysis, J.N.Ž., S.D., M.S., D.M. and B.F.; investigation, M.S., D.M., B.F., J.N.Ž., S.D. and T.M.; data curation, S.D., J.N.Ž. and N.A.; writing—original draft preparation, S.D., J.N.Ž. and N.A.; writing—review and editing, N.A., S.D., B.F., M.S. and D.M.; visualization, J.N.Ž., S.D. and B.F.; supervision, S.D.; funding acquisition, S.D. All authors have read and agreed to the published version of the manuscript.

Funding: This research was funded by the Ministry of Science, Technological Development and Innovation of the Republic of Serbia, Grant No. 451-03-47/2023-01/200007.

Institutional Review Board Statement: Not applicable.

Informed Consent Statement: Not applicable.

Data Availability Statement: The data presented in this study are available on request from the corresponding authors.

Acknowledgments: The authors would like to thank the Ministry of Science, Technological Development and Innovation of the Republic of Serbia for the financial support of this research. The authors would like to thank Danijela Mišić and Uroš Gašić (Institute for Biological Research "Siniša Stanković"—National Institute of the Republic of Serbia, University of Belgrade, Serbia) for valuable help during the process of the work and Milica Bogdanović (Institute for Biological Research "Siniša Stanković"—National Institute of the Republic of Serbia, University of Belgrade, Serbia) for help in photographing the dehydrated leaves.

Conflicts of Interest: The authors declare no conflict of interest.

References

1. Rhodes, D.; Nadolska-Orczyk, A. Stress factors, their influence on plant metabolism, and tolerance or resistance to stress. *Plant Stress Phys.* **2001**, *23*, 1–7.
2. Lamers, J.; van der Meer, T.; Testerink, C. How plants sense and respond to stressful environments. *Plant Physiol.* **2020**, *182*, 1624–1635. [CrossRef] [PubMed]
3. Zhu, J.K. Abiotic stress signaling and responses in plants. *Cell* **2016**, *167*, 313–324. [CrossRef]
4. Saijo, Y.; Loo, E.P.I. Plant immunity in signal integration between biotic and abiotic stress responses. *New Phytol.* **2020**, *225*, 87–104. [CrossRef]
5. Shao, H.B.; Chu, L.Y.; Jaleel, C.A.; Zhao, C.X. Water-deficit stress-induced anatomical changes in higher plants. *C. R. Biol.* **2008**, *331*, 215–225. [CrossRef]
6. Sah, S.K.; Reddy, K.R.; Li, J. Abscisic acid and abiotic stress tolerance in crop plants. *Front. Plant Sci.* **2016**, *7*, 571. [CrossRef] [PubMed]
7. Lee, S.C.; Luan, S. ABA signal transduction at the crossroad of biotic and abiotic stress responses. *Plant Cell Environ.* **2012**, *35*, 53–60. [CrossRef]
8. Radwan, A.; Kleinwächter, M.; Selmar, D. Impact of drought stress on specialised metabolism: Biosynthesis and the expression of monoterpene synthases in sage (*Salvia officinalis*). *Phytochemistry* **2017**, *141*, 20–26. [CrossRef]
9. Camejo, D.; Guzmán-Cedeño, Á.; Moreno, A. Reactive oxygen species, essential molecules, during plant–pathogen interactions. *Plant Phys. Biochem.* **2016**, *103*, 10–23. [CrossRef]
10. Sachdev, S.; Ansari, S.A.; Ansari, M.I.; Fujita, M.; Hasanuzzaman, M. Abiotic stress and reactive oxygen species: Generation, signaling, and defense mechanisms. *Antioxidants* **2021**, *10*, 277. [CrossRef]
11. Das, K.; Roychoudhury, A. Reactive oxygen species (ROS) and response of antioxidants as ROS-scavengers during environmental stress in plants. *Front. Environ. Sci.* **2014**, *2*, 53. [CrossRef]
12. Veitch, N.C. Structural determinants of plant peroxidase function. *Phytochem. Rev.* **2004**, *3*, 3–18. [CrossRef]

13. Tohidi, B.; Rahimmalek, M.; Arzani, A. Essential oil composition, total phenolic, flavonoid contents, and antioxidant activity of *Thymus* species collected from different regions of Iran. *Food Chem.* **2017**, *220*, 153–161. [CrossRef]
14. Abdallah, S.B.; Rabhi, M.; Harbaoui, F.; Zar-kalai, F.; Lachâal, M.; Karray-Bouraoui, N. Distribution of phenolic compounds and antioxidant activity between young and old leaves of *Carthamus tinctorius* L. and their induction by salt stress. *Acta Phys. Plant.* **2013**, *35*, 1161–1169. [CrossRef]
15. Bistgani, Z.E.; Hashemi, M.; Dacosta, M.; Craker, L.; Maggi, F.; Morshedloo, M.R. Effect of salinity stress on the physiological characteristics, phenolic compounds and antioxidant activity of *Thymus vulgaris* L. and *Thymus daenensis* Celak. *Ind. Crop. Prod.* **2019**, *135*, 311–320. [CrossRef]
16. Chen, Z.; Ma, Y.; Yang, R.; Gu, Z.; Wang, P. Effects of exogenous Ca^{2+} on phenolic accumulation and physiological changes in germinated wheat (*Triticum aestivum* L.) under UV-B radiation. *Food Chem.* **2019**, *288*, 368–376. [CrossRef]
17. Daniels, C.W.; Rautenbach, F.; Marnewick, J.L.; Valentine, A.J.; Babajide, O.J.; Mabusela, W.T. Environmental stress effect on the phytochemistry and antioxidant activity of a South African bulbous geophyte, *Gethyllis multifolia* L. Bolus. *S. Afr. J. Bot.* **2015**, *96*, 29–36. [CrossRef]
18. Vosoughi, N.; Gomarian, M.; Pirbalouti, A.G.; Khaghani, S.; Malekpoor, F. Essential oil composition and total phenolic, flavonoid contents, and antioxidant activity of sage (*Salvia officinalis* L.) extract under chitosan application and irrigation frequencies. *Ind. Crops Prod.* **2018**, *117*, 366–374. [CrossRef]
19. Šamec, D.; Karalija, E.; Šola, I.; Vujčić Bok, V.; Salopek-Sondi, B. The role of polyphenols in abiotic stress response: The influence of molecular structure. *Plants* **2021**, *10*, 118. [CrossRef]
20. Bezerra, G.S.N.; Pereira, M.A.V.; Ostrosky, E.A.; Barbosa, E.G.; Moura, M.D.F.V.D.; Ferrari, M.; Aragão, C.F.S.; Gomes, A.P.B. Compatibility study between ferulic acid and excipients used in cosmetic formulations by TG/DTG, DSC and FTIR. *J. Therm. Anal. Calorim.* **2016**, *127*, 1683–1691. [CrossRef]
21. Ruwizhi, N.; Aderibigbe, B.A. Cinnamic acid derivatives and their biological efficacy. *Int. J. Mol. Sci.* **2020**, *21*, 5712. [CrossRef]
22. Wang, S.-J.; Chen, Q.; Liu, M.-Y.; Yu, H.-Y.; Xu, J.-Q.; Wu, J.-Q.; Zhang, Y.; Wang, T. Regulation effects of rosemary (*Rosmarinus officinalis* Linn.) on hepatic lipid metabolism in OA induced NAFLD rats. *Food Funct.* **2019**, *10*, 7356–7365. [CrossRef]
23. Georgiev, M.; Pastore, S.; Lulli, D.; Alipieva, K.; Kostyuk, V.; Potapovich, A.; Panetta, M.; Korkina, L. Verbascum xanthophoeniceum-derived phenylethanoid glycosides are potent inhibitors of inflammatory chemokines in dormant and interferon-gamma-stimulated human keratinocytes. *J. Ethnopharmacol.* **2012**, *144*, 754–760. [CrossRef]
24. Zhao, L.; Zhang, Y.; Liu, G.; Hao, S.; Wang, C.; Wang, Y. Black rice anthocyanin-rich extract and rosmarinic acid, alone and in combination, protect against DSS-induced colitis in mice. *Food Funct.* **2018**, *9*, 2796–2808. [CrossRef] [PubMed]
25. Costa, P.; Sarmento, B.; Gonçalves, S.; Romano, A. Protective effects of *Lavandula viridis* L'Hér extracts and rosmarinic acid against H_2O_2-induced oxidative damage in A172 human astrocyte cell line. *Ind. Crops Prod.* **2013**, *50*, 361–365. [CrossRef]
26. Aničić, N.; Gašić, U.; Lu, F.; Ćirić, A.; Ivanov, M.; Jevtić, B.; Dimitrijević, M.; Anđelković, B.; Skorić, M.; Živković, J.N.; et al. Antimicrobial and immunomodulating activities of two endemic *Nepeta* species and their major iridoids isolated from natural sources. *Pharmaceuticals* **2021**, *14*, 414. [CrossRef] [PubMed]
27. Ferraro, V.; Madureira, A.R.; Sarmento, B.; Gomes, A.; Pintado, M.E. Study of the interactions between rosmarinic acid and bovine milk whey protein α-Lactalbumin, β-Lactoglobulin and Lactoferrin. *Food Res. Int.* **2015**, *77*, 450–459. [CrossRef]
28. Li, Z.; Henning, S.M.; Zhang, Y.; Zerlin, A.; Li, L.; Gao, K.; Lee, R.-P.; Karp, H.; Thames, G.; Bowerman, S.; et al. Antioxidant-rich spice added to hamburger meat during cooking results in reduced meat, plasma, and urine malondialdehyde concentrations. *Am. J. Clin. Nutr.* **2010**, *91*, 1180–1184. [CrossRef] [PubMed]
29. Li, P.; Yang, X.; Lee, W.J.; Huang, F.; Wang, Y.; Li, Y. Comparison between synthetic and rosemary-based antioxidants for the deep frying of French fries in refined soybean oils evaluated by chemical and non-destructive rapid methods. *Food Chem.* **2021**, *335*, 127638. [CrossRef] [PubMed]
30. Marchev, A.S.; Vasileva, L.V.; Amirova, K.M.; Savova, M.S.; Koycheva, I.K.; Balcheva-Sivenova, Z.P.; Vasileva, S.M.; Georgiev, M.I. Rosmarinic acid—From bench to valuable applications in food industry. *Trends Food Sci. Technol.* **2021**, *117*, 182–193. [CrossRef]
31. Tupas, G.D.; Otero, M.C.B.; Ebhohimen, I.E.; Egbuna, C.; Aslam, M. Antidiabetic lead compounds and targets for drug development. In *Phytochemicals as Lead Compounds for New Drug Discovery*; Elsevier: Amsterdam, The Netherlands, 2020; pp. 127–141.
32. Açıkgöz, M.A. Establishment of cell suspension cultures of *Ocimum basilicum* L. and enhanced production of pharmaceutical active ingredients. *Ind. Crops Prod.* **2020**, *148*, 112278. [CrossRef]
33. Gonçalves, S.; Mansinhos, I.; Rodríguez-Solana, R.; Pérez-Santín, E.; Coelho, N.; Romano, A. Elicitation improves rosmarinic acid content and antioxidant activity in *Thymus lotocephalus* shoot cultures. *Ind. Crops Prod.* **2019**, *137*, 214–220. [CrossRef]
34. Mišić, D.; Šiler, B.; Gašić, U.; Avramov, S.; Živković, S.; Živković, J.N.; Milutinović, M.; Tešić, Ž. Simultaneous UHPLC/DAD/(+/-)HESI-MS/MS analysis of phenolic acids and nepetalactones in methanol extracts of *Nepeta* species: A possible application in chemotaxonomic studies. *Phytochem. Anal.* **2015**, *26*, 72–85. [CrossRef] [PubMed]
35. Murashige, T.; Skoog, F. A revised medium for rapid growth and bio assays with tobacco tissue cultures. *Physiol. Plant.* **1962**, *15*, 473–497. [CrossRef]
36. Aničić, N.; Matekalo, D.; Skorić, M.; Živković, J.N.; Petrović, L.; Dragićević, M.; Dmitrović, S.; Mišić, D. Alterations in nepetalactone metabolism during polyethylene glycol (PEG)-induced dehydration stress in two *Nepeta* species. *Phytochemistry* **2020**, *174*, 112340. [CrossRef] [PubMed]

37. Michel, B.E. Evaluation of the water potentials of solutions of polyethylene glycol 8000 both in the absence and presence of other solutes. *Plant Physiol.* **1983**, *72*, 66–70. [CrossRef] [PubMed]
38. Dmitrović, S.; Perišić, M.; Stojić, A.; Živković, S.; Boljević, J.; Živković, J.N.; Aničić, N.; Ristić, M.; Mišić, D. Essential oils of two *Nepeta* species inhibit growth and induce oxidative stress in ragweed (*Ambrosia artemisiifolia* L.) shoots in vitro. *Acta Physiol. Plant.* **2015**, *37*, 64. [CrossRef]
39. Bradford, M.M. A rapid and sensitive method for the quantitation of microgram quantities of protein utilizing the principle of protein-dye binding. *Anal. Biochem.* **1976**, *72*, 248–254. [CrossRef] [PubMed]
40. Yang, X.; Lu, M.; Wang, Y.; Wang, Y.; Liu, Z.; Chen, S. Response mechanism of plants to drought stress. *Horticulturae* **2021**, *7*, 50. [CrossRef]
41. Zhao, J.; Davis, L.C.; Verpoorte, R. Elicitor signal transduction leading to production of plant secondary metabolites. *Biotechnol. Adv.* **2005**, *23*, 283–333. [CrossRef]
42. Zhao, J.; Fujita, K.; Sakai, K. Oxidative stress in plant cell culture: A role in production of β-thujaplicin by *Cupresssus lusitanica* suspension culture. *Biotechnol. Bioeng.* **2005**, *90*, 621–631. [CrossRef] [PubMed]
43. Petersen, M. Rosmarinic acid: New aspects. *Phytochem. Rev.* **2013**, *12*, 207–227. [CrossRef]
44. Liu, H.; Wang, X.; Wang, D.; Zou, Z.; Liang, Z. Effect of drought stress on growth and accumulation of active constituents in *Salvia miltiorrhiza* Bunge. *Ind. Crops Prod.* **2011**, *33*, 84–88. [CrossRef]
45. Szabó, K.; Radácsi, P.; Rajhárt, P.; Ladányi, M.; Németh, É. Stress-induced changes of growth, yield and bioactive compounds in lemon balm cultivars. *Plant Physiol. Biochem.* **2017**, *119*, 170–177. [CrossRef] [PubMed]
46. Chen, Y.; Guo, Q.; Liu, L.; Liao, L.; Zhu, Z. Influence of fertilization and drought stress on the growth and production of secondary metabolites in *Prunella vulgaris* L. *J. Med. Plants Res.* **2011**, *5*, 1749–1755. Available online: http://www.academicjournals.org/JMPR (accessed on 21 February 2011).
47. Zulfiqar, F.; Chen, J.; Finnegan, P.M.; Younis, A.; Nafees, M.; Zorrig, W.; Hamed, K.B. Application of trehalose and salicylic acid mitigates drought stress in sweet basil and improves plant growth. *Plants* **2021**, *10*, 78. [CrossRef]
48. Osmolovskaya, N.; Shumilina, J.; Kim, A.; Didio, A.; Grishina, T.; Bilova, T.; Keltsieva, O.A.; Zhukov, V.; Tikhonovich, I.; Tarakhovskaya, E.; et al. Methodology of drought stress research: Experimental setup and physiological characterization. *Int. J. Mol. Sci.* **2018**, *19*, 89. [CrossRef]
49. Hadi, N.; Sefidkon, F.; Shojaeiyan, A.; Šiler, B.; Jafari, A.A.; Aničić, N.; Mišić, D. Phenolics' composition in four endemic *Nepeta* species from Iran cultivated under experimental field conditions: The possibility of the exploitation of *Nepeta* germplasm. *Ind. Crops Prod.* **2017**, *95*, 475–484. [CrossRef]
50. Duda, S.C.; Mărghitaş, L.A.; Dezmirean, D.; Duda, M.; Mărgăoan, R.; Bobiş, O. Changes in major bioactive compounds with antioxidant activity of *Agastache foeniculum*, *Lavandula angustifolia*, *Melissa officinalis* and *Nepeta cataria*: Effect of harvest time and plant species. *Ind. Crops Prod.* **2015**, *77*, 499–507. [CrossRef]
51. Cvetković, J.; Milutinović, M.; Božunović, J.; Aničić, N.; Nestorović Živković, J.; Živković, S.; Mišić, D. Paraquat-mediated oxidative stress in *Nepeta pannonica* L. *Bot. Serbica* **2015**, *39*, 121–128.
52. Smiljković, M.; Dias, M.I.; Stojković, D.; Barros, L.; Bukvički, D.; Ferreira, I.C.; Soković, M. Characterization of phenolic compounds in tincture of edible *Nepeta nuda*: Development of antimicrobial mouthwash. *Food Funct.* **2018**, *9*, 5417–5425. [CrossRef] [PubMed]
53. Nestorović, J.; Mišić, D.; Šiler, B.; Soković, M.; Glamočlija, J.; Ćirić, A.; Maksimović, V.; Grubišić, D. Nepetalactone content in shoot cultures of three endemic *Nepeta* species and the evaluation of their antimicrobial activity. *Fitoterapia* **2010**, *81*, 621–626. [CrossRef] [PubMed]
54. Nestorovic-Zivkovic, J.; Zivkovic, S.; Siler, B.; Anicic, N.; Dmitrovic, S.; Divac-Rankov, A.; Giba, Z.; Misic, D. Differences in bioactivity of three endemic *Nepeta* species arising from main terpenoid and phenolic constituents. *Arch. Biol. Sci.* **2018**, *70*, 63–76. [CrossRef]
55. Hosseini, N.S.; Ghasimi Hagh, Z.; Khoshghalb, H. Morphological, antioxidant enzyme activity and secondary metabolites accumulation in response of polyethylene glycol-induced osmotic stress in embryo-derived plantlets and callus cultures of *Salvia leriifolia*. *Plant Cell. Tissue Organ Cult.* **2020**, *140*, 143–155. [CrossRef]
56. Mansinhos, I.; Gonçalves, S.; Rodríguez-Solana, R.; Duarte, H.; Ordóñez-Díaz, J.L.; Moreno-Rojas, J.M.; Romano, A. Response of *Thymus lotocephalus* in vitro cultures to drought stress and role of green extracts in cosmetics. *Antioxidants* **2022**, *11*, 1475. [CrossRef]
57. Pietta, P.G. Flavonoids as antioxidants. *J. Nat. Prod.* **2000**, *63*, 1035–1042. [CrossRef] [PubMed]
58. Sharma, A.; Shahzad, B.; Rehman, A.; Bhardwaj, R.; Landi, M.; Zheng, B. Response of phenylpropanoid pathway and the role of polyphenols in plants under abiotic stress. *Molecules* **2019**, *24*, 2452. [CrossRef]
59. Ibrahim, K.M.; Musbah, H.M. Increasing poly phenols in *Coleus blumei* at the cellular and intact plant levels using PEG stress. *Res. J. Pharm. Technol.* **2018**, *11*, 321–327. [CrossRef]
60. Razavizadeh, R.; Farahzadianpoor, F.; Adabavazeh, F.; Komatsu, S. Physiological and morphological analyses of *Thymus vulgaris* L. in vitro cultures under polyethylene glycol (PEG)-induced osmotic stress. *Vitr. Cell. Dev. Biol. Plant* **2019**, *55*, 342–357. [CrossRef]
61. Bettaieb, I.; Hamrouni-Sellami, I.; Bourgou, S.; Limam, F.; Marzouk, B. Drought effects on polyphenol composition and antioxidant activities in aerial parts of *Salvia officinalis* L. *Acta Physiol. Plant.* **2011**, *33*, 1103–1111. [CrossRef]
62. Krasensky, J.; Jonak, C. Drought, salt, and temperature stress-induced metabolic rearrangements and regulatory networks. *J. Exp. Bot.* **2012**, *63*, 1593–1608. [CrossRef] [PubMed]

63. Singh, H.P.; Kaur, S.; Mittal, S.; Batish, D.R.; Kohli, R.K. Essential oil of *Artemisia scoparia* inhibits plant growth by generating reactive oxygen species and causing oxidative damage. *J. Chem. Ecol.* **2009**, *35*, 154–162. [CrossRef] [PubMed]
64. Zhang, J.; Kirkham, M.B. Drought-stress-induced changes in activities of superoxide dismutase, catalase, and peroxidase in wheat species. *Plant Cell Physiol.* **1994**, *35*, 785–791. [CrossRef]
65. Baloğlu, M.C.; Kavas, M.; Aydin, G.; Öktem, H.A.; Yücel, A.M. Antioxidative and physiological responses of two sunflower (*Helianthus annuus*) cultivars under PEG-mediated drought stress. *Turk. J. Bot.* **2012**, *36*, 707–714. [CrossRef]
66. Patade, V.Y.; Bhargava, S.; Suprasanna, P. Salt and drought tolerance of sugarcane under iso-osmotic salt and water stress: Growth, osmolytes accumulation, and antioxidant defense. *J. Plant Interact.* **2011**, *6*, 275–282. [CrossRef]
67. Sarker, U.; Oba, S. Catalase, superoxide dismutase and ascorbate-glutathione cycle enzymes confer drought tolerance of *Amaranthus tricolor*. *Sci. Rep.* **2018**, *8*, 16496. [CrossRef]

horticulturae

MDPI

Article

Unlocking the Potential of *In Vitro* Photoautotrophy for *Eryngium foetidum*: Biomass, Morphophysiology, and Acclimatization

Darlyara Reis Silva [1], Marion Nayon Braga Soares [1], Maria Cristina Rocha Silva [1], Mariana Costa Lima [1], Vitória Karla de Oliveira Silva-Moraes [2], Givago Lopes Alves [2], Anyela Marcela Ríos-Ríos [3], Aldilene da Silva Lima [2], Marcos Vinícius Marques Pinheiro [4], Thais Roseli Corrêa [2,5], Diego Silva Batista [5,6], Fábio Afonso Mazzei Moura de Assis Figueiredo [5], Fabrício de Oliveira Reis [2], Tiago Massi Ferraz [2,5] and Sérgio Heitor Sousa Felipe [5,*]

1 Centro de Ciências Agrárias, Universidade Estadual do Maranhão, São Luís 65055-310, Brazil; darlyarareissilva.125@gmail.com (D.R.S.); mnbs2018sh@gmail.com (M.N.B.S.); maria.r.silva1508@gmail.com (M.C.R.S.); lcmarianalima@gmail.com (M.C.L.)

2 Programa de Pós-Graduação em Agroecologia, Universidade Estadual do Maranhão, São Luís 65055-310, Brazil; vitoriakarlaos@gmail.com (V.K.d.O.S.-M.); engivago@gmail.com (G.L.A.); aldilene29@gmail.com (A.d.S.L.); thaiscorrea@professor.uema.br (T.R.C.); fareoli@gmail.com (F.d.O.R.); ferraztm@gmail.com (T.M.F.)

3 Programa de Pós-Graduação em Biodiversidade e Biotecnologia da Amazônia Legal–BIONORTE, Universidade Estadual do Maranhão, São Luís 65055-310, Brazil; anyelamrios@gmail.com

4 Universidade Regional Integrada do Alto Uruguai e das Missões, Campus Frederico Westphalen (URI-FW), Frederico Westphalen 98400-000, Brazil; macvini@gmail.com

5 Programa de Pós-Graduação em Agricultura e Ambiente, Universidade Estadual do Maranhão, São Luís 65055-310, Brazil; diegoesperanca@gmail.com (D.S.B.); figueiredo.uema@gmail.com (F.A.M.M.d.A.F.)

6 Departamento de Agricultura, Universidade Federal da Paraíba, Campus III, Bananeiras 58220-000, Brazil

* Correspondence: sergio.h.s.felipe@gmail.com

Abstract: *Eryngium foetidum* L., a biennial herb with diverse applications in food and traditional medicine, holds economic and pharmacological significance. Given its growing commercial interest, implementing biotechnological approaches like plant tissue culture is vital for sustainable propagation and metabolite production. In this study, we investigated the *in vitro* photoautotrophic potential of *Eryngium foetidum*, examining growth, chlorophyll *a* fluorescence, photosynthetic pigments, and anatomical features under sucrose concentrations (0 and 30 g L^{-1}) and gas exchange rate (14 and 25 $\mu L\ L^{-1}\ s^{-1}\ CO_2$). Acclimatization and survival rates of plants after *ex vitro* transfer were also assessed. *Eryngium foetidum* exhibited robust growth in both photoautotrophic and photomixotrophic conditions, with natural ventilation significantly enhancing plant development. Chlorophyll *a* fluorescence and photosynthetic performance were influenced by sucrose and gas exchange, highlighting the importance of these factors in plant micropropagation. Moreover, the species demonstrated remarkable plasticity during acclimatization, with high survival rates and rapid inflorescence development. The research provides valuable insights into optimizing *in vitro* cultivation conditions for *Eryngium foetidum*, emphasizing the potential for large-scale clonal propagation and exploring secondary metabolites. The observed phenotypic plasticity underscores the adaptability of the species to diverse environments. These biotechnological strategies open avenues for future studies, including the application of elicitors for enhanced secondary metabolite production.

Keywords: culantro; gas exchange rate; natural ventilation; plant tissue culture; sucrose supplementation

1. Introduction

Eryngium foetidum L. (Apiaceae family), commonly known as culantro, is a biennial herb with an erect growth habit, featuring a rosette of long spatulate spiny-toothed leaves and reaching a height of 10–50 cm after inflorescence development. It is commonly found in tropical and subtropical regions [1,2]. This species serves a dual purpose, functioning both as a food source (unconventional seasoning vegetable) and a traditional medicine in Latin America, the Central and Pacific Islands, Africa, and Asia [3]. The leaves of *Eryngium foetidum* are notably rich in phosphorus, potassium, and dietary micronutrients, making them suitable for potential use in food fortification [4–6]. In the realm of medicine, *Eryngium foetidum* is recognized for its high biological activity as an antioxidant. It is utilized in the treatment of diseases related to the gastrointestinal tract and exhibits antibacterial, analgesic, anti-inflammatory, anthelmintic, anticonvulsant, and anticancer properties [7–9].

The pharmacological importance attributed to *Eryngium foetidum* is particularly notable due to the various classes of compounds produced in its secondary metabolism, including aldehydes, carotenoids, phenols, and anthraquinones [7]. Furthermore, it is important to note that essential oils are found in the leaves, including molecules such as (2*E*)-2-dodecenal, *trans*-2-tetradecenal, 1-dodecanal, 1-decanal, tetradecanal, γ-terpinene, mesitylene, ρ-cymene, α-pinene, 1-undecanal, carotol, and 1-tetradecene [10]. Collectively, this phytochemical profile adds economic value to international trade and in the cosmetic, pharmaceutical, and food industries [7–11].

Due to the increasing commercial interest in *Eryngium foetidum*, it becomes essential to implement biotechnological approaches, such as plant tissue culture, for the rational propagation of the species. This is crucial to mitigate the risk of genetic erosion resulting from the disorderly exploitation of the population [12]. Furthermore, apart from being a biotechnological strategy, plant tissue culture offers the potential for controlled production of metabolites of interest, in contrast to field conditions. It also enables the exploration of complex biosynthetic pathways that have not been achievable artificially until now [13,14].

Plant tissue culture serves as a platform for plant cloning under aseptic conditions and environmental control in the laboratory [15]. Traditionally, *in vitro* multiplication relies on three nutritional types of cultures: heterotrophic, where there are no photosynthetically active organs, and the carbon source is solely the carbohydrate supplied in the medium; photomixotrophic, where photosynthetically active organs use the carbohydrate in the medium, along with the consumption of CO_2 in the flask headspace; and photoautotrophic (or sugar-free medium), where photosynthetically active organs rely on CO_2 in the flask headspace as the carbon source [16,17]. The latter two cultivation systems, photoautotrophic and photomixotrophic, promote *in vitro* plant growth compared to the first (heterotrophic), which is more suitable for callus cultivation [16].

While photoautotrophic cultivation requires the use of gas-permeable films to enhance gas exchange between the flask headspace and the external environment, these films can also be employed in photomixotrophic (sugar-provided medium) cultivation to improve the ventilation rate of the culture flask [17–19]. Investigating the impact of natural ventilation on both photoautotrophy and photomixotrophy in the *in vitro* cultivation of *Eryngium foetidum* plants may offer valuable insights into the morphophysiological characteristics influencing the growth and development of plants in these systems. This research can contribute new knowledge for implementing strategies applicable to clonal propagation on a large scale and in the *in vitro* germplasm banks.

Considering the information presented above, our objective was to assess the *in vitro* photoautotrophic potential of *Eryngium foetidum*. In this study, we initially characterized the growth and development of *Eryngium foetidum* plants under *in vitro* photomixotrophic conditions (without and with natural ventilation) and photoautotrophic systems. Subsequently, we examined the acclimatization of these plants. Our research provides novel insights into how *Eryngium foetidum* responds to photomixotrophy and photoautotrophy, demonstrating its adaptability to different carbon sources.

2. Materials and Methods

2.1. Plant Material and Experimental Design

The *Eryngium foetidum* plants from a natural population utilized in the experiments were sourced from the Germplasm Bank at the Tissue Culture Laboratory (LCT), State University of Maranhão, MA, Brazil (2°34′00″ S and 44°12′00″ W). These plants are regularly maintained in 350 mL transparent glass flasks, each containing 60 mL of Murashige and Skoog (MS) medium with vitamins [20] (PhytoTechnology®, Lenexa, KS, USA), 3% sucrose (*w*/*v*; Dinâmica® Química Contemporânea Ltd.a, São Paulo, SP, Brazil), and solidified with 5.5 g L^{-1} agar (Dinâmica® Química Contemporânea Ltd.a, São Paulo, SP, Brazil). The maintenance conditions include a temperature of 25 ± 2 °C, an irradiance of 85 μmol m^{-2} s^{-1} (provided by four tubular white LED lamps, T8, 9 W, Avant, São Paulo, SP, Brazil), and a 16 h photoperiod.

Explants (approximately 2.0–2.5 cm) were inoculated into 350 mL transparent glass flasks containing 60 mL of MS culture medium [20] (PhytoTechnology®, Lenexa, KS, USA), following the conditions employed *in vitro* cultivation in the Germplasm Bank, with sucrose concentrations of 0 and 30 g L^{-1}. Two explants were inoculated per flask. We employed two flask-sealing systems, resulting in different gas exchange rates [21]: polypropylene lids without membranes (14 μL L^{-1} s^{-1} CO_2 exchange rate) and polypropylene lids with two 10 mm holes covered by microporous tape membranes, as proposed by Saldanha et al. [18] (25 μL L^{-1} s^{-1} CO_2 exchange rate) (Figure 1A).

Figure 1. Schematic representation of the experimental design. (**A**) Polypropylene lids without a membrane (14 μL L^{-1} s^{-1} CO_2 exchange rate) and with two membranes (25 μL L^{-1} s^{-1} CO_2 exchange rate). (**B**) Explants of *Eryngium foetidum* at the beginning of *in vitro* culture. (**C**) Experimental design. Further details can be found in the Materials and Methods section. Scale bars: (**A**) = 1 cm, (**B**) = 2 cm, and (**C**) = 5 cm.

The flasks were maintained for 45 days in a growth room under an irradiance of 85 μmol m^{-2} s^{-1}, a 16 h photoperiod, and a temperature of 25 ± 2 °C (Figure 1B). The design was completely randomized in a 2 × 2 factorial [sucrose concentrations (0 and 30 g L^{-1}) and gas exchange rate (14 and 25 μL L^{-1} s^{-1} CO_2)], consisting of twelve replications for each treatment (Figure 1C). The experiments were repeated twice.

2.2. Chlorophyll a Fluorescence

Chlorophyll fluorescence measurements were conducted on the third fully expanded leaf below the apex. The following parameters: initial fluorescence (F_0), maximum fluorescence (F_m), variable fluorescence (F_v), PSII maximum quantum yield (F_v/F_m), total number of active reaction centers per absorption (RC/ABS), maximum primary efficiency of PSII (F_v/F_0), and performance index (PI) were assessed using a portable nonmodulated fluorimeter (Pocket PEA, Hansatech Instrument Ltd., King's Lynn, Norfolk, UK). The instrument was calibrated to the leaf under dark conditions for 30 min using a leaf-clip system (Hansatech Instrument Ltd., King's Lynn, Norfolk, UK). This calibration ensured the complete opening of reaction centers with minimal heat loss and the complete oxidation of QA [22].

2.3. Determination of Photosynthetic Pigments

Pigments were extracted from leaf discs (5 mm in diameter), which were obtained from the third fully expanded leaf below the apex. The leaf discs were immersed in 5 mL of dimethyl sulfoxide (DMSO; Isofar® Ltd.a, Duque de Caxias, RJ, Brazil) and left in the dark for 48 h [23]. Absorbance measurements of the samples at 665 nm, 649 nm, and 480 nm were taken using a UV-visible spectrophotometer (mono-beam) (model UV-M51; BEL Engineering Company, Monza, Italy) in a 10 mm quartz cuvette. The calculation of chlorophyll *a*, *b*, and carotenoids followed the method outlined by Wellburn [24].

2.4. Quantification of Photosynthetic Performance

Gas exchange and quantification of the *in vitro* photosynthetic rate were conducted following the procedures outlined by Castro et al. [25]. A CO_2 gas analyzer (Model SBA-5; PP Systems International, Amesbury, MA, USA) and LoggerLite 1.9.3 software (Vernier Software & Technology Caliper, Beaverton, OR, USA) were used for these measurements. Gas exchange was determined by calculating the difference between the reference CO_2 and the CO_2 levels of plants exposed to atmospheric air. This calculation was based on the leaf area per plant (cm^2).

2.5. Growth Analyzes

To analyze the growth and development of plants grown *in vitro* for 45 d, various parameters were assessed, including shoot length (cm), root length (cm), base diameter of the rosette (mm), leaf area (cm^2), number of leaves, shoot dry mass (g), root dry mass (g), and shoot-to-root dry mass ratio. The leaf area was determined following the method outlined by Silva et al. [26]. To determine the dry mass, the samples were subjected to 60 °C in an oven with forced air circulation (AmbiKontrol, Arujá, SP, Brazil) until a constant weight was reached.

2.6. Leaf Anatomical Analysis

For an anatomical structural characterization of plants grown *in vitro* for 45 days, samples of the third pair of expanded leaves of the median region were fixed in FAA solution (formalin: acetic acid: 50% ethyl alcohol in 1:1:18 ratio) [27]. After fixation, the samples were dehydrated in increasing ethanolic series (10, 20, 30, 40, 50, 60, 70, 80, 85, 90, and 95%) and included in acrylic resin (Historesin; Leica Instruments, Jena, Germany). Cross-sections of 6 µm thickness were obtained on a semi-automated rotary microtome (model MRP2015, Lupetec Tecnologia Aplicada, São Carlos, SP, Brazil) and stained with toluidine blue (pH 3.2) [28]. Slides were mounted and images were captured using a light microscope (model B20T; Bioptika, Colombo, PR, Brazil) with the U-photo system, coupled to a digital photographic camera (model CMOS-5.0; Bioptika, Colombo, PR, Brazil) and computer with Capture V2.1 software.

2.7. Ex Vitro Acclimatization

The culture medium was removed from the root systems of 45-day-old *Eryngium foetidum* plants grown *in vitro* by gently rinsing them under running water. The plants of each *in vitro* treatment were then transplanted into 350 mL polystyrene pots (model KP-350 TR; Cristalcopo Descartaveis S/A, Içara, SC, Brazil) containing commercial substrate (Carolina soil®, Santa Cruz do Sul, RS, Brazil). The plants were grown in a greenhouse with an average temperature of 28.5 ± 2 °C; humidity relative of 80 ± 5%; natural light conditions with an intensity of ~1050 µmol photons m^{-2} s^{-1}, and a 12 h photoperiod.

2.8. Statistical Analyzes

The data were statistically analyzed by two-way analysis of variance (ANOVA), with the means compared by Tukey's test. All statistical analyses were performed with Genes software, version 1990.2018.39 [29].

3. Results

3.1. Gas Exchange Rates Improve Chlorophyll a Fluorescence and Photosynthetic Rate in Eryngium foetidum Grown In Vitro

Our results indicate that *Eryngium foetidum* plants exhibit phenotypic traits indicative of robust growth and development under photoautotrophic conditions (0 g L^{-1} sucrose and 25 μL L^{-1} s^{-1} CO_2) and photomixotrophic conditions (30 g L^{-1} sucrose) both without (14 μL L^{-1} s^{-1} CO_2; lower gas exchange rate) and with (25 μL L^{-1} s^{-1} CO_2; higher gas exchange rate) natural ventilation. However, plants under 0 g L^{-1} sucrose and 14 μL L^{-1} s^{-1} CO_2 were negatively impacted, likely due to the limiting condition of carbon availability—considered as the control treatment—(Figure 2).

Figure 2. Forty-five-day-old *Eryngium foetidum* plants under sucrose concentrations (0 and 30 g L^{-1}) and gas exchange levels (14 and 25 μL L^{-1} s^{-1} CO_2 exchange rates). Scale bars: 2 cm.

Eryngium foetidum plants grown under 0 g L^{-1} sucrose and 14 μL L^{-1} s^{-1} CO_2 displayed a significantly higher initial fluorescence compared to the other treatments (45% higher) (Figure 3A). In parallel, there were no significant differences in the maximum fluorescence for all treatments (ranged from 2112 to 2250) (Figure 3B). Plants under a higher gas exchange rate (25 μL L^{-1} s^{-1} CO_2), regardless of sucrose concentration, displayed better results for variable fluorescence, total number of active reaction centers per absorption, maximum primary efficiency of PSII, and performance index (Figure 3C,E–G). Lastly, only plants under the limiting condition of *in vitro* cultivation (0 g L^{-1} sucrose and 14 μL L^{-1} s^{-1} CO_2) exhibited an average of 0.61 for PS II maximum quantum yield (Figure 3D), which indicates that they are facing stress.

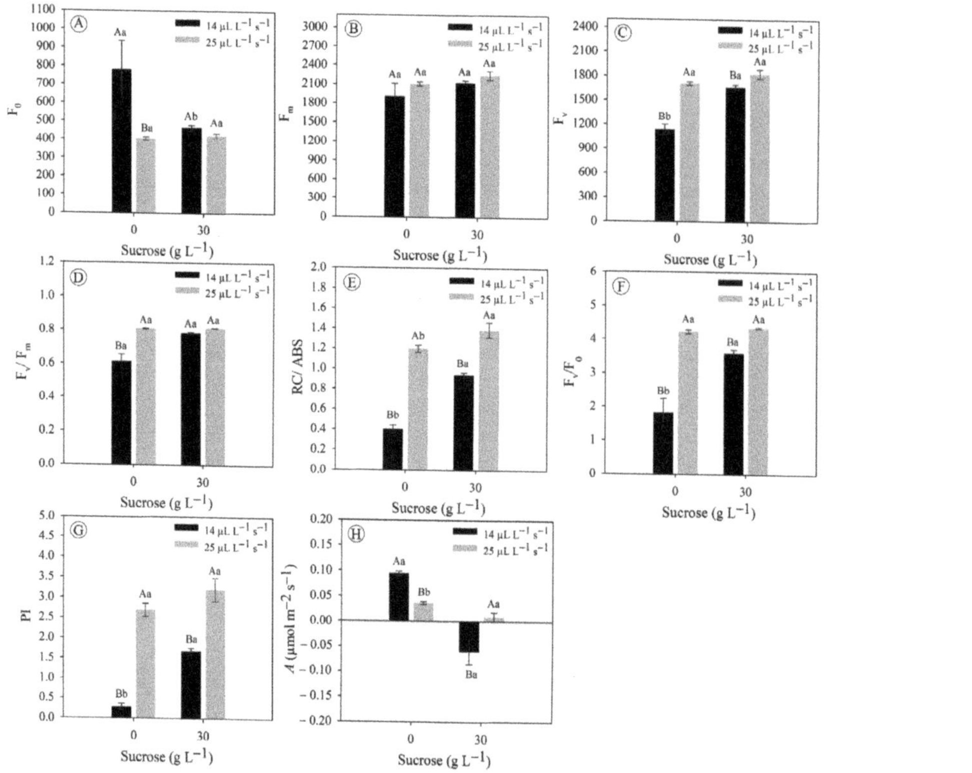

Figure 3. Chlorophyll fluorescence parameters and photosynthetic rate in *Eryngium foetidum* after 45 days of *in vitro* cultivation under sucrose concentrations (0 and 30 g L^{-1}) and gas exchange levels (14 and 25 μL L^{-1} s^{-1} CO_2 exchange rates). (**A**) initial fluorescence; (**B**) maximum fluorescence; (**C**) variable fluorescence; (**D**) PS II maximum quantum yield; (**E**) total number of active reaction centers per absorption; (**F**) maximum primary efficiency of PSII; (**G**) performance index; and (**H**) photosynthetic rate. Uppercase letters compare the means of plants under different CO_2 exchange rates at the same sucrose concentration, while lowercase letters compare plants under the same CO_2 exchange rate at different sucrose concentrations (Tukey's test; $p \leq 0.05$). The bars represent the standard error (n = 5).

The photosynthetic rate showed that *Eryngium foetidum* plants in the absence of sucrose exhibited better performance compared to treatments with the presence of this carbohydrate in the culture medium (3–8 times in sucrose-free with 14 and 25 μL L^{-1} s^{-1} CO_2) (Figure 3H). Notably, plants under cultivation with sucrose and low gas exchange (30 g L^{-1} sucrose and 25 μL L^{-1} s^{-1} CO_2) did not show photosynthetic carbon assimilation when their gas exchange was analyzed (Figure 3H). Collectively, the data indicate that sucrose negatively affects photosynthetic performance, but is mitigated when using a higher gas exchange rate (25 μL L^{-1} s^{-1} CO_2).

3.2. Sucrose Concentrations and Gas Exchange Rates Do Not Modulate Pigment Concentration in Eryngium foetidum Grown In Vitro

Eryngium foetidum plants grown under photoautotrophic (0 g L^{-1} sucrose and 25 μL L^{-1} s^{-1} CO_2), photomixotrophic (0 g L^{-1} sucrose and 14 μL L^{-1} s^{-1} CO_2), and photomixotrophic with ventilation natural (0 g L^{-1} sucrose and 25 μL L^{-1} s^{-1} CO_2) did not differ statistically from each other for the chlorophyll *a* (ranged from 37 to 49 μg cm^{-2}), total chlorophyll (ranged from 45 to 57 μg cm^{-2}), carotenoids (ranged from 10 to 12 μg cm^{-2}), and total chlorophyll/carotenoid ratio (~5 μg cm^{-2}); but these three cultivation systems differed significantly for cultivation under 0 g L^{-1} sucrose and 14 μL L^{-1} s^{-1} CO_2 (Figure 4A,D–F). Chlorophyll *b* concentrations and the chlorophyll *a*/*b* ratio did not show significant differences between all treatments, with averages of 8 and 6 μg cm^{-2}, respectively (Figure 4B,C).

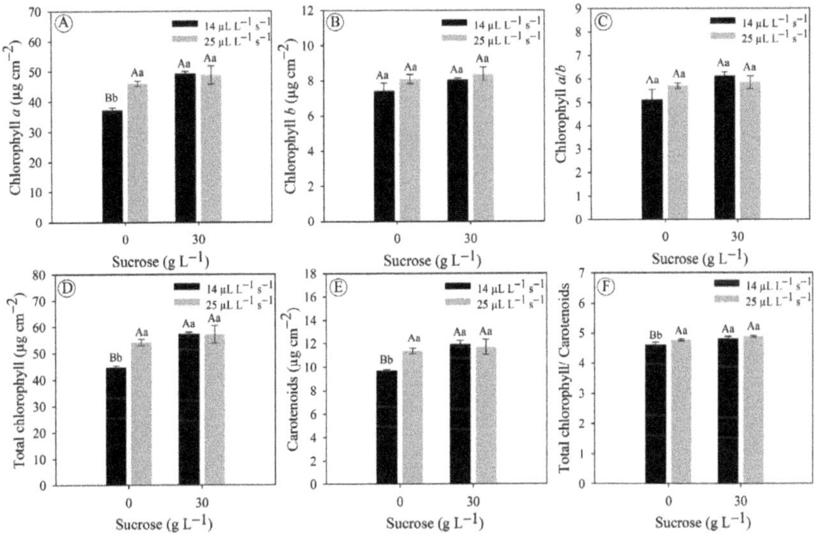

Figure 4. Concentration of photosynthetic pigments in *Eryngium foetidum* after 45 days of *in vitro* cultivation under sucrose concentrations (0 and 30 g L^{-1}) and gas exchange levels (14 and 25 μL L^{-1} s^{-1} CO_2 exchange rates). (**A**) chlorophyll *a*; (**B**) chlorophyll b; (**C**) ratio of chlorophyll *a* to chlorophyll *b*; (**D**) total chlorophyll; (**E**) carotenoids; and (**F**) ratio of total chlorophyll to carotenoids. Uppercase letters compare the means of plants under different CO_2 exchange rates at the same sucrose concentration, while lowercase letters compare plants under the same CO_2 exchange rate at different sucrose concentrations (Tukey's test; $p \leq 0.05$). The bars represent the standard error (n = 5).

3.3. *Sucrose Concentrations and Gas Exchange Rates Affect the In Vitro Growth and Development of Eryngium foetidum*

Eryngium foetidum plants grown under a higher gas exchange rate (25 μL L^{-1} s^{-1} CO_2) displayed an increase in shoot size of 74% and 11% under concentrations of 0 and 30 g L^{-1} sucrose, respectively, compared to the lower gas exchange rate (14 μL L^{-1} s^{-1} CO_2) (Figure 5A). Furthermore, in this same condition of greater gas exchange, there was a significant increase in root length by 382% and 39% at concentrations of 0 and 30 g L−1 sucrose, respectively (Figure 5B). Plants under a lower gas exchange rate (14 μL L^{-1} s^{-1} CO_2) displayed a larger mean rosette base diameter compared to the treatment with a higher gas exchange rate (25 μL L^{-1} s^{-1} CO_2) under the concentration of 30 g L^{-1} sucrose (10% larger diameter); while at the concentration of 0 g L^{-1} sucrose, the opposite occurred, an expected result since this treatment (0 g L^{-1} and 14 μL L^{-1} s^{-1} CO_2) is an extremely limiting cultivation condition for growth and plant development (Figure 5C).

Interestingly, plants grown in a photoautotrophic system (0 g L^{-1} sucrose and 25 μL L^{-1} s^{-1} CO_2) displayed as good an increase in leaf area as the photomixotrophic with natural ventilation (30 g L^{-1} sucrose and 25 μL L^{-1} s^{-1} CO_2; it ranged from 43 to 59 cm^2), both treatments being significantly superior to the system with the lowest gas exchange rate (14 μL L^{-1} s^{-1} CO_2) (Figure 5D). However, these differences in leaf area did not follow for the variable number of leaves, in which the plants under 0 g L^{-1} sucrose and 14 μL L^{-1} s^{-1} CO_2 and the plants under 30 g L^{-1} sucrose and 14 or 25 μL L^{-1} s^{-1} CO_2 displayed number of leaves with close values (~5 leaves) (Figure 5E).

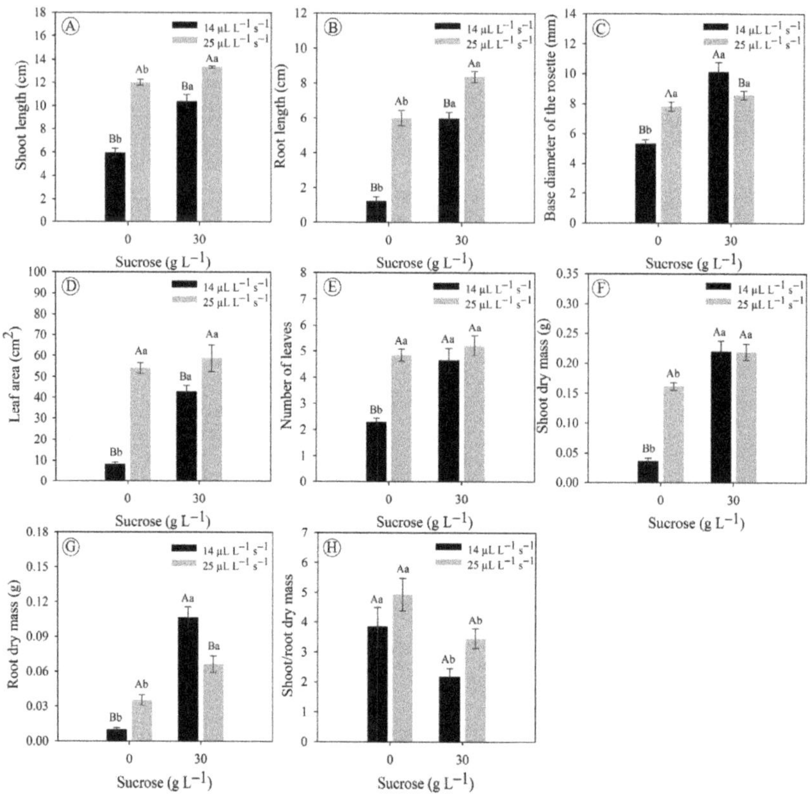

Figure 5. Growth parameters of *Eryngium foetidum* after 45 days of *in vitro* cultivation under sucrose concentrations (0 and 30 g L^{-1}) and gas exchange levels (14 and 25 μL L^{-1} s^{-1} CO_2 exchange rate). (**A**) shoot length; (**B**) root length; (**C**) base diameter of the rosette; (**D**) leaf area; (**E**) number of leaves; (**F**) shoot dry mass; (**G**) root dry mass; and (**H**) shoot-to-root dry mass ratio. Uppercase letters compare the means of plants under different CO_2 exchange rates at the same sucrose concentration, while lowercase letters compare plants under the same CO_2 exchange rate at different sucrose concentrations (Tukey's test; $p \leq 0.05$). The bars represent the standard error (n = 7).

Plants grown with sucrose supplementation displayed greater accumulation of dry mass in the shoot and root, independent of the gas exchange rate (Figure 5F,G). Concerning the shoot-to-root dry mass ratio, it was observed that in plants under 0 g L^{-1} sucrose, there was a greater carbon allocation in the shoot, while plants under 30 g L^{-1} sucrose displayed a more balanced carbon allocation between the shoot and root (Figure 5H).

3.4. Sucrose Supplementation and Gas Exchange Rates Affect Leaf Anatomy

Eryngium foetidum plants displayed the same leaf histological organization in all *in vitro* cultivation conditions accessed. The anatomy of the midrib cross-sections consisted of a uniseriate and amphistomatic epidermis, subepidermal collenchyma on the adaxial and abaxial surfaces; secretory canal and idioblasts distributed throughout the fundamental parenchyma; vascular bundles and fundamental parenchyma distributed throughout the central cavity (Figure 6). However, natural ventilation affected the histological differentiation of leaves compared to treatments with less gas exchange, leading to the formation of thicker and larger leaves (Figure 6).

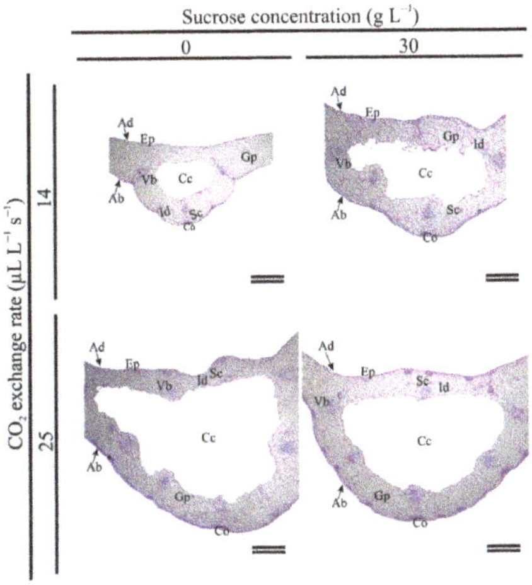

Figure 6. Leaf midrib cross-sections of *Eryngium foetidum* after 45 days of *in vitro* cultivation under sucrose concentrations (0 and 30 g L^{-1}) and gas exchange levels (14 and 25 μL L^{-1} s^{-1} CO_2 exchange rate). Ab: abaxial epidermis; Ad: adaxial epidermis; Cc: central cavity; Co: collenchyma; Ep: epidermis; Gp: ground parenchyma; Id: idioblast; Sc: secretory canal; and Vb: vascular bundle. Scale bars: 300 μm.

3.5. Ex Vitro Acclimatization of Eryngium foetidum Grown under Sucrose Concentrations and Gas Exchange Rates

Plants showed interesting plasticity to face the acclimatization phase (Figure 7A). *Eryngium foetidum* plants grown under photoautotrophic (0 g L^{-1} sucrose and 25 μL L^{-1} s^{-1} CO_2), photomixotrophic (30 g L^{-1} sucrose) without (14 μL L^{-1} s^{-1} CO_2) and with (25 μL L^{-1} s^{-1} CO_2) natural ventilation displayed 100% survival, while only the limiting (0 g L^{-1} sucrose and 14 μL L^{-1} s^{-1} CO_2) displayed a low survival (~33%) (Figure 7B). An interesting fact was the development of inflorescences in plants grown under these photoautotrophy and photomixotrophy treatments (Figure 7C).

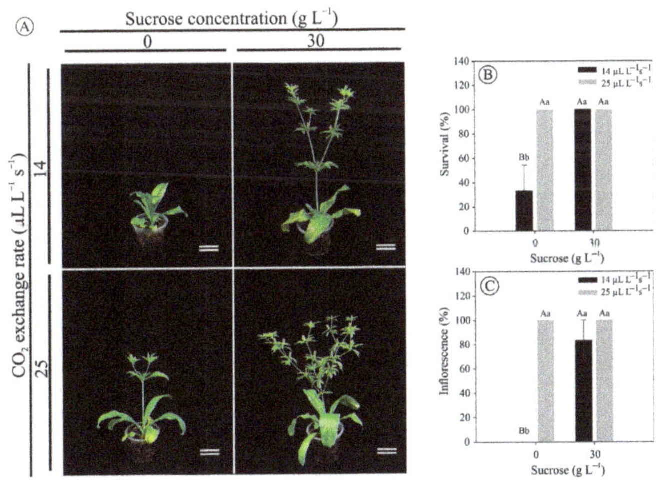

Figure 7. Acclimatization of *Eryngium foetidum* plants after 45 days of *in vitro* cultivation under sucrose concentrations (0 and 30 g L^{-1}) and gas exchange levels (14 and 25 μL L^{-1} s^{-1} CO_2 exchange

rate). (**A**) Sixty-five-day-old *Eryngium foetidum* plants during acclimatization; (**B**) Survival after 60 days of *ex vitro* cultivation; and (**C**) Inflorescence development after 60 days of *ex vitro* cultivation. Uppercase letters compare the means of plants under different CO_2 exchange rates at the same sucrose concentration, while lowercase letters compare plants under the same CO_2 exchange rate at different sucrose concentrations (Tukey's test; $p \leq 0.05$). Scale bars: (**A**) = 5 cm and (**B**,**C**) = the bars represent the standard error (n = 6).

4. Discussion

Plants coordinate their growth and development with carbon availability and source *in vitro* [30–32]. Understanding how sucrose levels (carbon source provided in the cultivation medium) and natural ventilation (carbon source provided in the flask headspace) can affect plant micropropagation offers new perspectives for producing plants with improved morphophysiological characteristics [19]. Our results here demonstrate that *Eryngium foetidum* plants exhibit *in vitro* photoautotrophic potential. However, the addition of sucrose in the photomixotrophic cultivation system also increased the accumulation of plant dry mass.

It is worth highlighting that the *Eryngium foetidum* plants under limiting cultivation conditions (14 $\mu L\ L^{-1}\ s^{-1}$ CO_2 and 0 g L^{-1} sucrose) were drastically affected in their physiological entirety, reducing growth and affecting plant development. This response was expected, since there is a strong limitation in the supply of carbon to plants, both in the cultivation medium due to only organic components making carbon available in small quantities (e.g., vitamins and myoinositol), and in the headspace of the flask due to the use of rigid lids making it difficult for CO_2 to diffuse [16,19]. Although this cultivation condition is an obstacle, the explants did not die, which reflects the ability of *Eryngium foetidum* to adapt to adverse environments. In contrast, the other three *in vitro* culture systems exemplified by photoautotrophic and photomixotrophic without and with natural ventilation provided better morphophysiological responses.

In the photomixotrophic system with a lower gas exchange rate (or without natural ventilation), the microenvironment of the flask is the factor with the greatest influence on the growth and development, since carbon is made available through the supply of sucrose in the medium [33–35]. Undeniably, sealing the bottles (rigid lids or without membranes) provides high humidity inside them, greater accumulation of the phytohormone ethylene, reduced gas exchange, and low CO_2 concentration [36–39]. These factors lead to the development of plants with morphophysiological disorders, such as leaf formation displaying mesophyll with large intercellular spaces, poorly developed vascular system, lack of epicuticular wax formation, and non-functional stomata [16,18,19,40]. In this system, the accumulation of ethylene is a key factor that can strongly affect the growth and development of plants [36].

Ethylene affects several processes in plant growth and development, in which its synthesis increases as the concentration of its precursor increases, suggesting that 1-aminocyclopropane-1-carboxylic acid can act as a signaling molecule in plants [41]. Depending on the plant species and type of flask sealing, there may be a high accumulation of ethylene in the flask headspace, causing a lower photosynthetic rate due to inducing leaf senescence pathways and, consequently, lower plant growth [36,42,43]. In the current study, although some symptoms of leaf chlorosis were seen in plants under systems with a lower gas exchange rate (or 14 $\mu L\ L^{-1}\ s^{-1}$ CO_2), there was no leaf abscission; thus, we believe that ethylene may indeed harm plants, but not to the level of being decisive for lower photosynthetic performance. Probably, the sucrose supplied in the cultivation medium is the main inhibitor of photosynthesis in these systems without natural ventilation [19,44].

Sucrose supplementation in the culture medium affects carbon assimilation by photosynthesis based on the activity reduction in the ribulose-1,5-bisphosphate carboxylase/oxygenase (RuBisCO) enzyme [44] since active carboxylation sites of Rubisco are blocked with phosphorylated sugars [45]. To face these morphophysiological disorders, there is the photoautotrophic system, which improves the environment in the flask headspace in line with the elimination of the carbon source (carbohydrate; e.g., sucrose) [16,40,42]. Despite this, we cannot consider the sucrose supplemented in the cultivation medium to be completely deleterious. In our study, sucrose supplemented in the *in vitro* culture medium increased the absolute values of plant dry weight.

Previous studies with *Solanum tuberosum* plants grown under 3% sucrose estimated that CO_2 assimilation contributed about 40% of the dry mass, while the other 60% came from sucrose absorbed from the culture medium [33]. This may be the explanation for the greater biomass of *Eryngium foetidum* plants in a photomixotrophic without and with natural ventilation compared to photoautotrophic, as, often, these plants grown in a photoautotrophic system may not be achieving maximum photosynthesis performance due to extrinsic factors such as quantity (irradiance) and quality (spectral) of light, in addition to the low concentration of CO_2 in the headspace of the bottle. Based on this, CO_2 enrichment—or forced ventilation—has been applied to plant tissue culture (see [30–32,46]).

Considering the chlorophyll *a* fluorescence and photosynthesis of *Eryngium foetidum* plants, it appears that the response of CO_2 assimilation by photosynthesis was not likely associated with photochemical limitations within cultivation systems. Instead, it suggests potential obstacles in terms of biochemical reactions to CO_2 fixation. The utilization of natural ventilation in the *in vitro* cultivation of *Eryngium foetidum* plants (both photoautotrophic and photomixotrophic with natural ventilation) significantly improved fluorescence aspects in correlation with the photosynthetic rate, compared to photomixotrophic cultivation without natural ventilation.

Photoautotrophic or photomixotrophic with natural ventilation systems also provided greater development of the leaf area of *Eryngium foetidum*, in addition to altering the cellular differentiation of the leaves. These results corroborate findings observed in other species, for example, *Vernonia condensate* Baker [19], *Pfaffia glomerata* (Spreng.) Pedersen [40], *Aechmea blanchetiana* (Baker) L.B.Sm. [47], *Lippia dulcis* Trev. [48], and *Brassavola tuberculata* Hook. [49], in which natural ventilation led to the formation of thicker and larger leaves. These response patterns are important in photosynthetic performance; however, even better, natural ventilation also leads to the formation of functional stomata even *in vitro* cultivation, capable of better controlling gas exchange and evapotranspiration during the acclimatization phase [16,42].

Eryngium foetidum plants grown *in vitro* showed excellent survival during post-culture acclimatization, indicating that not only the photoautotrophic system but also photomixotrophy without and with natural ventilation are effective micropropagation systems. An interesting fact was the rapid development of inflorescences in plants during acclimatization. We do not know for sure whether the cultivation conditions caused an effective stress capable of leading to this, or whether it could be related to the vegetative propagation that favored the rapid development of inflorescences in the photoautotrophic and photomixotrophic treatments without and with natural ventilation.

Our initial findings, based on biotechnological strategies, open avenues for future studies. These studies could explore the application of elicitors to develop methods promoting the exploration of secondary metabolites in the medicinal context of the species. The considerable phenotypic plasticity we observed in the *in vitro* cultivation this species points to emphasizes the potential for further advancements in this field. Furthermore, it is necessary to advance knowledge through studies that can evaluate different genotypes in light of the phenotypic plasticity of the species (see [50]).

5. Conclusions

Eryngium foetidum plants demonstrate *in vitro* photoautotrophic potential. Moreover, when supplemented with sucrose and provided with natural ventilation, there was a significant increase in plant biomass. The implications of this research extend to the mass propagation of this species, aligning with the exploration of secondary metabolites. This is particularly relevant in the phytochemical industry, especially concerning essential oils.

Author Contributions: Conceptualization, S.H.S.F. and M.V.M.P.; methodology, S.H.S.F. and D.R.S.; formal analysis, D.R.S., M.N.B.S., M.C.R.S., M.C.L., V.K.d.O.S.-M., G.L.A., A.M.R.-R. and A.d.S.L.; investigation, D.R.S., M.N.B.S., M.C.R.S., M.C.L., V.K.d.O.S.-M., G.L.A., A.M.R.-R. and A.d.S.L.; resources, T.R.C., F.A.M.M.d.A.F., F.d.O.R. and T.M.F.; data curation, S.H.S.F., D.R.S. and G.L.A.; writing—original draft preparation, S.H.S.F., D.S.B. and D.R.S.; writing—review and editing, S.H.S.F.; D.S.B.; M.V.M.P.; T.R.C., F.A.M.M.d.A.F., F.d.O.R. and T.M.F.; supervision, S.H.S.F. All authors have read and agreed to the published version of the manuscript.

Funding: Funding by the National Council for Scientific and Technological Development (CNPq, Brasília, DF, Brazil: Grant no. PQ 304214/2022-1 to Diego S Batista).

Data Availability Statement: The original contributions presented in the study are included in the article, further inquiries can be directed to the corresponding author.

Acknowledgments: The authors thank the Universidade Estadual do Maranhão (UEMA) and the Fundação de Amparo à Pesquisa e ao Desenvolvimento Científico e Tecnológico do Maranhão (FAPEMA), MA, Brazil.

Conflicts of Interest: The authors declare no conflict of interest.

References

1. Castro, T.F.D.; Carneiro, W.F.; Reichel, T.; Fabem, S.L.; Machado, M.R.F.; Souza, K.K.C.; Resende, L.V.; Murgas, L.D.S. The toxicological effects of *Eryngium foetidum* extracts on zebrafish embryos and larvae depend on the type of extract, dose, and exposure time. *Toxicol. Res.* **2022**, *11*, 891–899. [CrossRef] [PubMed]
2. Lucas, D.B.; Cardozo, A.L. Eryngium in Flora e Funga do Brasil. Jardim Botânico do Rio de Janeiro. Available online: https://floradobrasil.jbrj.gov.br/FB15529 (accessed on 5 December 2023).
3. POWO. Plants of the World Online. Available online: https://powo.science.kew.org/ (accessed on 30 November 2023).
4. Singh, S.; Singh, D.R.; Banu, S.; Salim, K.M. Determination of bioactives and antioxidant activity in *Eryngium foetidum* L.: A traditional culinary and medicinal herb. *Proc. Natl. Acad. Sci. India Sect. B Biol. Sci.* **2013**, *83*, 453–460. [CrossRef]
5. Anju, T.; Rai, N.K.S.; Uthirchamkavu, I.; Sreedharan, S.; Ndhlala, A.R.; Singh, P.; Kumar, A. Analysis of nutritional and antioxidant potential of three traditional leafy vegetables for food security and human wellbeing. *S. Afr. J. Bot.* **2022**, *145*, 99–110. [CrossRef]
6. Do Socorro Teixeira Costa Leitão, D.; Pereira Barbosa-Carvalho, A.P.; Campos de Siqueira, F.; Pontes s Sousa, R.; Santos Lopes, A.; Campos Chisté, R. Extracts of *Eryngium foetidum* Leaves from the Amazonia Were Efficient Scavengers of ROS and RNS. *Antioxidants* **2023**, *12*, 1112. [CrossRef]
7. Rodrigues, T.L.M.; Silva, M.E.P.; Gurgel, E.S.C.; Oliveira, M.S.; Lucas, F.C.A. *Eryngium foetidum* L. (Apiaceae): A literature review of traditional uses, chemical composition, and pharmacological activities. *Evid. Based Complement. Alternat Med.* **2022**, *14*, 2896895. [CrossRef] [PubMed]
8. Paw, M.; Gogoi, R.; Sarma, N.; Saikia, S.; Chanda, S.K.; Lekhak, H.; Lal, M. Anti-microbial, anti-oxidant, anti-diabetic study of leaf essential oil of *Eryngium foetidum* L. along with the chemical profiling collected from North East India. *J. Essent. Oil-Bear.* **2022**, *26*, 814–829. [CrossRef]
9. Zhang, X.; Chen, J.; Zhou, S.; Zhao, H. Ethanol extract of *Eryngium foetidum* leaves induces mitochondrial associated apoptosis via ROS generation in human gastric cancer cells. *Nutr. Cancer* **2022**, *74*, 2996–3006. [CrossRef]
10. Rodrigues, T.L.M. Desempenho Fisiológico e Perfil Químico do Óleo Essencial. Orientadora: Eloisa Helena de Aguiar Andrade. Masters's Thesis, Universidade Federal Rural da Amazônia, Museu Paraense Emílio Goeldi, Belém, Brazil, 2019; 60p. Available online: http://repositorio.ufra.edu.br/jspui/handle/123456789/812 (accessed on 10 December 2023).
11. Devi, P.B.; Deb, P.; Singh, H.B. Promotion, utilization, and commercial cultivation of local spices with special reference to eryngo (*Eryngium foetidum* L.) as a measure for livelihood improvement towards achieving the goal of sustainable development in the indo-burma biodiversity hotspot: A case study from manipur, North-East India. In *Climate Change Adaptation, Risk Management and Sustainable Practices in the Himalay*; Sharma, S., Kuniyal, J.C., Chand, P., Singh, P., Eds.; Springer International Publishing: Cham, Switzerland, 2023; pp. 253–266. [CrossRef]
12. Rajamohan, S. Harbouring the potential of medicinal and aromatic plants of India: Novel biotechnological approach and extraction technologies. In *Medicinal and Aromatic Plants of India Vol. 1. Medicinal and Aromatic Plants of the World, Vol. 8*; Máthé, Á., Khan, I.A., Eds.; Springer International Publishing: Cham, Switzerland, 2022; pp. 323–339. [CrossRef]
13. Karuppusamy, S. A review on trends in production of secondary metabolites from higher plants by *in vitro* tissue, organ and cell cultures. *J. Med. Plants Res.* **2009**, *3*, 1222–1239.
14. Fazili, M.A.; Bashir, I.; Ahmad, M.; Yaqoob, U.; Geelani, S.N. *In vitro* strategies for the enhancement of secondary metabolite production in plants: A review. *Bull. Natl. Res. Cent.* **2022**, *46*, 35. [CrossRef]
15. Nhut, D.T. General information: Some aspects of plant tissue culture. In *Plant Tissue Culture: New Techniques and Application in Horticultural Species of Tropical Region*; Nhut, D.T., Tung, H.T., Yeung, E.C.T., Eds.; Springer: Singapore, 2022, pp. 1–23. [CrossRef]

16. Kozai, T.; Kubota, C. Concepts, definitions, ventilation methods, advantages and disadvantages. In *Photoautotrophic (Sugar–Free Medium) Micropropagation as a New Micropropagation and Transplant Production System*; Kozai, T., Afreen, F., Zobayed, S., Eds.; Springer: Dordrecht, The Netherlands, 2005; pp. 19–30. [CrossRef]
17. Ramos, J.C.M.; Ribeiro, L.M.; Nunes, G.P.; Soares, J.S.; Sorgato, J.C. *In vitro* and *ex vitro* production of *Schomburgkia crispa*: Effect of flask sealing systems and different light sources. *Rodriguesia* **2023**, *74*, e01062022. [CrossRef]
18. Saldanha, C.W.; Otoni, C.G.; Azevedo, J.L.F.; Dias, L.L.C.; Rêgo, M.M.; Otoni, W.C. A low-cost alternative membrane system that promotes growth in nodal cultures of Brazilian ginseng [*Pfaffia glomerata* (Spreng.) Pedersen]. *Plant Cell Tiss. Organ. Cult.* **2012**, *110*, 413–422. [CrossRef]
19. Fortini, E.A.; Batista, D.S.; Mamedes-Rodrigues, T.C.; Felipe, S.H.S.; Correia, L.N.F.; Chagas, K.; Silva, P.O.; Rocha, D.I.; Otoni, W.C. Gas exchange rates and sucrose concentrations affect plant growth and production of flavonoids in *Vernonia condensata* grown *in vitro*. *Plant Cell Tiss. Organ. Cult.* **2021**, *144*, 593–605. [CrossRef]
20. Murashige, T.; Skoog, F. A revised medium for rapid growth and bio assays with tobacco tissue cultures. *Physiol. Plant* **1962**, *15*, 473–497. [CrossRef]
21. Batista, D.S.; Dias, L.L.C.; Rêgo, M.M.; Saldanha, C.W.; Otoni, W.C. Flask sealing on *in vitro* seed germination and morphogenesis of two types of ornamental pepper explants. *Ciênc. Rural.* **2017**, *47*, e20150245. [CrossRef]
22. Bolhar-Nordenkampf, H.R.; Long, S.P.; Baker, N.R.; Oquist, G.; Schreiber, U.L.E.G.; Lechner, E.G. Chlorophyll fluorescence as a probe of the photosynthetic competence of leaves in the field: A review of current instrumentation. *Funct. Ecol.* **1989**, *3*, 497–514. [CrossRef]
23. Santos, R.P.; Ferreira Da Cruz, A.C.; Iarema, L.; Kuki, K.N.; Campos Otoni, W. Protocolo para extração de pigmentos foliares em porta-enxertos de videira micropropagados. *Rev. Ceres* **2008**, *55*, 356–364.
24. Wellburn, A.R. The spectral determination of chlorophylls a and b, as well as total carotenoids, using various solvents with spectrophotometers of different resolution. *J. Plant Physiol.* **1994**, *144*, 307–313. [CrossRef]
25. Castro, K.M.; Batista, D.S.; Fortini, E.A.; Silva, T.D.; Felipe, S.H.S.; Fernandes, A.M.; Sousa, R.M.J.; Nascimento, L.S.Q.; Campos, V.R.; Grazul, R.M.; et al. Photoperiod modulates growth, morphoanatomy, and linalool content in *Lippia alba* L. (Verbenaceae) cultured *in vitro*. *Plant Cell Tiss. Organ. Cult.* **2019**, *139*, 139–153. [CrossRef]
26. Silva, T.D.; Batista, D.S.; Fortini, E.A.; Castro, K.M.; Felipe, S.H.S.; Fernandes, A.M.; Sousa, R.M.J.; Chagas, K.; Silva, J.V.S.; Correia, L.N.F.; et al. Blue and red light affects morphogenesis and 20-hydroxyecdisone content of *in vitro Pfaffia glomerata* accessions. *J. Photochem. Photobiol. B Biol.* **2020**, *203*, 111761. [CrossRef]
27. Johansen, D.A. *Plant Microtechnique*; Mac Graw-Hill Book Co., Inc.: New York, NY, USA, 1940; 523p.
28. O'Brien, T.P.; McCully, M.E. *The Study of Plant Structure: Principles and Selected Methods*; Termarcarphi Pty., Ltd.: Melbourne, Australia, 1981.
29. Cruz, C.D. Genes: A software package for analysis in experimental statistics and quantitative genetics. *Acta Sci. Agron.* **2013**, *35*, 271–276. [CrossRef]
30. Ferreira, P.R.B.; Cruz, A.C.F.; Batista, D.S.; Nery, L.A.; Andrade, I.G.; Rocha, D.I.; Felipe, S.H.S.; Koehler, A.D.; Nunes-Nesi, A. CO_2 enrichment and supporting material impact the primary metabolism and 20-hydroxyecdysone levels in Brazilian ginseng grown under photoautotrophy. *Plant Cell Tiss. Organ. Cult.* **2019**, *139*, 77–89. [CrossRef]
31. Louback, E.; Batista, D.S.; Pereira, T.A.R.; Mamedes-Rodrigues, T.C.; Silva, T.D.; Felipe, S.H.S.; Rocha, D.I.; Steinmacher, D.A.; Otoni, W.C. CO_2 enrichment leads to altered cell wall composition in plants of *Pfaffia glomerata* (Spreng.) Pedersen (Amaranthaceae). *Plant Cell Tiss. Organ. Cult.* **2021**, *145*, 603–613. [CrossRef]
32. Luis, S.J.; Jabín, B.J. CO_2–enriched air in a temporary immersion system induces photomixotrophism during in vitro multiplication in vanilla. *Plant Cell Tiss. Organ. Cult.* **2023**, *155*, 29–39. [CrossRef]
33. Wolf, S.; Kalman-Rotem, N.; Yakir, D.; Zrv, M. Autotrophic and heterotrophic carbon assimilation of in vitro grown potato (*Solanum tuberosum* L) plants. *J. Plant Physiol.* **1998**, *153*, 574–580. [CrossRef]
34. Shin, K.S.; Park, S.Y.; Paek, K.Y. Sugar metabolism, photosynthesis, and growth of in vitro plantlets of *Doritaenopsis* under controlled microenvironmental conditions. *In Vitro Cell Dev. Biol. Plant* **2013**, *49*, 445–454. [CrossRef]
35. Ribeiro, I.S.; Ribeiro, L.M.; Soares, J.S.; Ramos, J.C.M.; Sorgato, J.C. Light condition, flask sealing, and cultivation time on the germination and initial in vitro development of *Dendrobium nobile* Lindl. *Ornam. Hortic.* **2022**, *28*, 407–413. [CrossRef]
36. Vinh, B.V.T.; Tung, H.T.; Mai, N.T.N.; Khai, H.D.; Luan, V.Q.; Phong, T.H.; Phong, T.N.P.; Nhut, D.T. Enhanced efficient micropropagation and reduced abnormal phenomena in *Phyllanthus amarus* plantlets cultured on medium containing silver nanoparticles. *S. Afr. J. Bot.* **2023**, *163*, 217–225. [CrossRef]
37. Askari, N.; Aliniaeifard, S.; Visser, R.G.F. Low CO_2 Levels are detrimental for in vitro plantlets through disturbance of photosynthetic functionality and accumulation of reactive oxygen species. *Horticulturae* **2022**, *8*, 44. [CrossRef]
38. Silva, L.M.; Carvalho, V.S.; Generoso, A.L.; Miranda, D.P.; Costa Júnior, O.D.; Simioni, P.F.; Santana, D.B.; Cunha, M.; Oliveira, J.G.; Viana, A.P. Micropropagation of interspecific hybrids of *Vitis* spp. in microenvironments with different gas exchanges. *Sci. Hortic.* **2022**, *305*, 111413. [CrossRef]
39. Carrari-Santos, R.; Vettorazzi, R.G.; Pinto, V.B.; Pinto, V.B.; Sena, E.O.A.; Oliveira, J.G.; Campostrini, E.; Silveira, V.; Santa-Catarina, C. Microporous membrane and culture medium affect in vitro seedling development of *Dalbergia nigra* (Vell.) Ex Benth. (Fabaceae) by modulation of the protein profile and accumulation of ethylene and CO_2. *Plant Cell Tiss. Organ. Cult.* **2023**, *153*, 559–576. [CrossRef]

40. Iarema, L.; Cruz, A.C.F.; Saldanha, C.W.; Dias, L.L.C.; Vieira, R.F.; Oliveira, E.J.; Otoni, W.C. Photoautotrophic propagation of Brazilian ginseng [*Pfaffia glomerata* (Spreng.) Pedersen]. *Plant Cell Tiss. Organ. Cult.* **2012**, *110*, 227–238. [CrossRef]
41. Wang, K.L.C.; Li, H.; Ecker, J.R. Ethylene biosynthesis and signaling networks. *Plant Cell* **2002**, *14* (Suppl. S1), S131–S151. [CrossRef] [PubMed]
42. Kubota, C. Concepts and background of photoautotrophic micropropagation. In *Molecular Breeding of Woody Plants*; Morohoshi, N., Komamine, A., Eds.; Elsevier: Amsterdam, The Netherlands, 2001; pp. 325–334.
43. Cournac, L.; Dimon, B.; Carrier, P.; Lohou, A.; Chagvardieff, P. Growth and photosynthetic characteristics of *Solanum tuberosum* plantlets cultivated in vitro in different conditions of aeration, sucrose supply, and CO_2 enrichment. *Plant Physiol.* **1991**, *97*, 112–117. [CrossRef]
44. Desjardins, Y.; Hdider, C.; Riek, J. Carbon nutrition in vitro—Regulation and manipulation of carbon assimilation in micropropagated systems. In *Automation and Environmental Control in Plant Tissue Culture*; Aitken-Christie, J., Kozai, T., Smith, M.A.L., Eds.; Springer: Dordrecht, The Netherlands, 1995; pp. 441–471. [CrossRef]
45. Mayak, S.; Tirosh, T.; Ilan, A.; Duvdevani, A.; Khayat, E. Growth and development of pineapple (*Ananas comosus* L.) plantlets cultured in vitro at enriched and ambient CO_2 environments. In *International Symposium on Biotechnology of Tropical and Subtropical Species Part 2*; ISHS: Brisbane, Australia, 1998; Volume 461, pp. 225–230. [CrossRef]
46. Marques, I.; Fernandes, I.; Paulo, O.S.; Lidon, F.C.; DaMatta, F.M.; Ramalho, J.C.; Ribeiro-Barros, A.I. A transcriptomic approach to understanding the combined impacts of supra-optimal temperatures and CO_2 revealed different responses in the polyploid *Coffea arabica* and its diploid progenitor *C. canephora*. *Int. J. Mol. Sci.* **2021**, *22*, 3125. [CrossRef] [PubMed]
47. Martins, J.P.R.; Almeida-Rodrigues, L.C.; Santos, E.R.; Gontijo, A.B.P.L.; Falqueto, A.R. Impacts of photoautotrophic, photomixotrophic, and heterotrophic conditions on the anatomy and photosystem II of in vitro-propagated *Aechmea blanchetiana* (Baker) L.B. Sm. (Bromeliaceae). *In Vitro Cell. Dev. Biol. Plant* **2020**, *56*, 350–361. [CrossRef]
48. Rocha, T.T.; Araújo, D.X.; Silva, A.M.; Oliveira, J.P.V.; Carvalho, A.A.; Gavilanes, M.L.; Bertolucci, S.K.V.; Alves, E.A.; Pinto, J.E.B.P. Morphoanatomy and changes in antioxidant defense associated with the natural ventilation system of micropropagated Lippia dulcis plantlets. *Plant Cell Tiss. Organ. Cult.* **2022**, *151*, 467–481. [CrossRef]
49. Soares, J.S.; Ramos, J.C.M.; Sorgato, J.C.; Ribeiro, L.M.; Reis, L.C. *Brassavola tuberculata* Hook.: In vitro growth and *ex vitro* establishment as a function of the micropropagation system and sucrose. *Braz. J. Biol.* **2023**, *83*, e270892. [CrossRef]
50. Gonçalves, V.P.; Andrade, F.L.N.; Gomes, R.F.; Santos, L.S. Genetic diversity in creole genotypes of Amazon chicory. *Acta Amazon.* **2022**, *52*, 89–95. [CrossRef]

horticulturae

MDPI

Article

De Novo Regeneration of *Cannabis sativa* cv. Cheungsam and Evaluation of Secondary Metabolites of Its Callus

S. M. Ahsan [1,2], Da Bin Kwon [1,2], Md. Injamum-Ul-Hoque [3], Md. Mezanur Rahman [4], Inhwa Yeam [2,5] and Hyong Woo Choi [1,2,*]

1 Department of Plant Medicals, Andong National University, Andong 36729, Republic of Korea; smvahsan@gmail.com (S.M.A.); kdb0626@naver.com (D.B.K.)
2 Institute of Cannabis Biotechnology, Andong National University, Andong 36729, Republic of Korea
3 Department of Applied Biosciences, Kyungpook National University, Daegu 41566, Republic of Korea; mdinjamum92@knu.ac.kr
4 Institute of Genomics for Crop Abiotic Stress Tolerance, Department of Plant and Soil Science, Texas Tech University, Lubbock, TX 79409, USA; mdmerahm@ttu.edu
5 Department of Horticulture and Breeding, Andong National University, Andong 36729, Republic of Korea; iyeam@andong.ac.kr
* Correspondence: hwchoi@anu.ac.kr

Abstract: *Cannabis sativa* L. cv. 'Cheungsam' is an industrial hemp plant of Republic of Korea origin, primarily cultivated for fiber and seed production. In vitro seed germination and tissue culture are valuable tools for developing various biotechnological techniques. In the present study, we aimed to develop a tissue culture process for hemp plants using Cheungsam as a model plant and examine the secondary metabolites produced from its callus. We also developed a method to prepare pathogen-free seedlings from field-derived seeds using hydrogen peroxide (H_2O_2) solution as a liquid germination medium. Treating seedlings with removed seed coat in 3% H_2O_2 significantly reduced the contamination rate. Callus formation and de novo organogenesis of shoots and roots from callus were successfully achieved using cotyledon and leaf tissues prepared from the pathogen-free seedlings. The most effective in vitro regeneration results were obtained using the Murashige and Skoog (MS) medium supplemented with certain targeted growth regulators. An optimal combination of 0.5 mg/L thidiazuron (TDZ) and 1.0 mg/L 1-naphthalene acetic acid proved highly effective for callus induction. The addition of 0.5 mg/L TDZ in the MS medium significantly stimulated shoot proliferation, while robust root development was best supported by MS medium supplemented with 2.5 mg/L indole-3-butyric acid for both cotyledon and leaf explants. Finally, gas chromatography–mass spectrometry (GC–MS) analysis of ethanol extract from Cheungsam leaf callus revealed the presence of different secondary metabolites, including 9-octadecenamide, methyl salicylate, dodecane, tetradecane, and phenol, 2,4-bis-(1,1-dimethylethyl). This study provides a comprehensive de novo regeneration protocol for Cheungsam plants and insight into the secondary metabolite profiles of its callus.

Keywords: *Cannabis sativa* 'Cheungsam'; tissue culture; secondary metabolites; pathogen-free seedlings; GC–MS

1. Introduction

Cannabis (e.g., *Cannabis sativa* L.) is one of the world's oldest cultivated plants, valued for its wide range of applications. Typically, it is classified and recognized as either industrial hemp or marijuana, a distinction based primarily on the concentration of Δ9-tetrahydrocannabinol (THC) contents, the principal psychoactive compound [1]. Cannabis plants are gaining increasing attention due to their medicinal effects (anti-cancer, anti-inflammatory, anti-spastic, anti-pruritic, anticonvulsant, antimicrobial, immunomodulatory, neuro-protective, and anti-psychotic effects), which are derived from various secondary metabolites, such as phytocannabinoids (THC, CBD, CBN, CBC, CBG, THCV, CBDV, etc.), terpenoids, flavonoids, fatty acids, and more [1–9]. In addition to its medicinal value, Cannabis plants are also widely cultivated to produce various industrial products, such as fiber, oil seed, animal feed, biodegradable plastics, construction materials, biofuel, and textiles [1–8]. Thus, Cannabis plants provide numerous benefits across different industries, making them an important crop worldwide.

Explant sources, especially juvenile tissues from in vitro-grown seedlings, play a fundamental role in in vitro tissue culture. Under natural conditions, endophytic microbes within seeds provide plants with enhanced nutrient availability, beneficial metabolites, and enzymes that protect them from different stress factors [3,10–20]. However, endophytic microbes can also create serious contamination problems under in vitro culture conditions [3,10–20]. Notably, the generation of pathogen-free seedlings in vitro from seeds having a thick seed coat, such as Cannabis, is highly challenging, and surface sterilization alone is not usually enough [3,10–20]. Recently, the dual roles of H_2O_2 as a disinfectant and a germination-promoting signaling molecule have led to remarkable success in generating pathogen-free seedlings in vitro [10]. However, varying concentrations of H_2O_2 showed conflicting roles in the survival of seeds; thus, its use (or concentration) must be critically managed and optimized in relation to contamination frequency, germination rate, and seed viability [10].

Due to the totipotency of plants, plant tissues cultured in vitro can be regenerated into complete, fertile, and morphologically and physiologically true-to-type plants from explants derived from various tissues under appropriate cultivation conditions [21–24]. This process is called de novo shoot/root organogenesis (DNSO/DNRO). DNSO/DNRO can be induced in vitro by targeted manipulation of cytokinin (CK) and auxin signaling. By modifying the CK or auxin signaling pathways, callogenesis, shooting, and root regeneration can be considerably enhanced, even in recalcitrant tissues [22–24]. Plant tissue culture offers a controlled environment for the propagation of plants and serves as a basis for the genetic modification of plants. This technique allows for the mass production of disease-free, genetically uniform plants, therefore extensively used in research, agriculture, horticulture, and plant secondary metabolite production [22–24]. However, the tissue culture of Cannabis plants poses unique challenges due to their high recalcitrance, with significant variation among explant types and varieties creating a bottleneck for the application of in vitro tissue culture to the improvement of *C. sativa* [22–24]. Thus, a variety-specific optimization of tissue culture protocols is required to successfully regenerate specific genotypes (or cultivars) of Cannabis plants.

Beyond its pivotal role as a propagation method, plant tissue culture has garnered substantial industrial interest for its broader application in plant improvement and the large-scale production of valuable secondary metabolites [25]. Tissue culture techniques have become invaluable not only for the mass propagation of plants and the production of plant-based products at an industrial scale through different bioreactor systems and genetic manipulation but also as a tool for advancing core insights in biochemistry, plant physiology, embryology, cytology, and molecular biology [25]. Recently, plant biotechnology has increasingly utilized callus culture as a prominent approach for the in vitro synthesis of secondary metabolites and large-scale bioactive compounds compared with other existing techniques, primarily due to the efficiency of in vitro callus induction as a direct and rapid cell multiplication system [26,27]. Plant cells can be readily cultivated to produce secondary metabolites sustainably under controlled conditions, allowing their direct extraction from the callus without the need to sacrifice the whole plant [26,27].

Metabolomics is a powerful analytical technique for identifying a vast array of metabolites at specific time points, and it has become widely applicable in food and nutritional science research [28]. Among these techniques, GC–MS stands out as a rapid, reliable, and highly influential analytical method, particularly suited for analyzing volatile and thermally stable compounds and primary metabolites, such as alcohols, sugars, organic acids, amino acids, volatile oil, esters, fatty acids, amino acids, and nitro compounds.

It is also instrumental in tailoring biosynthetic pathway precursors across various plant species, including *Cannabis sativa* [25,28–34]. Consequently, it has been used to describe the response of carbon (C) and nitrogen (N) primary metabolism to major environmental cues, for example, herbivores, CO_2 mole fraction, drought, nutrient conditions, or stress/elicitor combinations [35]. Recently, numerous non-targeted metabolome studies employing GC–MS [36] have been conducted across diverse plant species, demonstrating the technique's capability to identify metabolite-level variations in various physiological and metabolic processes [28].

Cheungsam is an industrial hemp cultivar widely used in South Korea and recently studied for its richness in beneficial metabolites [2,4–6]. Several reports suggest that its extracts exhibit antioxidant, anti-inflammatory, anti-obesity, and anti-diabetic activities, thereby offering significant potential as a raw material in the food and medicinal industries [2,4–6]. Recently, one study established a micropropagation protocol for the Cheungsam variety, enabling the direct production of normal plantlets from in vitro-grown seedlings [37]. However, an optimized protocol for callogenesis and in vitro organogenesis from the Cheungsam plant is not yet available. Thus, in this study, we aimed to: (i) develop optimized conditions for callus formation from leaf and cotyledon explants, (ii) achieve in vitro regeneration from callus, and (iii) investigate the metabolite profile of the callus.

2. Materials and Methods

2.1. Optimization of Sterilization Methods for the Generation of Pathogen-Free Seedlings

A large-seeded variety of *Cannabis sativa* L., named Cheungsam, was collected from the field and used for this study. Seeds were surface sterilized with 70% ethanol for 3 min and 10% sodium hypochlorite containing 0.1% Tween 20 for 10 min and then washed with sterile distilled water until all bubbles were completely removed [9]. Surface-sterilized seeds were directly submerged in conical tubes containing different volumes (10, 15, 20, and 25 mL) of liquid germination medium (LGM: 1% H_2O_2). Sterile distilled water was used as a negative control in the LGM experiments. Fresh LGM was replaced twice daily, with the old LGM removed each time, for 4 days. Experiments were performed with three replicates per treatment group, and 25 seeds were used for each replicate. H_2O_2-treated seeds were stored at room temperature under dark conditions. The germination rate for each treatment was assessed by observing radicle emergence, while the survival rate was evaluated based on the development of light-green cotyledons.

As LGM alone proved insufficient to prevent contamination of in vitro seedlings, further sterilization was applied to the seedlings after their seed coats were removed. Germinated seedlings in LGM were detached from seed coats using sterilized forceps and submerged in 1% or 3% H_2O_2 solutions. For 1% H_2O_2 treatment, seedlings with removed seed coats were submerged for 1 day, with the solution changed twice daily. For 3% H_2O_2 treatment, seedlings with removed seed coat were submerged for different durations (1, 2, 3, and 4 h). Sterilized seedlings were washed with 1% sodium hypochlorite (1 min), 70% ethanol (1 min), and sterile distilled water (3 times) and then transferred to Murashige and Skoog (MS) agar medium to promote further growth [37]. Seedlings grown on MS medium were transferred to a growth chamber set at 24 °C with an 18 h light/6 h dark cycle and a light intensity of 200 $\mu mol \cdot m^{-2} \cdot s^{-1}$ for two weeks to assess contamination and survival rates. To prepare 1 L of MS medium, 4.43 g of MS salts (including vitamins) and 20 g of sucrose were dissolved in 1 L of distilled water. The pH of the medium was adjusted to 5.7 using KOH, followed by the addition of 8 g of agar before autoclaving [10]. The contamination rate was calculated by dividing the number of contaminated seedlings by the total number of seedlings and expressing the result as a percentage.

2.2. Callus, Shoot, and Root Induction from Cotyledon and Leaf Explants

Plant hormones and chemicals used in tissue culture study were purchased from Duchefa Biochemia (Haarlem, The Netherlands). To identify the optimal conditions for inducing callus formation from cotyledon and leaf explant, callus induction medium (CIM) containing full-strength MS medium with different concentrations of thiadizuron (TDZ; 0 and 1 mg/L) and naphthalene-acetic acid (NAA; 0, 0.5, 1.0, 1.5, and 2.0 mg/L) were tested. Each explant was precisely sectioned into 1 cm^2 using a sterile scalpel (Duchefa Biochemie, Haarlem, The Netherlands) and was obtained from the central axis along the mid-rib tissues, excluding petioles or apex areas from leaves and cotyledons. For optimal callus formation, the abaxial side of one explant was gently placed onto the CIM and carefully tapped to maximize surface contact, reducing any unintended movement during the handling process. Experiments were performed with three replicates, and each replicate included 10 explants of leaf or cotyledon tissues. The capped jars were securely double-wrapped with sealing tape (Kisan Bio, Seoul, Republic of Korea) and randomly arranged within a controlled growth chamber set to 24 °C with an 18 h light/6 h dark cycle and a light intensity of 200 $\mu mol \cdot m^{-2} s^{-1}$. Leaf and cotyledon explants were cultivated under these precisely regulated conditions to support callus induction over 8 and 3 weeks, respectively, with weekly evaluations conducted to monitor callus development. To avoid nutrient and hormone depletion, each explant and callus was transferred to fresh CIM with the same concentrations of hormones every week. After 8 weeks of culture on the CIM, the callus mass (g) and response percentage were recorded.

Callus derived from leaves and cotyledons were transferred to the shoot induction medium (SIM) after 4 and 3 weeks of incubation on the CIM, respectively. To determine the optimal SIM, full-strength MS medium was supplemented with varying concentrations of TDZ (0, 0.5, 1.0, 2.5, 5.0, and 10.0 mg/L) and evaluated. Eight-week-old calli previously cultured on the CIM were transferred into plastic containers containing 50 mL of SIM and kept under the same conditions as previously described. Calli cultures were maintained on SIM until shoots reached a height of over 2.5 cm; at this point, they were transferred to the root induction medium (RIM). Non-responsive calli, including those that had ceased growing or died, were discarded after a four-month cultivation period. Finally, to assess the de novo root organogenesis, RIM composed of full-strength MS medium was prepared with varying concentrations of IBA (0, 0.5, 1.0, 2.5, 5.0, and 10.0 mg/L). De Novo organogenesis of roots was induced in plastic jars containing 50 mL of RIM and maintained under the same conditions described above.

2.3. GC–MS Measurement of Secondary Metabolites in Callus Cultures

One-month-aged calli were transferred on fresh medium and subcultured every 3 weeks. Then, callus was separated from the medium. A 0.1 g sample of crushed callus was dried in an oven at 150 °C for 30 min to extract secondary metabolites. The dried callus was ground using a Silamat S6 device (Ivolar Vivadent Inc., New York, NY, USA) and glass beads, mixed with GC-grade EtOH (10 mL EtOH added per 0.1 g sample), thoroughly mixed by vortexing for 1 min, and incubated at room temperature for 1 h under vigorous shaking. Undissolved plant debris for each sample was removed by centrifugation at 12,000× *g* for 5 min. The supernatant of each sample was transferred to a new tube, filtered with a syringe filter (pore size of 0.2 µm, Pall Corporation, New York, NY, USA), and used for the GC–MS analysis.

A GC/MSD system (5977A Series, Agilent Technologies, Santa Clara, CA, USA) and DB-5 ms columns (Agilent Technologies, Santa Clara, CA, USA) were used for the GC–MS analysis. The oven temperature was held at 80 °C for 1 min and then increased to 240 °C at a rate of 10 $°C \cdot min^{-1}$, from 240 °C to 260 °C at a rate of 5 $°C \cdot min^{-1}$, and from 260 °C to 300 °C at a rate of 20 $°C \cdot min^{-1}$, and was finally held at 300 °C for 10 min. The injector and mass interface temperature was 300 °C. Helium was used as a carrier gas at a flow rate of 1 $mL \cdot min^{-1}$. The injection mode was split, and the scanned mass range was from 40 to 450 m/z. Two µL of each sample was injected for the analysis. The result was analyzed using the National Institute of Standards and Technology version 11 (NIST 11 spectral library).

2.4. Data Analysis

The average seed germination rate across various liquid germination media, contamination rate in the medium multiplied, and survival percentages on MS medium were determined and reported as means ± standard errors (SEs). Each experiment was performed in triplicate and provided consistent results, with 25 seeds in each replicate. Statistical analysis (One-way ANOVA) was performed using Minitab (Version 21.3.1), with significance tested at $p < 0.05$ using the post-hoc Tukey Honestly Significant Difference (HSD) test. GraphPad Prism (version 8.1, GraphPad Software, San Diego, CA, USA) was used for graphical visualization.

3. Results

3.1. Preparation of Pathogen-Free Seedlings

In the previous study, various hemp plant seeds showed the highest germination rate when a 1% H_2O_2 solution was used as liquid germination medium (LGM) [10]. To examine the optimal volume of H_2O_2 solution for the seed germination of Cheungsam plants, the same numbers of seeds (25 seeds) were incubated with different volumes of LGM (10 mL water for control and 10, 15, 20, and 25 mL of 1% H_2O_2). As expected, seeds treated with a 1% H_2O_2 solution as LGM showed a higher germination rate than those germinated in water (Figure 1A). Interestingly, the seeds incubated with 15 mL H_2O_2 showed significantly higher germination rates than the other conditions. This suggests that using the optimal volume of 1% H_2O_2 solution as LGM can maximize the germination of Cannabis seeds.

Unexpectedly, we observed a significant contamination in the seedlings treated with a 1% H_2O_2 solution as LGM (Figure 1B). This could be due to the seeds used in this study having more severe endophytic contamination compared to the previous study [10]. To further reduce the contamination rate, we performed additional sterilization of seedlings after removing the seed coat using different volumes of 1% H_2O_2 for 24 h. Although additional sterilization of seedlings with a removed seed coat with 25 mL of 1% H_2O_2 reduced the contamination rate to 40.0 ± 2.36%, it was not enough to fully remove the endophytic contamination (Figure 1C). Thus, we used a higher concentration of H_2O_2 for seedling sterilization for different time periods. Notably, sterilization of seedlings with a removed seed coat with 25 mL of 3% H_2O_2 further reduced the contamination rate of seedlings in a sterilization duration time-dependent manner (Figure 1D). Sterilization of seedlings with removed seed coat with 25 mL of 3% H_2O_2 for 4 h successfully reduced the contamination rate to 0.00%. However, as shown in Figure 1E, prolonged 3% H_2O_2 treatment induced seedling growth retardation. This suggests that the additional sterilization of seedlings with a removed seed coat using 3% H_2O_2 can be applied to generate pathogen-free seedlings from heavily contaminated seeds; however, the protocol needs to be optimized considering the seed contamination rate under the premise of normal growth conditions.

3.2. In Vitro Callogenesis from Leaf and Cotyledon Explants

To generate callus, excised leaf and cotyledon tissues were placed on the surface of the callus induction medium (CIM) containing a fixed concentration of TDZ (1.0 mg/L) and varying concentrations of NAA to determine the optimal auxin–cytokinin ratio for callus induction. Among the different combinations of TDZ and NAA, both leaf and cotyledon showed the highest callogenesis response on the CIM containing 1.0 mg/L TDZ and 0.5 mg/L NAA (Table 1). As expected, no callogenesis was observed from either leaf or cotyledon explants when hormone-free MS medium was used as CIM. Leaf tissue excided from the contamination-free seedlings started to show initial callus formation 14 days after placing it on CIM (Figure 2A,B). Four weeks after incubation, the callus formed a green nodular photosynthetic and friable callus (Figure 2C,D). Leaf explants generated the highest amount of callus (6.87 ± 0.12 g) and showed the highest callogenesis response (92.80 ± 1.46%) on CIM containing 1.0 mg/L TDZ and 0.5 mg/L NAA (Table 1). Increasing concentrations of NAA (≥1.0 mg/L) negatively affected both the callogenesis response and the amount of callus generated. For example, leaf explants on CIM containing

1.0 mg/L TDZ and 2.0 mg/L NAA showed a 74.8% reduction in the amount of callus generated and a 28.4% reduction in callogenesis response compared to those on CIM containing 1.0 mg/L TDZ and 0.5 mg/L NAA.

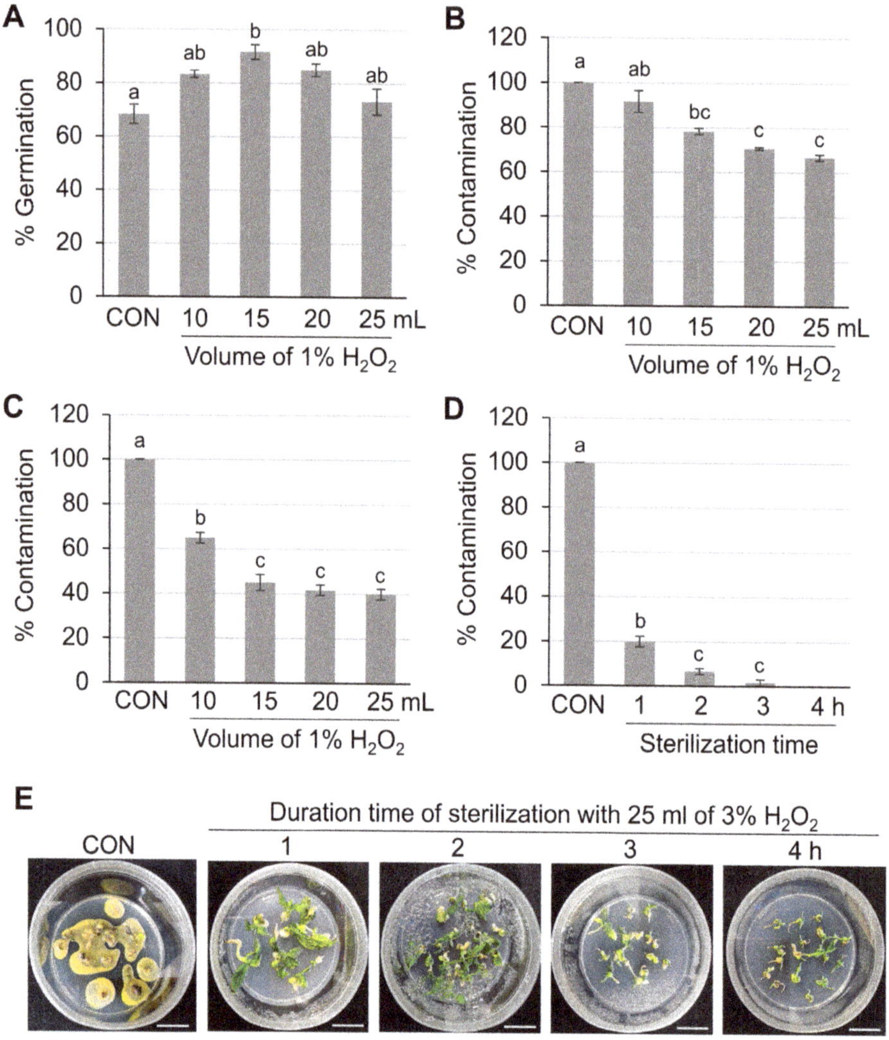

Figure 1. Pathogen-free in vitro seedling preparation from Cheungsam seeds. (**A**) Enhanced germination rates of Cheungsam seeds germinated in different volumes of 1% H_2O_2. (**B**) Reduced contamination rates of seedlings germinated in different volumes of 1% H_2O_2. Contamination rates were assessed 2 weeks after transferring the sterilized seedlings to MS medium. (**C**) Reduced contamination rates of seedlings additionally sterilized with different volumes of 1% H_2O_2 after removing the seed coat. (**D**) Reduced contamination rates of seedlings with removed seed coat, additionally sterilized with 3% H_2O_2 for different durations (1, 2, 3, and 4 h). (**E**) Representative images of seedlings with removed seed coat sterilized with 3% H_2O_2 for different durations (1, 2, 3, and 4 h). Pictures were taken 2 weeks after transferring the sterilized seedlings to MS medium. CON: water control (Scale bar in all panels = 2 cm). Different letters above the bars indicate a statistically significant difference at $p < 0.05$ according to the Tukey HSD test.

Figure 2. Callus formation and de novo organogenesis of the excised leaf of Cheungsam plants. (**A**) Pathogen-free seedlings germinated in H_2O_2 solution as a liquid germination medium and grown on MS medium. Red circle: Leaf tissue used for callus formation. Scale bar = 0.2 cm. (**B–D**) Callus induced from leaf explants. Scale bar = 0.2 cm. Initial swelling and formation of callus from leaf explants (4-week-old, (**B**)). Green nodular photosynthetic and friable callus (red arrows, 5-week-old, (**C**)). The initial stage of shoot regeneration from photosynthetic callus (red arrow, 6-week-old). (**D**). Scale bar = 0.2 cm. (**E**,**F**) De Novo shoot morphogenesis from leaf-induced callus on shoot induction medium (SIM) containing 0.5 mg/L TDZ. Scale bar = 0.2 cm. (**G**) De Novo root morphogenesis from leaf-induced callus on root induction medium (RIM) containing 2.5 mg/L IBA. Scale bar = 1 cm. Regenerated roots are indicated by the red arrow.

Cotyledon tissues started to show initial callus formation 5–7 days after placing them on CIM (Figure 3A,B). Similarly, cotyledon explants on CIM containing 1 mg/L TDZ and 0.5 mg/L NAA generated the highest amount of callus (4.87 $\pm$ 0.45 g) and showed the highest callus induction rate (88.37 $\pm$ 1.77%) (Table 1). Cotyledon explants on CIM containing more NAA ($\geq$1.0 mg/L) showed reduced callus and callogenesis responses compared to those on CIM containing 1.0 mg/L TDZ and 0.5 mg/L NAA. Notably, cotyledon explants generated green, nodular, photosynthetic, and friable callus faster than leaf explants (within 3 weeks; Figure 3B).

Table 1. Effect of different concentrations of TDZ and NAA on callogenesis from leaf and cotyledon explants of Cheungsam.

Tissue Used	TDZ (mg/L)	NAA (mg/L)	Callogenesis Response (%)	Amount of Callus (g)
Leaf	0.0	0.0	0.00 ± 0.00 [d]	0.00 ± 0.00 [d]
	1.0	0.5	92.80 ± 1.46 [a]	6.87 ± 0.12 [a]
	1.0	1.0	80.03 ± 0.90 [b]	4.90 ± 0.61 [b]
	1.0	1.5	73.67 ± 1.18 [c]	3.23 ± 0.30 [c]
	1.0	2.0	66.40 ± 5.73 [c]	1.73 ± 0.32 [c]
Cotyledon	0.0	0.0	0.00 ± 0.00 [d]	0.00 ± 0.00 [c]
	1.0	0.5	88.37 ± 1.77 [a]	4.87 ± 0.45 [a]
	1.0	1.0	78.67 ± 1.45 [b]	2.93 ± 0.30 [b]
	1.0	1.5	71.43 ± 4.77 [b]	1.67 ± 0.44 [b]
	1.0	2.0	57.93 ± 4.97 [c]	1.50 ± 0.12 [b]

TDZ: thidiazuron; NAA: naphthalene-acetic acid. Data were obtained 4 weeks after the incubation of leaf explants on CIM with the indicated concentrations of TDZ and NAA. Different letters indicate statistically significant differences at $p < 0.05$ according to the Tukey HSD test.

Figure 3. Callus formation and de novo organogenesis of excised cotyledon tissues from Cheungsam plants. (**A**) Pathogen-free seedlings germinated in H_2O_2 solution as a liquid germination medium and grown on MS medium. Red circle: Cotyledon tissue used for callus formation. Scale bar = 0.2 cm. (**B**) Callus induced from the cotyledon explants 7 days after incubation on CIM. Initial swelling and formation of callus from cotyledon explant. Green nodular photosynthetic and friable calli are indicated by red arrows. Scale bar = 0.2 cm. (**C**,**D**). De Novo shoot morphogenesis from cotyledon-induced callus on SIM containing 0.5 mg/L TDZ. Scale bar = 0.2 cm. (**E**) De Novo root morphogenesis from cotyledon-induced callus on RIM containing 2.5 mg/L IBA. Regenerated roots are indicated by the red arrow. Scale bar = 1 cm.

3.3. De Novo Shoot Morphogenesis from Callus

To induce de novo shoot morphogenesis from the callus, the fresh and green photosynthetic components of the callus masses were carefully excised with a sharp knife and placed on a shoot induction medium (SIM) containing different concentrations of TDZ. Among the SIMs with varying TDZ concentrations, the SIM with 0.5 mg/L TDZ elicited the highest de novo shoot morphogenesis responses from the calli derived from both leaf and cotyledon tissues (Table 2). De Novo shoot morphogenesis from leaf-derived photosynthetic callus was first observed 2 weeks after incubation on SIM. Leaf explant-induced callus generated the highest shooting response (15.67 ± 0.27%) with the highest number of shoots (11.60 ± 0.59) and the most extended shoot length (9.03 ± 0.39 cm) after subculturing for another 2 weeks on fresh SIM containing 0.5 mg/L TDZ. Similarly, cotyledon-derived callus showed the highest de novo shoot morphogenesis response (22.67 ± 1.19%) on SIM with 0.5 mg/L TDZ (Table 2; Figure 3C,D). Increasing concentrations of TDZ above 0.5 mg/L reduced de novo shoot morphogenesis responses from both leaf- and cotyledon-derived calluses.

Table 2. Effect of different concentrations of TDZ on de novo shoot morphogenesis from the calli derived from leaf or cotyledon explants of Cheungsam.

Origin of Callus	TDZ (mg/L)	Response (%)	Average Number of Shoots	Average Length of Shoots (cm)
Leaf	0.0	0.00 ± 0.00 ^{e}	0.00 ± 0.00 ^{d}	0.00 ± 0.00 ^{e}
	0.5	15.67 ± 0.27 ^{a}	11.60 ± 0.59 ^{a}	9.03 ± 0.39 ^{a}
	1.0	12.33 ± 0.98 ^{b}	7.90 ± 0.42 ^{b}	6.13 ± 0.84 ^{b}
	2.5	9.67 ± 0.98 ^{c}	7.47 ± 0.49 ^{b}	4.63 ± 0.44 ^{b,c}
	5.0	8.17 ± 0.49 ^{c}	4.33 ± 0.24 ^{c}	3.30 ± 0.26 ^{c,d}
	10.0	4.97 ± 0.80 ^{d}	2.80 ± 0.17 ^{c}	2.43 ± 0.29 ^{d}
Cotyledon	0.0	0.00 ± 0.00 ^{f}	0.00 ± 0.00 ^{d}	0.00 ± 0.00 ^{d}
	0.5	22.67 ± 1.19 ^{a}	8.33 ± 1.20 ^{a}	6.53 ± 0.48 ^{a}
	1.0	10.67 ± 0.98 ^{b}	5.67 ± 0.88 ^{a,b}	4.57 ± 0.50 ^{a,b}
	2.5	8.37 ± 0.62 ^{c}	3.33 ± 0.88 ^{b,c}	3.73 ± 0.39 ^{b,c}
	5.0	6.03 ± 0.78 ^{d}	2.83 ± 0.17 ^{b,c}	3.17 ± 0.69 ^{b,c}
	10.0	4.03 ± 0.42 ^{e}	2.00 ± 0.29 ^{c}	2.37 ± 0.19 ^{c}

Data were obtained 4 weeks after incubation of callus on SIM with indicated concentrations of TDZ. Different letters indicate statistically significant difference at $p < 0.05$ according to the Tukey HSD test.

3.4. De Novo Root Morphogenesis from Callus-Derived Shoots

To induce de novo root morphogenesis, shoots with callus portions reaching a height of more than 5 cm were placed on the root induction medium (RIM) with different concentrations of IBA. Among the RIMs with varying IBA concentrations, the RIM with 2.5 mg/L IBA induced the highest de novo root morphogenesis responses from the callus-derived shoots of both leaf and cotyledon explant tissues (Table 3; Figures 2G and 3E). De Novo root morphogenesis from leaf-derived callus was first observed 2 weeks after incubation on RIM. After subculturing for another 2 weeks on fresh RIM with 2.5 mg/L IBA, the highest rooting response (95.30 ± 1.14%) was achieved, with the highest number of roots (12.77 ± 1.43) and the longest root length (7.20 ± 0.64 cm). Similarly, shoots from cotyledon-derived calli showed the highest de novo root morphogenesis response (92.83 ± 2.31%) on RIM with 2.5 mg/L IBA. Increasing concentrations of IBA above 2.5 mg/L significantly reduced de novo root morphogenesis responses from both leaf- and cotyledon-derived calluses.

Table 3. Effect of different concentrations of IBA on de novo root morphogenesis from the calli derived from leaf or cotyledon explants of Cheungsam.

Origin of Callus	IBA (mg/L)	Response (%)	Average Number of Roots	Average Length of Roots (cm)
Leaf	0.0	0.00 ± 0.00^{d}	0.00 ± 0.00^{d}	0.00 ± 0.00^{c}
	0.5	86.27 ± 0.44^{b}	$5.70 \pm 0.92^{b,c}$	4.83 ± 0.32^{b}
	1.0	82.50 ± 0.64^{b}	$6.70 \pm 0.59^{b,c}$	5.23 ± 0.32^{b}
	2.5	95.30 ± 1.14^{a}	12.77 ± 1.43^{a}	7.20 ± 0.64^{a}
	5.0	77.10 ± 0.99^{c}	8.70 ± 0.42^{b}	$5.33 \pm 0.60^{a,b}$
	10.0	71.97 ± 0.42^{c}	4.57 ± 0.29^{c}	4.23 ± 0.12^{b}
Cotyledon	0.0	0.00 ± 0.00^{d}	0.00 ± 0.00^{d}	0.00 ± 0.00^{d}
	0.5	68.07 ± 2.87^{c}	4.47 ± 0.55^{c}	$4.60 \pm 0.95^{a,b}$
	1.0	75.43 ± 1.53^{b}	7.03 ± 0.62^{b}	$4.83 \pm 0.49^{a,b}$
	2.5	92.83 ± 2.31^{a}	9.70 ± 0.59^{a}	7.47 ± 0.71^{a}
	5.0	65.37 ± 1.98^{c}	4.70 ± 0.21^{c}	4.17 ± 1.05^{b}
	10.0	69.70 ± 4.14^{c}	3.70 ± 0.42^{c}	$2.80 \pm 0.17^{b,c}$

IBA: indole-3-butyric acid. Data were obtained 4 weeks after the incubation of callus on RIM with the indicated concentrations of IBA. Different letters indicate statistically significant differences at $p < 0.05$ according to the Tukey HSD test.

3.5. Gas Chromatography–Mass Spectrometry (GC–MS) Analysis

For the analysis of secondary metabolites in the leaf-derived callus cultures of Cheungsam, ethanol was used as the extraction solvent. GC–MS analysis revealed distinct compound profiles, with nineteen unique peaks detected in the ethanol extract. The total ion chromatograph (TIC) corresponding to the ethanol-extracted compounds from the callus culture is shown in Figure 4. Among the compounds detected, the top five most abundant chemical compounds are listed in Table 4. From the analysis, 9-Octadecenamide (Z) is identified as the most abundant compound. However, none of the cannabinoids are detected.

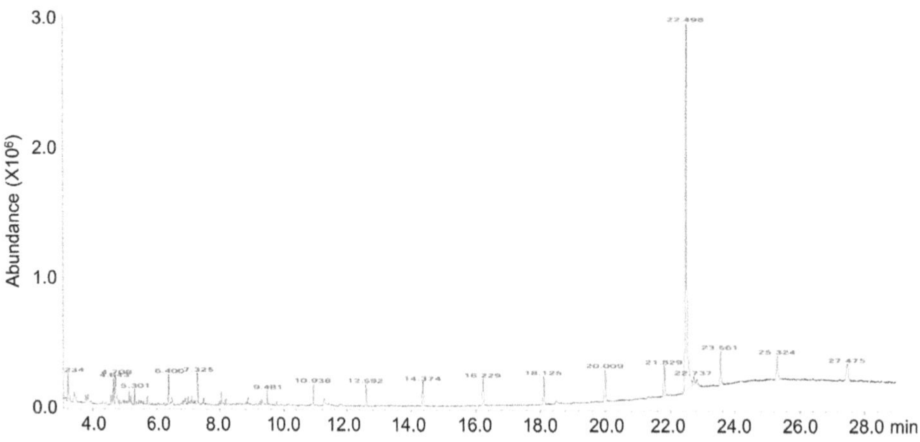

Figure 4. GC–MS total ion chromatogram (TIC) obtained from the analyses of ethanol extract of Cheungsam callus. TIC was generated from the analysis of ethanol extract of one-month-old Cheungsam leaf callus.

Table 4. List of five of the most abundant chemical compounds found in leaf-derived callus cultures of Cheungsam by gas chromatography–mass spectrometry (GC–MS) analysis.

No.	Compound	Structure	Peak No.	Formula	Molecular Weight	Retention Time (min)
1	9-Octadecenamide, (Z)-	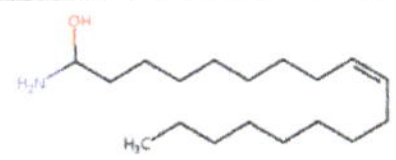	15	$C_{18}H_{35}NO$	281.4766	22.497
2	Methyl salicylate	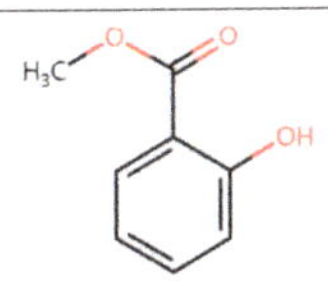	3	$C_8H_8O_3$	152.15	4.706
3	Dodecane		2	$C_{12}H_{26}$	170.33	4.644
4	Tetradecane		5	$C_{14}H_{30}$	198.39	6.401
5	2,4-bis(1,1-dimethylethyl)phenol	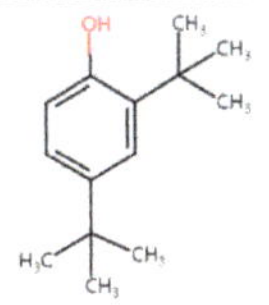	6	$C_{17}H_{30}O$	278.5	7.325

4. Discussion

In this study, we successfully developed optimized conditions for the in vitro regeneration of Cheungsam. The findings of the study are as follows: (i) the additional sterilization of seedlings with removed seed coats using H_2O_2 can further reduce the contamination rate of in vitro seedlings germinated from heavily contaminated seeds; (ii) optimal hormone concentrations that can be used for the preparation of Cheungsam-specific CIM, SIM, and RIM were determined; and (iii) secondary metabolites in callus derived from Cheungsam leaf explants were identified through GC–MS analysis. As previously reported, using a 1% H_2O_2 solution as an LGM significantly improved seed germination rates. However, unlike our previous report [10], this treatment still resulted in considerable contamination, likely due to severe endophytic contamination in the seeds. Subsequent sterilization of seedlings with removed seed coats using higher concentrations of H_2O_2 effectively reduced the contamination rates. However, the observed growth retardation caused by prolonged H_2O_2 exposure highlights the need for the precise optimization of both H_2O_2 concentration and exposure time during sterilization to ensure seedling health [10,38–42].

In plants, de novo organogenesis refers to the regeneration of new organs (e.g., roots or shoots) from wounded or detached explants through the formation of callus under in vitro conditions. When plants experience wounding stress, they activate responses to repair the damage. This includes forming an unorganized callus mass and, in some cases, generating new organs through vascular differentiation and de novo organogenesis [23,24,43]. These processes involve complex hormonal interactions between auxin and cytokinin, where (i) intermediate auxin-to-cytokinin ratio induces callus formation, (ii) high auxin-to-cytokinin ratio promotes root formation, and (iii) low auxin-to-cytokinin ratio promotes shoot regeneration. These hormonal signals are regulated by intricate gene networks, including transcription factors, hormonal transport/metabolism genes, and cell cycle-related regulators, ensuring successful organogenesis.

The selection of appropriate explants is critical for the success of in vitro de novo organogenesis. In dicotyledonous plants, leaves and cotyledons act as natural sources of auxin, with auxin transport occurring from the tip toward the stem/hypocotyl and roots [43,44]. This characteristic makes leaves and cotyledons particularly suitable for inducing cell division and serving as effective explants for de novo organogenesis. Interestingly, cotyledon explants often exhibit a higher callogenesis response compared to leaf explants in many dicotyledonous species, including Cannabis [45]. This is likely due to their greater capacity for auxin biosynthesis. In our experiments, the optimal callus induction medium (CIM) for leaf and cotyledon explants contained 1.0 mg/L TDZ and 0.5 mg/L NAA. Under these conditions, cotyledon explants demonstrated faster callus formation than leaf explants. However, higher concentrations of synthetic auxins, such as NAA (≥1.0 mg/L), are well-known for their phytotoxic effects [45–47]. These auxins, often referred to as "auxinic herbicides", can disrupt the auxin homeostasis, leading to the overproduction of ethylene, abscisic acid, reactive oxygen species (ROS), and other harmful metabolites. In our study, higher doses of NAA negatively impacted both the callogenesis response and callus production, underscoring the critical importance of maintaining hormonal balance in plant tissue culture.

Both leaf- and cotyledon-derived calluses exhibited optimal de novo shoot morphogenesis on shoot induction medium (SIM) containing 0.5 mg/L TDZ. TDZ, a potent plant growth regulator, is known to stimulate cytokinin-associated pathways by interacting with cytokinin receptors [48,49]. It also modulates the activities of enzymes involved in auxin and cytokinin regulation, enabling the induction of both the auxin and cytokinin signaling pathways. This dual hormonal regulation makes TDZ particularly effective in inducing morphogenesis, even in recalcitrant species such as Cannabis [48,49]. However, TDZ-driven morphogenesis is highly sensitive to its concentration and the duration of exposure. Excessive concentrations can inhibit growth and development, likely due to the disruption of hormonal balance. In our study, while callus induction medium (CIM) produced satisfactory results for both leaf and cotyledon explants, the morphogenic response on SIM was suboptimal for both explant types. This reduced response might be attributed to the highly recalcitrant nature of the tissues, as reported in earlier studies [23,24,43,45–47]. Additionally, shoot morphogenesis decreased significantly at higher TDZ concentrations (≥1.0 mg/L). This emphasizes the importance of precise hormonal regulation in plant tissue culture. Higher concentrations of TDZ have been associated with adverse effects, including tissue necrosis, loss of rooting ability, hyperhydricity (excessive water retention), and cytogenetic abnormalities, as highlighted in the recent reviews [50]. These findings further underscore the need to carefully optimize TDZ concentrations to balance effective morphogenesis with minimal phytotoxic effects.

Rooting from de novo-formed shoots is a crucial step in plant tissue culture to complete the micropropagation process [45]. Physiologically, roots play a vital role in channeling the shoot-derived auxin, which is essential for coordinated plant growth and development. Successful rooting depends on three key factors [23,45]: (i) High auxin production in specific xylem-associated cells within the stem. (ii) Effective delivery of auxin to the target tissues. (iii) Optimal hormonal balance to induce root formation without causing phytotoxicity. Several de novo root induction protocols have been developed, including treatments with relatively high auxin concentrations, such as IBA [23,24,44,51–54]. In our study, the best de novo root morphogenesis was observed on RIM containing 2.5 mg/L IBA. Both leaf- and cotyledon-derived calluses showed excellent responses, with high rooting percentages (>92%), significant root numbers (>9.7 roots per explant), and substantial root lengths (>7.2 cm). However, a reduction in root morphogenesis was observed at both lower and higher IBA concentrations, highlighting the importance of maintaining a precise balance in auxin application for successful root induction.

Our GC–MS analysis identified 19 different compounds; however, unexpectedly, no cannabinoids or their related compounds were detected. Instead, 9-octadecenamide (Z), also known as oleamide, was identified as the most abundant compound. This bioactive lipid belongs to the fatty acid amides (FAAs) class, a group of molecules found in both the central nervous system (CNS) and peripheral tissues [55]. FAAs, which are composed primarily of "orphan" ligands, often interact with the "orphan" or unidentified receptors, playing key roles in various biological processes [56]. Within this lipid class, oleamide functions as a chemical signaling molecule and has been shown to interact with cannabinoid receptors [55–57]. Through this interaction, oleamide can induce diverse biological effects, such as promoting sleep and modulating immune responses. Additionally, oleamide acts as a neurotransmitter and has been implicated in processes such as thermoregulation and pain modulation (antinociception). These findings highlight its potential role as a multifunctional signaling molecule with a broad spectrum of physiological effects.

Methyl salicylate (MeSA or methyl 2-hydroxybenzoate) was identified as the second most abundant compound in the GC–MS analysis. MeSA is a volatile organic compound (VOC) and serves as a methyl ester derivative of salicylic acid (SA). MeSA is widely emitted by various plant species, and it functions as a signal-inducing systemic acquired resistance (SAR), helping plants defend against pests, pathogens, and competing antagonists [58–62]. In plants, MeSA is synthesized from SA via the isochorismate and the phenylalanine ammonia-lyase pathways. However, beyond its role in plant defense, recent studies have highlighted the pharmacological significance of MeSA for human health [62]. Naturally occurring salicylates, including MeSA, are known to target multiple regulatory pathways, contributing to beneficial health effects. These include anti-inflammatory, anticancer, neuroprotective, and anti-diabetic properties, making MeSA a molecule of interest in both plant and medical sciences.

Our GC–MS analysis identified two specific alkanes, dodecane and tetradecane, as the third- and fourth-most abundant compounds, respectively. Alkanes, also referred to as paraffins (from Latin, meaning "with little chemical affinity"), are non-polar, stable, and minimally reactive organic compounds [63]. Structurally, they are linear, saturated hydrocarbons with the general formula C_nH_{2n+2} and are widely used as lipid biomarkers for characterizing essential and volatile oils [63–66]. In plants, alkanes are synthesized as part of the protective wax layer on leaf surfaces, serving critical physiological and ecological roles, such as reducing water loss, providing UV protection, and acting as a barrier against pests and pathogens [67–70]. Long-chain alkanes are also considered intermediate VOCs and may function as precursors for the synthesis of secondary metabolites, such as terpenes [71]. Terpenes, prominent secondary metabolites, exhibit potent bioactivity and have therapeutic potential for treating infectious and chronic diseases. Their properties include anti-inflammatory, antioxidant, antimicrobial, antitumor, and antidiabetic effects. Terpenes are synthesized from acetyl-CoA via the mevalonic acid pathway in the cytoplasm. Additionally, integrative genomic analyses have revealed co-expression networks of genes involved in the biosynthesis of both cannabinoids and terpenoids, highlighting the shared origin of the cannabinoid and terpenoid biosynthetic pathways [72]. Furthermore, the degradation and oxidation of alkanes can yield fatty acids and acetyl-CoA, which are essential intermediates for various biosynthetic pathways, including secondary metabolite production [67–70,73]. These findings underscore the multifaceted role of alkanes in the biosynthesis of bioactive compounds with significant ecological and therapeutic implications.

Finally, 2,4-bis(1,1-dimethylethyl)phenol (2,4-DTBP) was identified as the fifth-most abundant compound, highlighting its prominence among the detected metabolites. Also known as 2,4-Di-tert-butylphenol, 2,4-DTBP is a secondary metabolite produced by a wide range of organisms, including plants, fungi, and bacteria [74]. 2,4-DTBP is frequently detected as a major constituent in volatile and essential oils and is well-known for its pronounced toxicity against various organisms, including its own producers. This toxicity is mediated through multiple mechanisms, such as enzyme inhibition, DNA structure alterations, disruption of ribosomal translation, and cellular membrane destabilization [74,75]. Beyond its toxicity, 2,4-DTBP displays a wide range of bioactivities, including antimicrobial, herbicidal, antioxidant, insecticidal, nematicidal, and cytotoxic effects [74,75]. These properties make it a compound of interest in various fields, including agriculture, medicine, and environmental science.

5. Conclusions

This study successfully optimized the in vitro regeneration conditions for the Cheungsam cultivar, providing a robust framework for the mass propagation of pathogen-free, genetically uniform Cannabis plants. Through de novo organogenesis, this study demonstrates the importance of hormonal balance, particularly the auxin-to-cytokinin ratio, in regulating callus formation, root development, and shoot regeneration. Moreover, the GC–MS analysis revealed a diverse range of secondary metabolites, providing valuable insights into the chemical composition of Cheungsam-derived calluses. Key compounds such as oleamide, MeSA, dodecane, tetradecane, and 2,4-DTBP were identified, each possessing significant ecological and therapeutic potential. Collectively, this study lays the groundwork for future research into optimizing tissue culture techniques and exploring the molecular mechanisms underlying in vitro regeneration and secondary metabolite biosynthesis. The integration of advanced omics approaches, including transcriptomics, proteomics, and metabolomics, could further unravel the genetic and biochemical pathways involved, paving the way for metabolic engineering and the development of high-value compounds for agricultural and pharmaceutical applications.

Author Contributions: Concept and experimental design: S.M.A. and H.W.C.; experiment conduction: S.M.A.; analysis: S.M.A. and M.I.-U.-H.; data curation: S.M.A. and M.I.-U.-H.; writing—original draft: S.M.A.; writing—review and editing: S.M.A., I.Y., D.B.K., M.M.R. and H.W.C.; visualization: M.I.-U.-H. and H.W.C.; project implementation: H.W.C. All authors have read and agreed to the published version of the manuscript.

Funding: This research was funded by the Promotion (NO: P0023710) of Innovative Businesses for Regulation-Free Special Zones funded by the Ministry of SMEs and Startups (MSS, Korea).

Data Availability Statement: The original contributions presented in the study are included in the article, further inquiries can be directed to the corresponding author.

Conflicts of Interest: The authors declare no conflicts of interest.

References

1. Hesami, M.; Pepe, M.; Baiton, A.; Jones, A.M.P. Current status and future prospects in cannabinoid production through in vitro culture and synthetic biology. *Biotechnol. Adv.* **2023**, *62*, 108074. [CrossRef] [PubMed]
2. Aloo, S.O.; Kwame, F.O.; Oh, D.-H. Identification of possible bioactive compounds and a comparative study on in vitro biological properties of whole hemp seed and stem. *Food Biosci.* **2023**, *51*, 102329. [CrossRef]
3. Eliwa, G.I.; El-Dengawy, E.-R.F.; Gawish, M.S.; Yamany, M.M. Comprehensive study on in vitro propagation of some imported peach rootstocks: In vitro explant surface sterilization and bud proliferation. *Sci. Rep.* **2024**, *14*, 5586. [CrossRef] [PubMed]
4. Okomo Aloo, S.; Park, S.; Oh, D.-H. Impacts of germination and lactic acid bacteria fermentation on anti-nutrients, bioactive compounds, and selected functional properties of industrial hempseed (*Cannabis sativa* L.). *Food Chem.* **2023**, *428*, 136722. [CrossRef] [PubMed]
5. Okomo Aloo, S.; Park, S.; Martins Oyinloye, T.; Oh, D.-H. Rheological properties, biochemical changes, and potential health benefits of dehulled and defatted industrial hempseeds after fermentation. *Food Chem.* **2024**, *439*, 138086. [CrossRef]
6. Jeon, Y.; Kim, T.; Kwon, H.; Kim, J.-K.; Park, Y.-T.; Ham, J.; Kim, Y.-J. Cannabidiol Enhances Cabozantinib-Induced Apoptotic Cell Death via Phosphorylation of p53 Regulated by ER Stress in Hepatocellular Carcinoma. *Cancers* **2023**, *15*, 3987. [CrossRef]
7. Fordjour, E.; Manful, C.F.; Sey, A.A.; Javed, R.; Pham, T.H.; Thomas, R.; Cheema, M. Cannabis: A multifaceted plant with endless potentials. *Front. Pharmacol.* **2023**, *14*, 1200269. [CrossRef] [PubMed]
8. Gabotti, D.; Locatelli, F.; Cusano, E.; Baldoni, E.; Genga, A.; Pucci, L.; Consonni, R.; Mattana, M. Cell Suspensions of *Cannabis sativa* (var. Futura): Effect of Elicitation on Metabolite Content and Antioxidant Activity. *Molecules* **2019**, *24*, 4056. [CrossRef] [PubMed]
9. Walsh, K.B.; McKinney, A.E.; Holmes, A.E. Minor Cannabinoids: Biosynthesis, Molecular Pharmacology and Potential Therapeutic Uses. *Front. Pharmacol.* **2021**, *12*, 777804. [CrossRef] [PubMed]
10. Ahsan, S.M.; Shin, J.H.; Choi, H.W. Availability of Hydrogen Peroxide Solutions as a Germination Liquid Medium for Contamination-free in vitro Seedling Development of *Cannabis sativa*. *Hortic. Sci. Technol.* **2022**, *40*, 605–613. [CrossRef]
11. Hesami, M.; Pepe, M.; Monthony, A.S.; Baiton, A.; Phineas Jones, A.M. Modeling and optimizing in vitro seed germination of industrial hemp (*Cannabis sativa* L.). *Ind. Crops Prod.* **2021**, *170*, 113753. [CrossRef]
12. Hesami, M.; Adamek, K.; Pepe, M.; Jones, A.M.P. Effect of Explant Source on Phenotypic Changes of In Vitro Grown Cannabis Plantlets over Multiple Subcultures. *Biology* **2023**, *12*, 443. [CrossRef]

13. Pepe, M.; Hesami, M.; Jones, A.M.P. Machine Learning-Mediated Development and Optimization of Disinfection Protocol and Scarification Method for Improved In Vitro Germination of Cannabis Seeds. *Plants* **2021**, *10*, 2397. [CrossRef] [PubMed]
14. Leifert, C.; Cassells, A.C. Microbial hazards in plant tissue and cell cultures. *In Vitr. Cell. Dev. Biol.-Plant* **2001**, *37*, 133–138. [CrossRef]
15. Srivastava, V.; Nerwal, D.K.; Kandan, A.; Akhtar, J.; Sharma, N.; Kiran, R.; Bansal, S.; Agrawal, A. Management of microbial contaminants in the In Vitro Gene Bank: A case study of taro [*Colocasia esculenta* (L.) Schott]. *In Vitr. Cell. Dev. Biol.-Plant* **2021**, *57*, 152–163. [CrossRef]
16. Niedz, R.P.; Bausher, M.G. Control of In vitro contamination of explants from greenhouse- and field-grown trees. *In Vitr. Cell. Dev. Biol.-Plant* **2002**, *38*, 468–471. [CrossRef]
17. Taghinasab, M.; Jabaji, S. Cannabis Microbiome and the Role of Endophytes in Modulating the Production of Secondary Metabolites: An Overview. *Microorganisms* **2020**, *8*, 355. [CrossRef] [PubMed]
18. Dumigan, C.R.; Deyholos, M.K. Cannabis Seedlings Inherit Seed-Borne Bioactive and Anti-Fungal Endophytic Bacilli. *Plants* **2022**, *11*, 2127. [CrossRef] [PubMed]
19. Comeau, D.; Novinscak, A.; Joly, D.L.; Filion, M. Spatio-Temporal and Cultivar-Dependent Variations in the Cannabis Microbiome. *Front. Microbiol.* **2020**, *11*, 491. [CrossRef] [PubMed]
20. Punja, Z.K.; Scott, C. Organically grown cannabis (*Cannabis sativa* L.) plants contain a diverse range of culturable epiphytic and endophytic fungi in inflorescences and stem tissues. *Botany* **2023**, *101*, 255–269. [CrossRef]
21. Monthony, A.S.; Kyne, S.T.; Grainger, C.M.; Jones, A.M.P. Recalcitrance of *Cannabis sativa* to de novo regeneration; a multi-genotype replication study. *PLoS ONE* **2021**, *16*, e0235525. [CrossRef]
22. Hesami, M.; Pepe, M.; de Ronne, M.; Yoosefzadeh-Najafabadi, M.; Adamek, K.; Torkamaneh, D.; Jones, A.M.P. Transcriptomic Profiling of Embryogenic and Non-Embryogenic Callus Provides New Insight into the Nature of Recalcitrance in Cannabis. *Int. J. Mol. Sci.* **2023**, *24*, 14625. [CrossRef]
23. Raspor, M.; Motyka, V.; Kaleri, A.R.; Ninković, S.; Tubić, L.; Cingel, A.; Ćosić, T. Integrating the Roles for Cytokinin and Auxin in De Novo Shoot Organogenesis: From Hormone Uptake to Signaling Outputs. *Int. J. Mol. Sci.* **2021**, *22*, 8554. [CrossRef]
24. Lardon, R.; Geelen, D. Natural Variation in Plant Pluripotency and Regeneration. *Plants* **2020**, *9*, 1261. [CrossRef]
25. Chun, S.C.; Gopal, J.; Iyyakannu, S.; Muthu, M. An analytical retrospection of mass spectrometric tools established for plant tissue culture: Current endeavours and future perspectives. *Trends Anal. Chem.* **2020**, *126*, 115843. [CrossRef]
26. Carla Guimarães Sobrinho, A.; Silva Corpes, R.; Isabel Portilho dos Santos, K.; Maria Menezes Barra, I.; Kiyoshi Miyagawa, H.; Silva Santos, A. Untargeted GC-MS Metabolomics applied to wild leaves and callus produced by plant tissue culture of *Hibiscus sabdariffa* L. *Arab. J. Chem.* **2022**, *15*, 104103. [CrossRef]
27. El-Naggar, H.M.; Shehata, A.M.; Morsi, M.-A.A. Micropropagation and GC–MS analysis of bioactive compounds in bulbs and callus of white squill. *In Vitr. Cell. Dev. Biol.-Plant* **2023**, *59*, 154–166. [CrossRef]
28. Rashid, A.; Ali, V.; Khajuria, M.; Faiz, S.; Gairola, S.; Vyas, D. GC–MS based metabolomic approach to understand nutraceutical potential of Cannabis seeds from two different environments. *Food Chem.* **2021**, *339*, 128076. [CrossRef]
29. Liu, Y.; Liu, H.-Y.; Li, S.-H.; Ma, W.; Wu, D.-T.; Li, H.-B.; Xiao, A.-P.; Liu, L.-L.; Zhu, F.; Gan, R.-Y. *Cannabis sativa* bioactive compounds and their extraction, separation, purification, and identification technologies: An updated review. *Trends Anal. Chem.* **2022**, *149*, 116554. [CrossRef]
30. Micalizzi, G.; Vento, F.; Alibrando, F.; Donnarumma, D.; Dugo, P.; Mondello, L. *Cannabis sativa* L.: A comprehensive review on the analytical methodologies for cannabinoids and terpenes characterization. *J. Chromatogr. A* **2021**, *1637*, 461864. [CrossRef]
31. Nahar, L.; Gavril, G.-L.; Sarker, S.D. Application of gas chromatography in the analysis of phytocannabinoids: An update (2020–2023). *Phytochem. Anal.* **2023**, *34*, 903–924. [CrossRef] [PubMed]
32. Nahar, L.; Guo, M.; Sarker, S.D. Gas chromatographic analysis of naturally occurring cannabinoids: A review of literature published during the past decade. *Phytochem. Anal.* **2020**, *31*, 135–146. [CrossRef] [PubMed]
33. Salam, U.; Ullah, S.; Tang, Z.-H.; Elateeq, A.A.; Khan, Y.; Khan, J.; Khan, A.; Ali, S. Plant Metabolomics: An Overview of the Role of Primary and Secondary Metabolites against Different Environmental Stress Factors. *Life* **2023**, *13*, 706. [CrossRef]
34. Delgado-Povedano, M.M.; Sánchez-Carnerero Callado, C.; Priego-Capote, F.; Ferreiro-Vera, C. Untargeted characterization of extracts from *Cannabis sativa* L. cultivars by gas and liquid chromatography coupled to mass spectrometry in high resolution mode. *Talanta* **2020**, *208*, 120384. [CrossRef]
35. Abadie, C.; Lalande, J.; Tcherkez, G. Exact mass GC-MS analysis: Protocol, database, advantages and application to plant metabolic profiling. *Plant Cell Environ.* **2022**, *45*, 3171–3183. [CrossRef]
36. Oliveira, J.P.S.; Hakimi, O.; Murgu, M.; Koblitz, M.G.B.; Ferreira, M.S.L.; Cameron, L.C.; Macedo, A.F. Tissue culture and metabolome investigation of a wild endangered medicinal plant using high definition mass spectrometry. *Plant Cell Tissue Organ Cult.* **2018**, *134*, 153–162. [CrossRef]
37. Phillips, G.C.; Garda, M. Plant tissue culture media and practices: An overview. *In Vitr. Cell. Dev. Biol.-Plant* **2019**, *55*, 242–257. [CrossRef]
38. Jones, I.A.; Joshi, L.T. Biocide Use in the Antimicrobial Era: A Review. *Molecules* **2021**, *26*, 2276. [CrossRef] [PubMed]
39. Katalinić, I.; Smojver, I.; Morelato, L.; Vuletić, M.; Budimir, A.; Gabrić, D. Evaluation of the Photoactivation Effect of 3% Hydrogen Peroxide in the Disinfection of Dental Implants: In Vitro Study. *Biomedicines* **2023**, *11*, 1002. [CrossRef] [PubMed]

40. Nazir, F.; Fariduddin, Q.; Khan, T.A. Hydrogen peroxide as a signalling molecule in plants and its crosstalk with other plant growth regulators under heavy metal stress. *Chemosphere* **2020**, *252*, 126486. [CrossRef]
41. Wojtyla, Ł.; Lechowska, K.; Kubala, S.; Garnczarska, M. Different Modes of Hydrogen Peroxide Action During Seed Germination. *Front. Plant Sci.* **2016**, *7*, 66. [CrossRef] [PubMed]
42. Xuan, L.; Li, J.; Wang, X.; Wang, C. Crosstalk between Hydrogen Sulfide and Other Signal Molecules Regulates Plant Growth and Development. *Int. J. Mol. Sci.* **2020**, *21*, 4593. [CrossRef]
43. Xu, C.; Hu, Y. The molecular regulation of cell pluripotency in plants. *Abiotech* **2020**, *1*, 169–177. [CrossRef] [PubMed]
44. Pasternak, T.P.; Steinmacher, D. Plant Growth Regulation in Cell and Tissue Culture In Vitro. *Plants* **2024**, *13*, 327. [CrossRef] [PubMed]
45. Chaohua, C.; Gonggu, Z.; Lining, Z.; Chunsheng, G.; Qing, T.; Jianhua, C.; Xinbo, G.; Dingxiang, P.; Jianguang, S. A rapid shoot regeneration protocol from the cotyledons of hemp (*Cannabis sativa* L.). *Ind. Crops Prod.* **2016**, *83*, 61–65. [CrossRef]
46. Galán-Ávila, A.; Gramazio, P.; Ron, M.; Prohens, J.; Herraiz, F.J. A novel and rapid method for Agrobacterium-mediated production of stably transformed *Cannabis sativa* L. plants. *Ind. Crops Prod.* **2021**, *170*, 113691. [CrossRef]
47. Monthony, A.S.; Page, S.R.; Hesami, M.; Jones, A.M.P. The Past, Present and Future of *Cannabis sativa* Tissue Culture. *Plants* **2021**, *10*, 185. [CrossRef]
48. Erland, L.A.E.; Giebelhaus, R.T.; Victor, J.M.R.; Murch, S.J.; Saxena, P.K. The Morphoregulatory Role of Thidiazuron: Metabolomics-Guided Hypothesis Generation for Mechanisms of Activity. *Biomolecules* **2020**, *10*, 1253. [CrossRef] [PubMed]
49. Ali, H.M.; Khan, T.; Khan, M.A.; Ullah, N. The multipotent thidiazuron: A mechanistic overview of its roles in callogenesis and other plant cultures in vitro. *Biotechnol. Appl. Biochem.* **2022**, *69*, 2624–2640. [CrossRef] [PubMed]
50. Dewir, Y.H.; Nurmansyah; Naidoo, Y.; Teixeira da Silva, J.A. Thidiazuron-induced abnormalities in plant tissue cultures. *Plant Cell Rep.* **2018**, *37*, 1451–1470. [CrossRef]
51. Quareshy, M.; Prusinska, J.; Li, J.; Napier, R. A cheminformatics review of auxins as herbicides. *J. Exp. Bot.* **2017**, *69*, 265–275. [CrossRef] [PubMed]
52. Hesami, M.; Baiton, A.; Alizadeh, M.; Pepe, M.; Torkamaneh, D.; Jones, A.M.P. Advances and Perspectives in Tissue Culture and Genetic Engineering of Cannabis. *Int. J. Mol. Sci.* **2021**, *22*, 5671. [CrossRef]
53. Efferth, T. Biotechnology Applications of Plant Callus Cultures. *Engineering* **2019**, *5*, 50–59. [CrossRef]
54. Bielach, A.; Hrtyan, M.; Tognetti, V.B. Plants under Stress: Involvement of Auxin and Cytokinin. *Int. J. Mol. Sci.* **2017**, *18*, 1427. [CrossRef] [PubMed]
55. Tripathi, R.K.P. A perspective review on fatty acid amide hydrolase (FAAH) inhibitors as potential therapeutic agents. *Eur. J. Med. Chem.* **2020**, *188*, 111953. [CrossRef]
56. Merkler, D.J.; Leahy, J.W. Binding-based proteomic profiling and the fatty acid amides. *Trends Res.* **2018**, *1*, 10. [CrossRef] [PubMed]
57. Syed, R.U.; Moni, S.S.; Alfaisal, R.H.; Alrashidi, R.H.; Alrashidi, N.F.; Wadeed, K.M.; Alshammary, F.N.; Habib, A.M.; Alharbi, F.M.; ur Rehman, Z.; et al. Spectral characterization of the bioactive principles and antibacterial properties of cold methanolic extract of *Olea europaea* from the Hail region of Saudi Arabia. *Arab. J. Chem.* **2022**, *15*, 104006. [CrossRef]
58. Gondor, O.K.; Pál, M.; Janda, T.; Szalai, G. The role of methyl salicylate in plant growth under stress conditions. *J. Plant Physiol.* **2022**, *277*, 153809. [CrossRef]
59. Singewar, K.; Fladung, M.; Robischon, M. Methyl salicylate as a signaling compound that contributes to forest ecosystem stability. *Trees* **2021**, *35*, 1755–1769. [CrossRef]
60. Ashrafi, A.M.; Bytešníková, Z.; Cané, C.; Richtera, L.; Vallejos, S. New trends in methyl salicylate sensing and their implications in agriculture. *Biosens. Bioelectron.* **2023**, *223*, 115008. [CrossRef]
61. Ren, Y.; McGillen, M.R.; Daële, V.; Casas, J.; Mellouki, A. The fate of methyl salicylate in the environment and its role as signal in multitrophic interactions. *Sci. Total Environ.* **2020**, *749*, 141406. [CrossRef]
62. Yeasmin, F.; Choi, H.W. Natural Salicylates and Their Roles in Human Health. *Int. J. Mol. Sci.* **2020**, *21*, 9049. [CrossRef] [PubMed]
63. Srbinovska, A.; Conchione, C.; Menegoz Ursol, L.; Lucci, P.; Moret, S. Occurrence of n-Alkanes in Vegetable Oils and Their Analytical Determination. *Foods* **2020**, *9*, 1546. [CrossRef] [PubMed]
64. Mitra, P.; Das, S.; Barik, A. Leaf waxes from Lathyrus sativus: Short-range attractant and stimulant for nymph laying in a viviparous insect. *Chemoecology* **2020**, *30*, 117–129. [CrossRef]
65. Shi, M.; Han, J.; Wang, G.; Wang, J.; Han, Y.; Cui, L. A long-term investigation of the variation in leaf wax n-alkanes responding to climate on Dongling Mountain, north China. *Quat. Int.* **2021**, *592*, 67–79. [CrossRef]
66. Ceccopieri, M.; Scofield, A.L.; Almeida, L.; Araújo, M.P.; Hamacher, C.; Farias, C.O.; Soares, M.L.G.; Carreira, R.S.; Wagener, A.L.R. Carbon isotopic composition of leaf wax n-alkanes in mangrove plants along a latitudinal gradient in Brazil. *Org. Geochem.* **2021**, *161*, 104299. [CrossRef]
67. Rojas-Vargas, J.; Castelán-Sánchez, H.G.; Pardo-López, L. HADEG: A Curated Hydrocarbon Aerobic Degradation Enzymes and Genes Database. *Comput. Biol. Chem.* **2023**, *107*, 107966. [CrossRef]
68. Antonelli, M.; Donelli, D.; Barbieri, G.; Valussi, M.; Maggini, V.; Firenzuoli, F. Forest Volatile Organic Compounds and Their Effects on Human Health: A State-of-the-Art Review. *Int. J. Environ. Res. Public Health* **2020**, *17*, 6506. [CrossRef] [PubMed]

69. Liu, J.; An, Z. Leaf wax n-alkane carbon isotope values vary among major terrestrial plant groups: Different responses to precipitation amount and temperature, and implications for paleoenvironmental reconstruction. *Earth-Sci. Rev.* **2020**, *202*, 103081. [CrossRef]
70. Aguiar, J.; Gonçalves, J.L.; Alves, V.L.; Câmara, J.S. Relationship between Volatile Composition and Bioactive Potential of Vegetables and Fruits of Regular Consumption—An Integrative Approach. *Molecules* **2021**, *26*, 3653. [CrossRef] [PubMed]
71. Chacon, F.T.; Raup-Konsavage, W.M.; Vrana, K.E.; Kellogg, J.J. Secondary Terpenes in *Cannabis sativa* L.: Synthesis and Synergy. *Biomedicines* **2022**, *10*, 3142. [CrossRef]
72. Zager, J.J.; Lange, I.; Srividya, N.; Smith, A.; Lange, B.M. Gene Networks Underlying Cannabinoid and Terpenoid Accumulation in Cannabis. *Plant Physiol.* **2019**, *180*, 1877–1897. [CrossRef] [PubMed]
73. Li, J.; Li, K.; Li, H.; Wang, X.; Wang, W.; Wang, K.; Ge, M. Long-chain alkanes in the atmosphere: A review. *J. Environ. Sci.* **2022**, *114*, 37–52. [CrossRef] [PubMed]
74. Zhao, F.; Wang, P.; Lucardi, R.D.; Su, Z.; Li, S. Natural Sources and Bioactivities of 2,4-Di-Tert-Butylphenol and Its Analogs. *Toxins* **2020**, *12*, 35. [CrossRef]
75. Lobiuc, A.; Pavăl, N.-E.; Mangalagiu, I.I.; Gheorghiță, R.; Teliban, G.-C.; Amăriucăi-Mantu, D.; Stoleru, V. Future Antimicrobials: Natural and Functionalized Phenolics. *Molecules* **2023**, *28*, 1114. [CrossRef]

horticulturae

MDPI

Article

Cold Plasma Treatment Alters the Morphology, Oxidative Stress Response and Specialized Metabolite Content in Yellow Iris (*I. reichenbachii*) Callus

Slađana Jevremović [1,*], Milica Milutinović [1], Ksenija Veličković [2], Uroš Gašić [1], Nikola Škoro [3], Nevena Puač [3] and Suzana Živković [1,*]

[1] Department of Plant Physiology, Institute for Biological Research "Siniša Stanković"—National Institute of the Republic of Serbia, University of Belgrade, Bulevar Despota Stefana 142, 11108 Belgrade, Serbia; milica.milutinovic@ibiss.bg.ac.rs (M.M.); uros.gasic@ibiss.bg.ac.rs (U.G.)

[2] Faculty of Biology, University of Belgrade, Studentski Trg 16, 11158 Belgrade, Serbia; ksenija@bio.bg.ac.rs

[3] Center for Non-Equilibrium Processes, Institute of Physics, National Institute of Republic of Serbia, University of Belgrade, Pregrevica 118, 11080 Belgrade, Serbia; nikola.skoro@ipb.ac.rs (N.Š.); nevena.puac@ipb.ac.rs (N.P.)

* Correspondence: sladja@ibiss.bg.ac.rs (S.J.); suzy@ibiss.bg.ac.rs (S.Ž.)

Abstract: The application of non-thermal (cold) plasmas is considered an environmentally friendly method that could affect plant metabolism and cellular development or can be used for the commercial production of natural products that cannot be chemically synthesized. In the present study, the non-embryogenic callus of iris (*Iris reichenbachii* Heuff.) was treated with a Radio Frequency (RF) plasma needle device using He as a working gas. We investigated short-term (up to seven days) and long-term (up to one year) changes on morphological, physiological and biochemical levels. An increased production of O_2^- and H_2O_2 was observed in the callus tissue after plasma treatment. The enzymes SOD and CAT represented the frontline in the antioxidant defense against reactive oxygen species (ROS) produced during the first hour of treatment, while POX was the leading antioxidant enzyme seven days after plasma treatment. Significant long-term morphological changes were observed in the calli due to the increased mitotic activity of the plant cells. In addition, three flavonoids (naringenin, apigenin and acacetin) and two isoflavonoids (irisolidone and irilone) were detected only in the plasma-treated tissue even one year after plasma treatment. The present study emphasizes the application of the plasma technique to promote meristematic activity and stimulate the production of specialized metabolites in iris calli.

Keywords: callus; iris; cold plasma; antioxidant activity; isoflavonoids

1. Introduction

Cold plasma (CP), also known as non-thermal or low temperature plasma, is an innovative technology that has wide range application in agriculture including seed germination, cultivation, surface sterilization, microorganism decontamination, food manufacturing processing and storage, wound healing, plant growth as well as secondary metabolite production [1]. Increasing interest for the application of CP lies in the fact that no expensive vacuum systems are required, the geometries of electrode systems are versatile and can be adapted depending on the type of samples and the desired characteristics of the plasma [2–4]. The main disadvantage of CP is the need for high-voltage power supplies and, in most cases, the need for the addition of noble gases (Ar, He, etc.) that are used to lower the breakdown voltage. Both can increase the cost of plasma applications, but even with this, CP remains competitive with classical methods that involve the usage of additional chemical compounds or complicated processes. This is due to the most important feature of CPs—their rich chemistry [5,6]. There is some evidence that chemical species in the plasma, with ample amounts of reactive oxygen and nitrogen species (RONS), are responsible for triggering various mechanisms that affect the plant cells [6,7]. The demand for precise and localized in vivo treatments of living cells and tissues resulted in the fast development of various plasma devices that operate at atmospheric pressure [1,8,9]. The plasma sources have a variety of designs and operating procedures such as corona discharges, spark discharges, dielectric barrier discharges, or underwater discharges, with the plasma needle as one of the atmospheric pressure devices that meet the requirements for the precise and localized treatments of living cells and tissues [1,10].

Numerous studies have demonstrated that this modern eco-agricultural technology is effective in improving agronomic seed quality and has the potential to be used for seed decontamination, germination activation, and seedling growth [11–16]. A few reports have also investigated stressor-induced changes in physiological and biochemical activities, such as the amount of pigments and secondary metabolites, antioxidant capacity, and the regulation of photosynthesis [11,17,18]. There is a wealth of data on CP treatment that has been carried out on seeds and seedlings ex vitro (summarized in 1,7,19), and, in all cases, the results showed that CP treatment can increase the content of valuable secondary metabolites [19–23]. On the other hand, there are not as many reports on the effect of CP on plant material during cultivation in vitro, like calli culture [24,25], synthetic seeds [26], liverworts [27], or genome editing [28]. Direct plasma treatment has been shown to have significant long-term effects on callus growth in some ornamental plants with plasma-treated callus growing better than the untreated one [24,25]. Callus, as a growing mass of unorganized plant parenchyma cells, can also produce bioactive compounds. Over the last few decades, significant progress has been made in the production of these compounds using plant tissue culture techniques. There are many studies reporting the production of specialized metabolites in callus and/or differentiated tissue cultures with callus cultures being the preferred in vitro culture system for achieving high yields and consistent material production for commercial use [29–33]. Recently, it has been reported that the CP treatment of calli can trigger the production of valuable secondary metabolites [10,34]. The authors showed that corona discharge plasma enhanced callogenesis and stimulated the production of cannabinoids in cannabis calli [34] and atropine in the callus of *Datura inoxia* [10]. It has also been suggested that, considering the necessity of keeping sterile conditions, the development of new specialized devices is necessary to facilitate the application of plasma treatment in the field of plant tissue culture [10].

Irises, a group of plants from the family Iridaceae, have immense medicinal importance in addition to their ornamental value, as they contain a variety of valuable secondary metabolites [35], which are used in the treatment of cancer, inflammation, bacterial and viral infections summarized in the studies of [36–38]. Characteristic phytochemicals of irises are flavones, flavone C-glycosides, isoflavones, terpenoids, xanthones, phenols, stilbenes, and quinones [39,40]. Of particular pharmaceutical interest are isoflavonoids, which are also characteristic of plants from the family Leguminosae. They represent the typical ecophysiological active compounds that act as defense substances and are mainly accumulated in the underground part of the plants. To date, 85 isoflavonoid aglicones and glycosides have been found in 18 different species belonging to the genus iris [41–44]. Most of them were isolated from the rhizome and leaves of the plant material collected in nature and showed antioxidant, anti-inflammatory, antiviral, antibacterial and estrogen-like effects [38,45,46].

Plant tissue culture methods, including callus and in vitro shoot and root regeneration, can ensure the production of valuable secondary metabolites or raw materials in large quantities without destroying the natural habitat [47]. In the case of irises, plant regeneration in in vitro culture has been reported for many endemic, endangered, and threatened species [48–51], as well as for some ornamental [52–55] and pharmaceutically or perfumery valuable iris species [56–59]. The production of secondary metabolites in in vitro tissue culture has been investigated in several species of the genus iris (Table 1). In the culture of iris callus, cell suspensions or adventive roots, the production of flavones [60,61], isoflavones [62–64], irones [49,65], iridals [66,67], essential oils [68] and xanthones [63,69] was investigated. It should be emphasized that the yellow color of the iris callus is due to the accumulation of carotenoids and not xanthones [52,69]. So far, the isoflavonoid production ability of irises has been evaluated in adventive root cultures of *I. germanica* [62] and *I. pseudacorus* [64], shoot cultures of *I. sibirica* [63] and callus cultures of *I. pseudacorus* [70] and [*I. domestica* [71]. There are some data in the literature on elicitation of the production of secondary metabolites in iris during regeneration in tissue culture by varying different plant growth regulators in the culture medium [60,61,63] or by adding various abiotic elicitors, such as $CuCl_2$ [62], phenylalanine [70], jasmonic acid [72] or quinone [49]. In addition, some flavonoids were isolated from the callus tissue of *I. ensata* [61], which are uncharacteristic of intact plants collected in nature (Table 1).

Table 1. A list of specialized metabolites recorded during in vitro growth of irises.

Plant Species	Culture Type	Compound	Elicitation	References
Iris atrofusca	Callus	Irones Essential oils	Quinine	[49] [68]
Iris domestica	Callus	Isoflavonoids	$CuCl_2$	[71]
Iris ensata	Callus	Flavonoids Flavones		[60] [61]
Iris germanica	Root cultures Callus	Isoflavonoids Carotenoids	$CuCl_2$	[62] [73]
Iris pallida	Cell suspensions Leaves	Iridals		[66] [59]
Iris petrana	Callus	Irones Essential oils	Quinine	[49] [68]
Iris pseudacorus	Cell suspensions Callus Shoots, roots Root culture	Iridals Isoflavonoids Phenolics Isoflavones	 Phenilalanine Jasmonic acid	[67] [70] [72] [64]
Iris reichenbachii	Callus, Shoots	Xanthone (Mangiferin)		[69]
Iris sibirica	Callus Cell suspension	γ irone Volatile oils Flavonoids		[65] [74] [75]
	Shoots	Isoflavonoids Mangiferin	Auxin (NAA)	[63] [76]

The subject of this study was *Iris reichenbachii* (Heuffel), dwarf bearded or yellow iris (Syn: *I. bosniaca* G. Back 1960, *I. balkana* Janka, 1960, *I. skorpilii* Velenovsky, *I. serbica*

Pančić, 1856), which is an endemic plant and can be found in several locations on the Balkan Peninsula [77] (Figure 1a). So far, the synthesis of valuable xanthones (mangiferin, isomangiferin), flavones (apigenin, acacetin, chrysoeriol) and C-glycosylflavones (isoorientin, luteolin di-C-glycoside) was recorded mainly in leaves and rhizomes of this species [44,78–81]. Šavikin and coworkers showed that the isoflavonoid iridin was present in the rhizomes of *I. reichenbachii* collected in the wild [80]. In addition, Jevremović et al. [77,82,83] established a protocol for the in vitro regeneration of mature zygotic embryos of *I. reichenbachii* and investigated the production of mangiferin (C-flavone glycoside) during tissue culture induction and the plantlet regeneration of *I. reichenbachii* in vitro [69]. The authors stated that the synthesis of mangiferin was recorded during shoot regeneration, while, in callus culture, mangiferin was detected only in traces [69]. According to preliminary data, cold plasma may have some effects on callus growth as well as the production of secondary metabolites in iris calli, but the mechanism of plasma treatment-induced changes is not yet clear [84].

Figure 1. Dwarf iris, *Iris reichenbachii* (Heuffel): (**a**) Plants in natural habitat. (**b**) Three types of calli formed in the culture of zygotic embryos on a medium supplemented with 2,4-D: white, embryogenic (EC), green organogenic (OC) and yellow non-embryogenic (NEC) calli. Bars: (**a**) 10 mm; (**b**) 1 mm.

When in vitro cultures are induced on a 2,4-D containing medium, three types of calli can be distinguished in most iris species, including *I. reichenbachii* (Figure 1b) based on their color and consistency: (i) white calli, referred to as embryogenic calli (EC), (ii) green, organogenic calli (OC), and (iii) yellow referred to as non-embryogenic calli (NEC) [50,55,56,59,85,86]. After prolonged culture, yellow callus is often the most common and frequent, showing no morphogenetic response with very low regeneration potential, and therefore represents one of the greatest challenges in this field of research.

Accordingly, the present study aimed to complete the knowledge on the short- and long-term effects of CP on the yellow non-embryonic callus of *I. reichenbachii* at different levels: (i) morphological, (ii) histological and (iii) physiological, including the assessment of oxidative stress and antioxidant defense of calli immediately after plasma treatment. The long-term effects of plasma will focus on the ability of plasma priming to be used as an alternative tool/elicitor for the production of specialized metabolites in yellow iris calli.

2. Materials and Methods

2.1. Plant Material and Growth Conditions

For the experiments, a nodular, yellow non-embryogenic callus (NEC) of *I. reichenbachii* Heuff was used. Callus induction was performed in the culture of mature zygotic embryos as described in the study of [77]. Three types of calli were initially induced: white embryogenic calli (EC), green organogenic calli (OC) and yellow non-embryogenic calli

(NEC). All calli were grown on solid MS medium consisting of MS mineral solution [87], 3% sucrose, 250 mg of casein hydrolysate and 250 mg of proline, supplemented with 2,4-dichlorophenoxyacetic acid (2,4-D) and kinetin, 1.0 mg L^{-1} each and subcultured on fresh medium every four weeks. After several subcultures, the embryogenic potential decreased, and most calli became yellow nodular, slow-growing NEC. The yellow *I. reichenbachii* callus used in this work was maintained for several months prior to the experiments by transferring a portion of calli to the same medium. All cultures were grown at a temperature of 24 ± 2 °C and under a 16 h light/8 h dark photoperiod (photon flux density of 40 μmol m^{-2} s^{-1} provided by cool white fluorescent lamps).

2.2. Non-Thermal Plasma System and Treatment Conditions

A cold (non-thermal) atmospheric plasma source—plasma needle, which operates at a frequency of 13.56 MHz, was used for the treatment of iris calli (Figure 2a,b). It consisted of a central tungsten electrode (diameter 0.5 mm) enclosed in a ceramic tube and placed inside a glass tube (inner diameter 4 mm, outer diameter 6 mm). The body of the plasma needle was made of Teflon. Further details on the system can be found in [88]. The working gas was helium, and the flow was kept constant at 1 slm (standard liter per minute) in all experiments. At the end of the tube in the mixture of helium and ambient air, plasma was generated and enabled the formation of RONS important for the treatment of biological samples (Figure 2c). The power introduced into the plasma was monitored and did not exceed a few watts, which is another important factor for the treatment of sensitive samples.

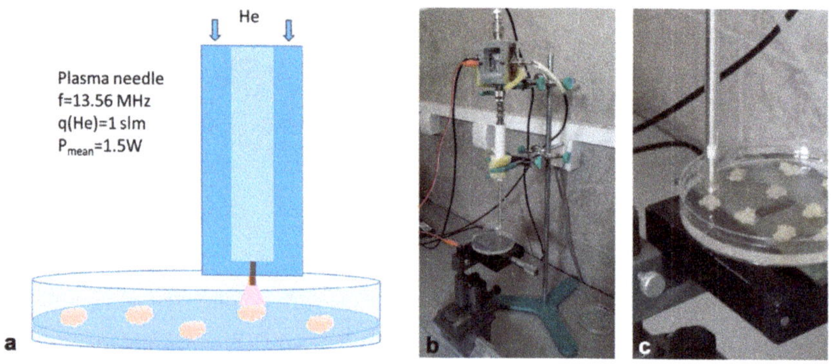

Figure 2. Plasma needle setup for the treatment of iris callus: (**a**) schematic illustration of cold atmospheric plasma (CP) treatment used in the experiments; (**b**) plasma needle device; (**c**) treatment of callus with plasma.

Ten calli samples (30–60 mg) were placed in a Petri dish containing 20 mL growth medium (MS medium with 2,4-D and kinetin, 1.0 mg L^{-1} each) before treatment with plasma. Plasma treatment was performed directly on the upper surface of callus using needle after uncovering the Petri dish. The treatment times were 10 s, 30 s and 60 s, at power deposited to the plasma of 1.5 W. All treatments were performed in triplicate. Untreated (control) calli samples were grown in a Petri dish filled with culture medium in the same way as the plasma-treated samples during the experimental procedure to ensure equal treatment conditions. After treatment, calli were either immediately frozen in liquid nitrogen and stored at −80 °C until use or continuously cultured in Petri dishes before use. The whole experiment was repeated twice.

2.3. Histological Analysis

The calli samples were fixed in FAA fixative (10% formalin, 5% acetic acid, 50% ethanol) at room temperature for 24 h. The fixed samples were dehydrated through a graded series of ethanol (50%, 70%, 96%, 100%), cleared in xylene and embedded in paraffin [89]. Tissue sections (10 μm) were routinely deparaffinized with xylene, rehydrated in graded ethanol, and stained with a 1% toluidine blue solution (Sigma-Aldrich, St. Louis, MO, USA). All sections were mounted in DPX mounting medium (Sigma-Aldrich, St. Louis, MO, USA) and examined using a DMLB microscope (Leica Microsystems, Wetzlar, Germany).

2.4. Histochemical Localization of O_2^- and Production of H_2O_2

The localization of superoxide anion radicals (O_2^-) and H_2O_2 was performed according to the modified method described in [90]. The formation of O_2^- was visualized using the reduction staining method of nitro blue tetrazolium (NBT, Sigma-Aldrich, St. Louis, MO, USA) as blue formazan precipitates due to the reduction in NBT in the presence of O_2^-. The control and treated calli samples were incubated in 2.5 mM NBT solution in 50 mM sodium phosphate buffer (pH 7.8) containing 10 nM of NaN_3 in the dark at room temperature for 2 h. Accumulation of H_2O_2 was detected by formation of a brown polymerization product after incubation of the calli with a 3,3′-diaminobenzidine (DAB, Sigma-Aldrich, St. Louis, MO, USA) solution (5 mM DAB-HCl, pH 3.8) for 2 h in the dark at room temperature. All samples were stored in a glycerol/ethanol solution (1:4; *v*/*v*) prior to light microscopic analysis and visualized with a Leica DMLB 2900 light microscope (Leica Microsystems, Wetzlar, Germany) using the LAS V4.11 program.

2.5. Protein Extraction

Yellow callus samples (NEC, approximately 50 mg FW) were grounded in liquid nitrogen using a mortar and pestle and homogenized in 100 mM potassium phosphate buffer (pH 6.5) containing 1 mM of phenylmethylsulfonyl fluoride (PMSF), 2 mM of EDTA, 5 mM of ascorbic acid, 0.5% Tween-20 and 2% (*w*/*v*) polyvinylpolypyrrolidone (PVPP). The homogenates were centrifuged at 14,000 g for 20 min, at +4 °C. The supernatant was used for the assays. The total protein content in the extracts was determined using a Qubit 3.0 Fluorimeter (Life Technologies, Thermo Fisher Scientific, Waltham, MA, USA) according to the instructions of the Qubit protein assay kit (Thermo Fisher Scientific, Waltham, MA, USA).

2.5.1. Native-PAGE Electrophoresis and Determination of Enzyme Activity

Separation by native polyacrylamide gel electrophoresis (native-PAGE) and activity staining of SOD, CAT and POX isoforms were performed according to procedure reported in the study of [11,91].

2.5.2. SDS-PAGE Electrophoresis and Immunoblotting

Sodium dodecyl sulfate-polyacrylamide gel electrophoresis (SDS-PAGE) and subsequent transfer of proteins to nitrocellulose membranes were performed as reported by the authors of [91]. The primary antibodies used in this work were CSD2 antibody (Anti-Rabbit chloroplastic CuZn-Superoxide Dismutase, AS06170; Agrisera Antibodies, Vännäs, Sweden), Anti-Rabbit Mn-Superoxide Dismutase antibody (AS09524, Agrisera Antibodies, Sweden), Anti-Catalase antibody (1:1000, AS09501; Agrisera Antibodies, Sweden), Anti-Peroxidase (AS09548, Agrisera Antibodies, Sweden) and goat Anti-Rabbit IgG peroxidase (A0545, Sigma-Aldrich, Saint Louis, MI, USA) as secondary antibodies. After intensive washing in the PBS buffer, the protein signals were visualized using an enhanced chemiluminescence detection (ECL) system. Densitometry analysis of band intensities was performed using ImageJ 1.32j software (W. Rasband, National Institute of Health, Bethesda, MD, USA). The signal intensities obtained were normalized to the value of the control sample, and the results are presented as relative abundances.

2.6. Preparation of the Extracts

Control (NEC and EC) and plasma-treated calli samples (30–50 mg) were ground to a fine powder with liquid nitrogen and extracted using methanol (1:10, *w:v*) at room temperature. After a short 10 min sonication in ultrasonic water bath (Sonorex Bandelin Electronic, Berlin, Germany), the extracts were centrifuged at 12,000 g for 20 min. The supernatants were filtered with 0.2 μm cellulose filters (Agilent Technologies, Santa Clara, CA, USA) and stored at +4 °C until use.

2.7. UHPLC-LTQ Orbitrap MS Analysis

The compounds of interest were separated using an Accela 600 ultra-high-performance liquid chromatography (UHPLC) system coupled to a linear ion trap—Orbitrap mass spectrometer (LTQ OrbiTrap MS; ThermoFisher Scientific, Bremen, Germany). A Syncronis C18 column (100 × 2.1 mm, 1.7 μm particle size) thermostatted at 40 °C was used to separate the compounds. The flow rate was set to 300 μL min^{-1}, and the mobile phase consisted of 0.1% formic acid in water (A) and acetonitrile + 0.1% formic acid (B). The injection volumes were 5 μL. The linear gradient program used for sample analysis was described in detail in the study of [92]. The MS was equipped with a heated electrospray ionization probe (HESI-II, Thermo Fisher Scientific, Bremen, Germany) operating in negative ionization mode. The MS spectra were acquired by full-range acquisition over 100–1000 m/z. The parameters of the ion source and the other MS settings were the same as previously described by the study of [93]. Xcalibur software (version 2.1) was used for instrument control, data acquisition, and analysis. Identification of unknown compounds was performed by exact mass search of their deprotonated molecule ($[M-H]^-$) and their MS^4 fragmentation, as well as by literature search of available chromatographic and MS data [41,94–97].

2.8. Statistical Analysis

Statistical analyses were performed using STATGRAPHICS software, v. 4.2 (Statgraphics Technologies, Inc., The Plains, VA, USA). Data were subjected to analysis of variance (ANOVA), and comparisons between the mean values of treatments were performed using Fisher's Least Significant Difference (LSD) post hoc test at a confidence level of $p < 0.05$ and $p < 0.01$.

3. Results

We evaluated the morphological (Section 3.1) and histological (Section 3.2) changes, as well as the oxidative stress assessment (Section 3.3) of yellow iris calli immediately after the plasma treatment of 10 s, 30 s, and 60 s (marked as 0 h time point), as well as one hour (1 h) and seven days (7d) after treatment. In addition, pronounced morphological changes and the development of white calli were observed only on samples after plasma treatment for 60 s. Therefore, long-term changes in phytochemical profile in plasma-treated (60 s) in comparison to untreated iris calli were evaluated after one week, three weeks and one year of cultivation following plasma treatment (Section 3.4).

3.1. Morphological Changes in Iris Calli After Plasma Treatment

After plasma treatment, all treated calli samples continued to grow on the same culture medium as untreated calli (Figure 3). It should be emphasized that there were no signs of plant material contamination, although the plasma needle setup and treatments were not performed under aseptic conditions (Figure 2). The first clearly visible morphological changes after the CP treatment were observed on the surface of treated calli (Figure 3b–h). Untreated calli had a compact nodular structure with a smooth surface (Figure 3a). Morphological change was recorded 7 days after plasma treatment, when a mucilaginous envelope around the plasma-treated calli samples was observed (Figure 3 b–d). After three weeks, the nodular structure disorganized, and protrusions of white calli appeared on the surface of the nodules (Figure 3e–g). It was also noted that the plasma-treated iris calli grew better after prolonged culture in comparison to untreated calli (Figure 3i).

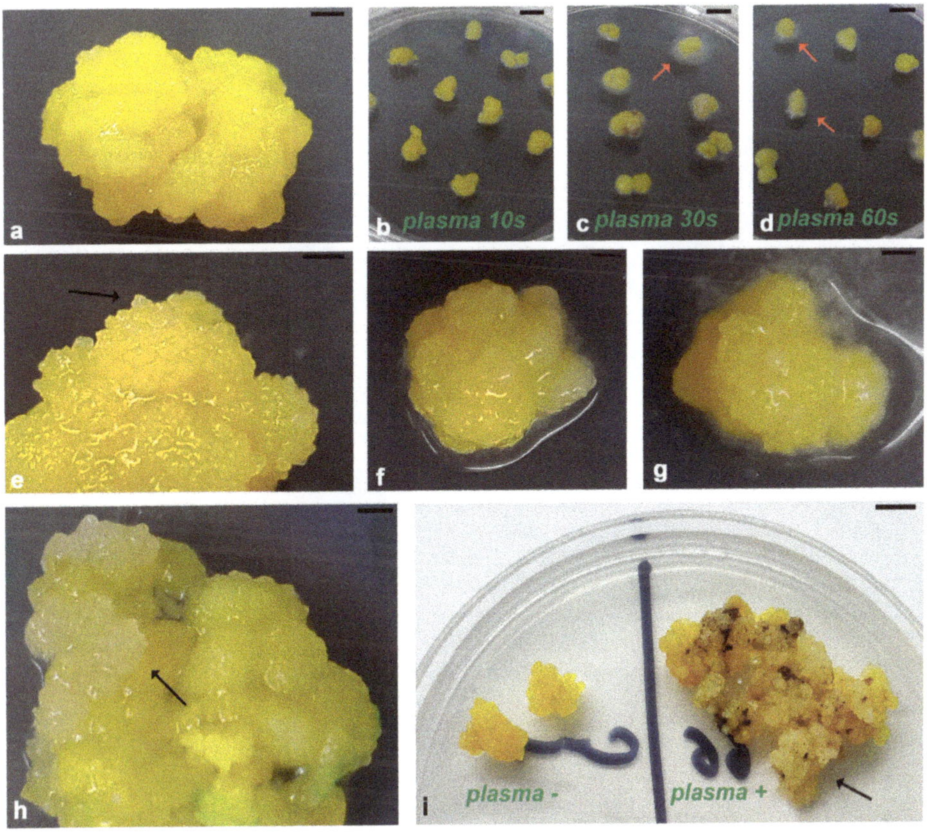

Figure 3. Morphological changes in iris calli after plasma treatment: (**a**) Untreated iris callus. (**b–d**) Calli samples 7 days after plasma treatment for 10 s (**b**), 30 s, (**c**) and 60 s (**d**). (**e–h**) Development of white calli on plasma-treated samples (60 s) three weeks after the treatment. (**i**) Calli cultures ten weeks after plasma treatment for 60 s (left—untreated calli; right—plasma-treated calli; note: growth of plasma-treated calli). The red arrows indicate the formation of mucilaginous "shell" around the callus; the black arrows indicate the formation of white callus on the surface of the treated calli. Bars: (**a,e,f,g,h**) 1 mm; (**b,c,d**) 5 mm; (i) 10 mm.

3.2. Histological Study of Iris Callus After PLASMA Treatment

Light microscopic examination revealed different cell types in the iris callus (Figure 4). The uniform cells were found in the untreated iris callus, where numerous parenchymatous cells (PCs) with a large vacuole, a thin cytoplasmic ring, and nuclei close to the cell wall were predominant (Figure 4a,d). Small regions in the inner parts of the untreated callus were characterized by tiny cells with a poorly developed vacuole system, dense cytoplasm, and a prominent nucleus and were designated as meristematic centers (MCs). In comparison to the untreated callus, the most prominent change observed in the CP-treated callus was related to meristematic cells. The areas with small and cytosol-rich cells in the plasma-treated calli appeared to be larger and more visible than those of the untreated callus (Figure 4b,e,f). In addition, the changes observed in the plasma-treated callus were also noted after one year of cultivation, with several regions of meristem initiation appearing (Figure 4c,g). Taken together, these results clearly showed that the cellular organization of the plasma-treated calli was significantly different from that of the untreated callus.

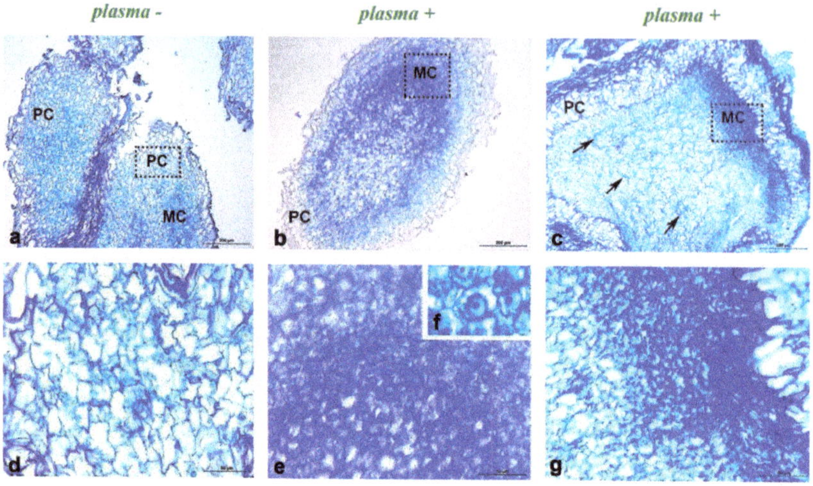

Figure 4. Histological study of *Iris reichenbachii* yellow callus treated with CP for 60 s: (**a–c**) Toluidine blue-stained cross-sections of untreated callus (**a**) and plasma-treated (**b**,**c**). (**d**) Detail of parenchymatous tissue of untreated callus. (**e**,**f**) Detail of meristematic center with typical meristematic cell (**f**) in plasma-treated calli. (**g**) Detail of cross-section of treated callus one year after plasma treatment for 60 s. Arrows indicate meristem initiation. Enlarged black dotted squares show typical histological features of parenchymatous cells (PCs) and meristematic cells in the meristematic center (MC). Bars: (**a–c**) 200 μm, (**d–g**) 50 μm.

3.3. *Oxidative Stress Response of Iris Calli to Plasma Treatment*

In order to investigate the oxidative stress response of iris callus to plasma treatment we evaluated the localization of reactive oxygen species (O_2^- and H_2O_2) production on the surface of the callus immediately after plasma treatment, one hour and one week after plasma treatment (Section 3.3.1). In addition, the activity (Section 3.3.2) and protein abundance of antioxidant enzymes (Section 3.3.3) were examined in parallel, during the same time periods.

3.3.1. Histochemical Localization of O_2^- Production and H_2O_2 Accumulation in Iris Calli After Plasma Treatment

Immediately after plasma treatment, an oxidative burst was observed on the surface of the iris yellow callus (Figure 5a–h). The production of oxygen anion (O_2^-) was localized mainly in cell walls and the extracellular matrix, outside of cells on untreated callus samples (Figure 5a,e), while in the plasma-treated callus, all surfaces of the callus were intensively stained with NBT (Figure 5b). In the cross-section, it was obvious that the increased production of oxygen anion was localized on the surface of the callus nodule (Figure 5f). In parallel, the accumulation of H_2O_2 was also recorded on the surface of the iris callus after plasma treatment (Figure 5d). In untreated calli, the accumulation of H_2O_2 was recorded sporadically around individual cells inside the callus (Figure 5g), while, in the plasma-treated callus, significant accumulation of brown DAB staining, indicating increased accumulation of H_2O_2, was observed at the surface of the callus (Figure 5h).

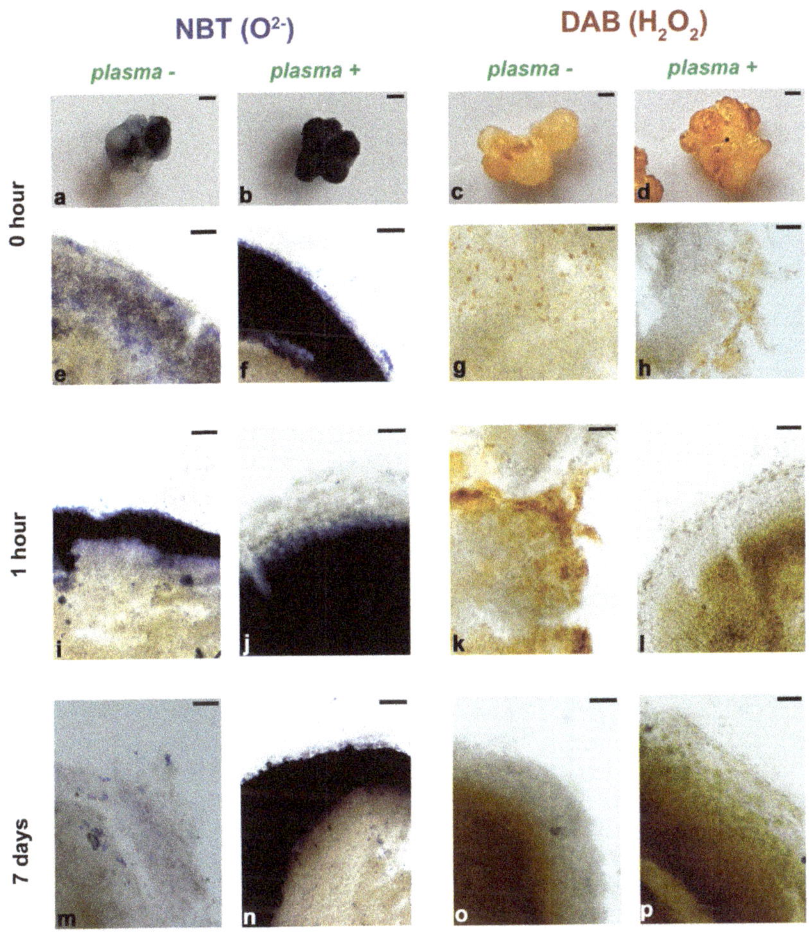

Figure 5. Histochemical localization of oxygen anion (O_2^-) and hydrogen peroxide (H_2O_2) production in *Iris reichenbachii* callus treated with plasma. Plasma-untreated and treated calli samples stained with NBT (**a**,**b**) and DAB (**c**,**d**) immediately after plasma treatment (0 h); cross-sections of untreated and CP-treated calli immediately after plasma treatment, 1 h and 7 days after plasma treatment. NBT staining (**e**,**f**,**i**,**j**,**m**,**n**); DAB staining (**g**,**h**,**k**,**l**,**o**,**p**). Bars: (**a**–**d**) 1 mm; (**e**–**p**) 200 µm.

One hour after plasma treatment, a significant increase in ROS production and accumulation of H_2O_2 was observed inside the callus (Figure 5j,l) in comparison to the untreated callus, where ROS production and H_2O_2 accumulation were detected on the surface of the iris callus (Figure 5i,k). Seven days after plasma treatment, an oxidative burst was noticed on the surface of plasma-treated callus (Figure 5n) compared to the untreated samples (Figure 5m). In contrast to O_2^- production, the H_2O_2 accumulation 7 days after plasma treatment was localized inside the callus nodule of both the untreated (Figure 5o) and treated callus (Figure 5p). However, cell layers with different DAB staining intensities (light and strong) were detected from the surface to the center of the callus nodule in the plasma-treated calli (Figure 5p).

3.3.2. Activity of Antioxidant Enzymes in Iris Calli After Plasma Treatment

Our results demonstrated that SOD and CAT in yellow iris calli were considered as the main enzymatic frontline in the antioxidant defense against ROS generated by plasma treatment (Figure 6). These enzymes showed a significant increase in total activity in response to CP treatment immediately and one hour after the plasma treatment (Figure 6a–c).

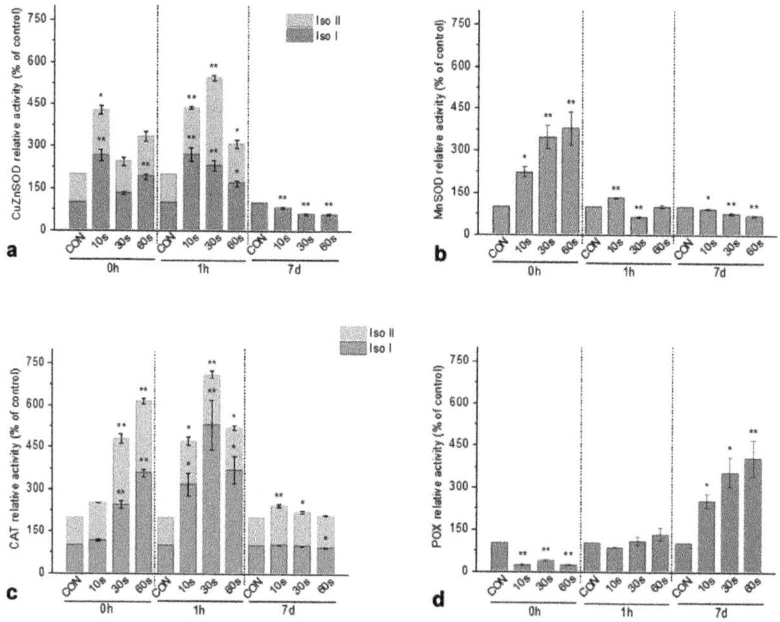

Figure 6. Activity of the antioxidant enzymes in *I. reichenbachii* callus after plasma treatment: (**a**,**b**) SOD activity, CuZnSOD (**a**), MnSOD (**b**); (**c**) CAT activity; (**d**) POX activity. Enzyme activities are measured immediately (0 h), one hour (1 h), or seven days (7d) of control (CON) and after CP treatment for 10 s, 30 s and 60 s. The band volume is determined by densitometric analysis. Data are normalized using the control signal. Values are means ± SE from two independent experiments. The asterisk indicates statistical significance compared to the control (* $p < 0.05$ and ** $p < 0.01$). Abbreviations: Iso I—Isoform I and Iso II—Isoform II.

Electrophoretic analyses of the SOD isoenzymes that were active and reacted after plasma treatment revealed three bands characterized as two CuZnSOD isoforms (Iso I and Iso II, Figure 6a) and one isoform of MnSOD (Figure 6b). The total CuZnSOD activity increased 1.5- to 3-fold in all plasma-treated samples collected immediately (0 h) and one hour (1 h) after treatment, respectively, compared to the untreated control sample (CON). Seven days (7d) after plasma treatment, total CuZnSOD activity decreased below baseline levels in all treated samples (Figure 6a). Contrary to CuZnSOD activity, a 2-4-fold increase in MnSOD activity was observed only immediately after plasma treatment (Figure 6b), which was below baseline one hour after plasma treatment. Seven days (7d) after plasma treatment, MnSOD isoforms showed a significant decrease in activity in comparison to untreated callus samples (Figure 6b).

A statistically significant increase in the activity of both detected CAT isoforms (Iso I and Iso II) was observed in all plasma-treated samples compared to the control (Figure 6c). Contrary to SOD and CAT, POX activity was significantly reduced immediately after treatment (0 h) compared with the control. In addition, POX activity in calli samples collected one hour (1 h) after treatment was not affected by low-temperature plasma treatment. Seven days (7d) after plasma treatment, POX becomes the leading antioxidant enzyme that protects the iris callus from excessive ROS production in callus cells. POX activity increased with prolonged plasma treatment and reached the highest value after 60s of plasma treatment where POX activity was increased more than fourfold compared to the control untreated samples (Figure 6c).

3.3.3. Immunoblot Analysis of Antioxidant Enzymes in Iris Calli After Plasma Treatment

The protein level of the antioxidant enzyme isoforms was evaluated by immunoblot analysis with specific antibodies (Figure 7). Immunoblot analysis revealed the presence of a CuZnSOD band with a molecular weight of 19 kDa in all samples (Figure 7a). The intensity of the band corresponding to the CuZnSOD isoform appeared to decrease significantly in all treatments compared to the control untreated samples. A sharp decline in CuZnSOD protein level (up to three-fold) was observed immediately and 1 h after the 60 s plasma treatment. Immunoblot analysis revealed a band of 50 kDa in iris calli that corresponded to MnSOD (Figure 7b). In contrast to CuZnSOD, MnSOD protein levels showed a similar pattern to CAT in all treatments (Figure 7c), characterized by the positive correlation between total enzyme activity and abundance in samples collected immediately after plasma treatment (Figure 7b,c). MnSOD protein levels (Figure 7b) 1 h and 7d after plasma treatment resulted in an increase in protein level, but activity analysis showed a statistically significant decrease compared to control samples (Figure 7b).

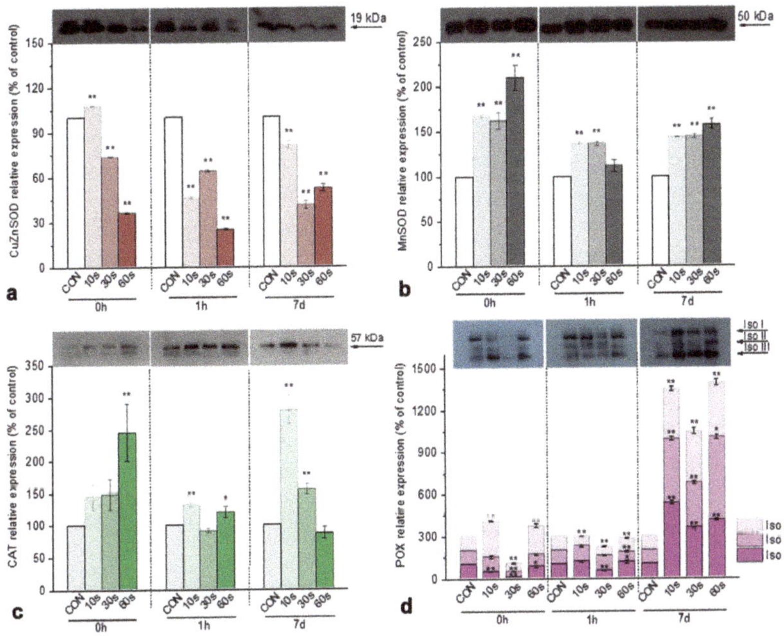

Figure 7. Immunoblot analysis of *I. reichenbachii* callus after plasma treatment: (**a**,**b**) SOD, (**a**) CuZnSOD, (**b**) MnSOD, (**c**) CAT, (**d**) POX measured immediately (0 h), one hour (1 h), or seven days (7d) of control (CON) and CP treatment (10 s, 30 s and 60 s). The band volume was determined by densitometric analysis. Data are normalized using the control signal. Values are means ± SE obtained from two independent experiments. The asterisk indicates statistical significance compared to the control (* $p < 0.05$ and ** $p < 0.01$). Abbreviations: Iso I—Isoform I, Iso II—Isoform II and Iso III—Isoform III.

A single band of approximately 57 kDa, corresponding to CAT, was present in all samples. Excessive ROS accumulation leading to oxidative stress resulted in an increase in CAT protein content in all plasma-treated samples immediately after the treatment, with the highest values after the 60 s treatment compared to the control untreated samples (Figure 6c). Furthermore, the enhanced CAT protein content at 0 h, 1 h and 7d was consistent with the increase in enzyme activity at the same time points, as shown in Figure 5c. The physiological response of plasma-treated iris calli measured immediately after plasma treatment (0 h), 1 h and 7d after plasma treatment was characterized by an increase in POX protein content of all three observed isoforms compared to the control untreated callus (Figure 7d).

3.4. *Secondary Metabolite Profile of Iris Calli Before and After Plasma Treatment*

The UHPLC-Orbitrap-MS characterization of extracts from iris callus in negative ionization mode resulted in the detection of a total of 40 compounds. The identified compounds can be categorized into four structurally distinct groups: flavonoids and their derivatives (4 compounds), isoflavone glycosides (20 compounds), isoflavone aglycones (13 compounds) and 3 other compounds (Table 2). All detected compounds were identified by an exact mass search of their deprotonated molecule $[M–H]^-$, MS^{2-}, MS^{3-} and MS^4-fragmentation behavior and by comparison with the available literature. The peak numbers, compound names, retention times (t_R, min), molecular formulas, calculated and exact masses ($[M–H]^-$, m/z), and the presence of selected compounds in various cold plasma-treated calli samples are summarized in Table 2, while parent and major MS^2, MS^3, and MS^4 fragmentations are given in Supplementary Table S1.

Isoflavones and their glycosides and aglycones are the major classes of polyphenolic compounds found in the yellow callus of *Iris reichenbachii*. One flavonoid, six isoflavone aglycones, and eleven isoflavone glycosides were detected in the untreated yellow callus (NEC). On the other hand, the secondary metabolite profile of the untreated EC callus was characterized by eight isoflavonoid aglycones, ten isoflavonoid glycozides, three acetylated sucrose derivatives and other compounds, such Shengasu C, its isomer and diferuloylsucrose. The overall profile of secondary metabolites in iris callus changed with plasma treatment. De novo synthesis of three flavonoids (narigenin, apigenin, and acacetin) and two isoflavonoid aglycones (irisolidone and irilone) was detected only in plasma-treated iris calli (Table 2, purple). These compounds were not present in the untreated iris callus (NEC and EC) but were also recorded in CP-treated calli subcultured for one year after plasma treatment (Table 2). In addition, two isoflavone glycosides, irisolone 4′-O-(6″-rhamnosyl)-hexoside and iriflogenin 4′-O-(6″-hexosyl)-hexoside, were not recorded in plasma-treated or in untreated EC callus (Table 2, light blue). One year after plasma treatment, most of the isoflavone glycosides and aglycones that were present in the untreated callus were not detected in CP-treated calli (Table 2).

Table 2. Comparative analysis of high-resolution MS recorded secondary metabolite compounds in the untreated and treated callus of *I. reichenbachii* 7 days (1w), 3 weeks and 1 year (1y) after plasma treatment. Marked in purple—de novo synthetized compounds in plasma-treated callus; marked in green: compounds present in plasma-untreated and plasma-treated callus; and marked in blue: compounds present only in non-embryonic calli.

No	Compound name	t_R * (min)	Molecular Formula [M–H]⁻	Calculated Mass [M–H]⁻	Exact Mass [M–H]⁻	Δ ppm	NEC **	PLASMA			EC
								1W	3W	1Y	
	Flavonoids and derivatives										
1	Apigenin 6-C-hexoside-8-C-pentoside	5.91	$C_{26}H_{27}O_{14}^-$	563,14063	563,13940	2.18	+	+	+	–	–
2	Naringenin	9.46	$C_{15}H_{11}O_5^-$	271,06120	271,06000	4.43	–	+	+	+	–
3	Apigenin	9.49	$C_{15}H_9O_5^-$	269,04554	269,0,474	2.97	–	+	+	+	–
4	Acacetin	11.80	$C_{16}H_{11}O_5^-$	283,06119	283,06110	0.32	–	+	+	+	–
	Isoflavone glycosides										
5	Tectoridin (Tectorigenin 7-O-hexoside)	5.93	$C_{22}H_{21}O_{11}^-$	461,10894	461,10894	−0.01	+	+	+	–	+
6	Iristectorin B (Iristectorigenin B 7-O-hexoside)	6.14	$C_{23}H_{23}O_{12}^-$	491,11950	491,11961	−0.23	+	+	+	–	+
7	Irislactin A	6.21	$C_{40}H_{47}O_{22}^-$	879,25645	879,25719	−0.84	–	–	–	–	+
8	Acetylated sucrose derivative 1	6.26	$C_{41}H_{49}O_{23}^-$	909,26701	909,26698	0.04	–	–	–	–	+
9	Isotectorigenin 7-O-hexoside	6.37	$C_{22}H_{21}O_{11}^-$	461,10894	461,10919	−0.55	–	–	–	–	+
10	Iridin (Irigenin 7-O-hexoside)	6.42	$C_{24}H_{25}O_{13}^-$	521,13007	521,13006	0.01	+	+	+	–	+
11	Homotectoridin (Homotectorigenin 7-O-glucoside)	6.52	$C_{23}H_{23}O_{12}^-$	491,11950	491,11987	−0.75	–	–	–	–	+
12	Isoridin (Isoirigenin 7-O-hexoside)	6.82	$C_{24}H_{25}O_{13}^-$	521,13007	521,13007	0.00	–	–	–	–	+
13	Acetylated sucrose derivative 2	7.05	$C_{45}H_{53}O_{25}^-$	993,28814	993,28742	0.73	–	–	–	–	+
14	Iristectorin A (Iristectorigenin A 7-O-hexoside)	7.27	$C_{23}H_{23}O_{12}^-$	491,11950	491.11764	3.79	+	+	+	–	+
15	Iridin S (Irisjaponin B 7-O-hexoside)	7.34	$C_{25}H_{27}O_{13}^-$	535,14572	535.14559	0.24	+	+	+	–	+
16	Irisolone 4′-O-(6″-rhamnosyl)-hexoside	7.52	$C_{29}H_{31}O_{15}^-$	619,16684	619,16681	0.05	+	–	–	–	–
17	Irilone 4′-O-(6″-hexosyl)-hexoside	7.54	$C_{28}H_{29}O_{16}^-$	621,14611	621,14423	3.03	–	–	–	–	–

Table 2. *Cont.*

No	Compound name	t_R * (min)	Molecular Formula [M–H]⁻	Calculated Mass [M–H]⁻	Exact Mass [M–H]⁻	Δ ppm	NEC **	PLASMA			EC
								1W	3W	1Y	
18	Iriflogenin 4′-O-(6″-hexosyl)-hexoside	7.60	$C_{29}H_{31}O_{17}^-$	651,15667	651,15625	0.65	+	–	–	–	–
19	Acetylated sucrose derivative 3	7.87	$C_{46}H_{53}O_{25}^-$	1005,28,814	1005,28768	0.46	–	–	–	–	+
20	Iristectorigenin A/B 7-O-(acetyl)hexoside	7.88	$C_{25}H_{25}O_{13}^-$	533,13006	533,13006	0.00	+	+	+	–	–
21	Acetylated sucrose derivative 4	7.90	$C_{47}H_{55}O_{26}^-$	1035,29871	1035,29881	–0.10	–	–	–	–	+
22	Irisolone 4′-O-[6″-(3-hydroxy-3-methylglutaryl)]-hexoside	8.30	$C_{29}H_{29}O_{15}^-$	617,15119	617,15070	0.79	+	+	+	–	–
23	Iriskashmirianin 4′-O-[6″-(3-hydroxy-3-methylglutaryl)]-hexoside	8.32	$C_{30}H_{31}O_{16}^-$	647,16176	647,16138	0.59	+	+	+	–	–
24	Iriflogenin 4′-O-[6″-(3-hydroxy-3-methylglutaryl)]-hexoside	8.90	$C_{29}H_{29}O_{16}^-$	633,14611	633,14349	4.14	+	+	+	–	–
	Isoflavone aglycones										
25	Iristectorigenin A	6.34	$C_{17}H_{13}O_7^-$	329,06668	329,06688	–0.62	–	–	–	–	+
26	5,7,4′-Trihydroxy-6,3′,5′-trimethoxyisoflavone	6.47	$C_{18}H_{15}O_8^-$	359,07724	359,07718	0.17	–	–	–	–	+
27	Irisolone	7.71	$C_{17}H_{11}O_6^-$	311,05611	311,05545	2.12	+	+	+	–	–
28	Noririsflorentin	7.88	$C_{19}H_{15}O_8^-$	371,07724	371,07657	1.81	+	+	+	–	–
29	Tectorigenin	8.01	$C_{16}H_{11}O_6^-$	299,05611	299,05605	0.20	+	+	+	–	+
30	Dalspinosin	8.17	$C_{18}H_{15}O_7^-$	343,308233	343,08127	3.09	+	+	+	–	–
31	Iristectorigenin B	8.19	$C_{17}H_{13}O_7^-$	329,06668	329,06667	0.02	–	–	–	–	+
32	Irigenin	8.33	$C_{18}H_{15}O_8^-$	359,07724	359,07715	0.24	+	+	+	+	+
33	8-Hydroxyirigenin	8.33	$C_{18}H_{15}O_9^-$	375,07216	375,07214	0.05	–	–	–	–	+
34	Iristectorigenin A/B	9.93	$C_{17}H_{13}O_7^-$	329,06668	329,06592	2.31	+	+	+	+	+
35	Isoirigenin	9.97	$C_{18}H_{15}O_8^-$	359,07724	359,07731	–0.18	–	–	–	–	+
36	Irisolidone	10.89	$C_{17}H_{13}O_6^-$	313,07176	313,07166	0.32	–	+	+	+	–
37	Irilone	11.19	$C_{16}H_9O_6^-$	297,04046	297,04046	0.00	–	+	+	+	–
	Other compounds										
38	Diferuloylsucrose	6.04	$C_{32}H_{37}O_{17}^-$	69,320,362	69,320,357	0.07	–	–	–	–	+
39	Shegansu C	8.19	$C_{44}H_{47}O_{21}^-$	91,126,153	91,126,037	1.28	–	–	–	–	+
40	Shegansu C isomer	8.47	$C_{44}H_{47}O_{21}^-$	91,126,153	91,126,104	0.55	–	–	–	–	+

* tR–retention time; ** NEC–untreated non-embryonic calli of *I. reichenbachii* (control); 1W–plasma-treated NEC 7 days (one week) after plasma treatment for 60 s; 3W–plasma-treated NEC three weeks after plasma treatment for 60 s; 1Y-newly formed white calli of *I. reichenbachii* one year after plasma treatment for 60 s; and EC–white non-treated embryogenic calli of *I. reichenbachii*.

4. Discussion

To date, a variety of cold plasma devices with different designs and operating procedures have been developed that can generate and deliver plasma-reactive species to seeds, plants, soil, or water [1]. However, not as many devices have been used for in vitro cultures, especially for callus tissue grown under aseptic conditions [10,34]. In some recently published reports, corona discharge plasma was used to treat samples placed in covered Petri dishes [10], or calli simples were treated under sterile conditions [34]. The effects of plasma treatment with a corona device were evaluated ten days or two weeks after treatment for *Cannabis* and *Datura inoxia* calli, respectively [10]. In the present study, we exposed *Iris reichenbachii* calli to a low-temperature plasma treatment by using a plasma needle. The plasma needle generates a gas-phase chemistry rich in RONS [2,6,25,98]. During the experiments, the plasma comes into direct contact with the surface of the plant calli, where both short- and long-lived RONS can reach the plant cells and trigger various mechanisms of RONS defense response in callus cells. We presented here changes recorded immediately after plasma treatment, one hour and seven days after plasma treatment, as well as long-term effects that include changes in morphology and secondary metabolite content up to one year after plasma treatment.

According to the available research data, the plasma priming of calli efficiently improved callus biomass production [24,99]. The only long-term effects observed so far in plasma-treated calli were the increase in fresh weight of *Fritillaria imperialis* calli [24]. Similar results regarding the increase in callus growth were recorded in our study for the yellow non-embryogenic calli *of I. reichenbachi* [99]. Recently it was shown that all newly formed calli of *Datura inoxia* and *Cannabis sativa* exhibited vigorous growth after plasma treatment, but no change in calli color was observed [10,34].

Yellow callus types with a nodular structure can be formed in many iris species and are usually characterized as non-embryogenic calli (NEC) [85,100–102]. These types of calli are poorly granulated and water-stained compared to the embryogenic callus (EC), which is pale yellow or white and has a friable structure [102]. The non-embryogenic callus of *I. reichenbachii* is characterized by irregularly shaped cells containing minimal cytoplasm with light nuclear staining. After plasma treatment, we observed regions of smaller, densely packed meristematic cells that had a regular shape and a centrally located nucleus, suggesting that these cells have a strong meristematic potential. This type of cell was found only in friable, whitish calli, which consist of compact, glassy-like, embryogenic nodules that give rise to somatic embryos from the region of pro-embryonic meristematic cells. This type of calli is referred to in the literature as EC [85,100,102]. In the tissue culture of *I. germanica*, yellow callus covered with viscous substances is commonly referred to as callus, on which somatic embryos usually develop and become embryogenic when transferred to growth regulator-free MS medium [58]. In the iris, whitish plasma-derived calli obtained in this work did not further participate in embryogenesis and the development of somatic embryos. However, in our previous reports, we have shown that CP treatment induces the accumulation of arabinogalactan proteins (AGPs) on the surface of the nodules and in the extracellular mucilagenous part of the yellow iris calli [103]. AGPs are proteins that occur in the cell wall and in the extracellular space and represent specific markers for the transition of cells from the somatic to the embryogenic state [104]. The question remains whether the further development of somatic embryos in plasma-treated calli was only blocked because the calli were constantly grown on 2,4-D and KIN medium after CP treatment, which promotes callus development but not somatic embryo formation. Results obtained in this study showed that the surface modifications on the iris callus can be attributed to enhanced cell division (mitotic activity) and the development of new meristematic centers after CP treatment [105,106]. One of the reasons for the observed changes in the plasma-treated iris calli could be related to the ability of plasma treatment to trigger the production of endogenous plant hormones. Therefore, the question of whether CP treatment also affects the hormonal status of iris calli needs to be further investigated.

To date, there is ample evidence that cold plasma-generated ROS and RONS play a crucial role in cell response [105–107]. Plasma treatment acts as an oxidative enhancer that disturbs the redox balance in plants, causing metabolic reactions and signaling cascades that consequently affect physiological and biochemical processes in plants [105,107]. In our study, we used two high-affinity probes (NBT and DAB) in iris calli samples after plasma treatment to provide information on the in situ site of formation and accumulation of ROS (O_2^- and H_2O_2). In the histochemical localization of oxygen radicals with NBT, the blue insoluble formazan and brownish DAB precipitates were larger in all plasma-treated calli than in the control samples. This difference indicated higher O_2^- production and H_2O_2 accumulation at the surface of the iris callus immediately after treatment, which shifts deep into the callus over the next hour. According to our results, plasma-derived RONS acted as an abiotic stressor upon contact with the surface of the callus, causing an oxidative burst and stress that impairs redox homeostasis. A similar accumulation of intracellular ROS immediately after plasma treatment was detected in *Arabidopsis* seedlings [108]. Generally, oxidative stress occurs when there is an imbalance between the availability of RONS and the ability of cells to neutralize or repair the damage caused by these reactive species. The content of H_2O_2 in plant material may be increased after plasma treatment or may not change significantly, as its level depends not only on production but also on the efficiency of the plant to remove toxic RONS from the system [34,108]. For example, H_2O_2 content in plasma-treated cannabis callus was not significantly different from untreated control tissue ten days after plasma treatment [34].

Direct plasma treatment represents exposure to a particular state of matter with high-energy particles that are chemically active and can have short- and long-term effects on biological processes in plant tissue [1]. According to the results obtained in this study, in response to plasma treatment, the iris calli react with complex constitutive and inducible defense mechanisms, such as the activation of key antioxidant defense enzymes and the synthesis of different types of compounds in response to physical stress. Plants have developed a number of mechanisms to protect themselves from RONS-mediated damage. There are two main pathways of ROS degradation in plants: non-enzymatic and enzymatic. Non-enzymatic systems include various secondary metabolites (antioxidants) such as phenols and flavonoids, which are capable of directly quenching ROS. Enzymatic systems, on the other hand, consist of antioxidant enzymes, including superoxide dismutase (SOD), catalase (CAT) and peroxidase (POX). SOD is the key enzyme that catalyzes the dismutation of O_2^- to H_2O_2, which is then further degraded to H_2O by CAT and POX [109]. SOD is the first step of the defense system against ROS generated in iris calli after plasma treatment. The activity of these enzymes, especially the CuZnSOD isoform, depended on the duration of plasma treatment and persisted during the first hour after plasma treatment. Increased SOD activity was found to be associated with an overproduction of ROS under various stress conditions, resulting in increased plant tolerance to environmental stress [110]. The observed changes in CuZnSOD and CAT activity in yellow iris callus corresponded with the rate of peroxide anion radical formation. Many authors have noted changes in the activity of antioxidant enzymes after CP treatment [10,34,111]. However, the changes were not uniform as they were influenced by the type of plasma device, the duration of the treatment, the working gas and the type of biological samples treated. In experiments conducted by various researchers, plasma-treated seeds and seedlings showed an increase in the activity of SOD, POX and CAT [108,112,113]. On the other hand, Rahman et al. [114] found no changes in SOD and POX activity in wheat seeds treated with plasma. SOD and POX activity in *Arabidopsis* seeds was initially increased and then decreased with the plasma treatment time, while CAT activity was significantly decreased in plasma-treated seeds compared to untreated seeds [115]. Our previous work on meristematic cells of *Daucus carota* also showed that cold plasma induced both higher SOD and CAT activity immediately and two weeks after treatment, clearly indicating a species-specific effect of cold plasma treatments [25]. In addition, the protein level of the antioxidant enzyme isoforms was detected by immunoblot analysis with specific antibodies correlating with the CAT protein content in all plasma-treated samples immediately after treatment, with the highest values measured after the 60 s treatment compared to the control samples. The physiological response of plasma-treated iris calli measured 7d after plasma treatment was characterized by an increase in POX activity and protein content compared to control samples. It should be noted that, although immunoblot analysis provides evidence for the synthesis of the enzyme subunits, this does not necessarily correlate with the activity of the enzyme in its native form [116]. This may explain why the slight difference between the protein content (apparent decrease or increase) and the changes in enzyme activity was observed in the same CP-treated calli samples. The precise determination of the specific isoforms and their respective activity would help to clarify this point. It should also be kept in mind that the different levels of enzymatic activity are not exclusively dependent on enzyme synthesis but also on post-translational modifications [117]. In addition, the non-enzymatic antioxidant system in plants includes small molecules that control ROS homeostasis by removing, converting or neutralizing the oxidant pool, which could explain the discrepancy observed in our results [118].

There is ample evidence that the cold plasma treatment of seeds resulted in increased amounts of different specialized metabolites, such as cannabinoids in *Cannabis indica* [119], isoflavones in *Triticum pratensae* [120], and various secondary metabolites in buckwheat [121] or natural sweeteners in *Stevia rebaudiana* [122]. In nature, the biosynthesis of specialized metabolites is generally limited to certain developmental stages and periods of stress [123]. Plant callus, an unorganized, fast-growing cell mass, can be produced in almost all plant species in response to wounding or plant growth regulators [124]. Callus tissue culture has been used extensively in plant biotechnology as a tool for the genetic manipulation of plants, for studying plant metabolism and cellular development and more recently for the large-scale commercial production of valuable compounds, i.e., plant secondary metabolites, as an alternative to the direct tissue extraction of naturally occurring products that cannot be chemically synthesized. However, not all secondary metabolites present in intact plants are produced in callus cultures [69]. The synthesis of some metabolites, depending on the levels of cellular differentiation and a certain level of organization in the cell culture, is very often required for metabolite synthesis, such as for the production of mangiferin in callus cultures of *Iris raichenbachii*. In addition, the production of these metabolites depended on the type of plant growth regulators (i.e., presence of 2,4-D) in culture medium [69]. In our work, we presented for the first time the complete metabolic profile of secondary metabolites produced in *I. rechenbachii* calli. In yellow iris callus, 40 phytochemicals were identified. Among them, one flavonoid glycoside (Apigenin 6-C-hexoside-8-C-pentoside) together with eight isoflavone glycosides and six corresponding aglycones were synthesized in yellow NEC. In addition, in this work we presented the metabolic profile of the white EC of the iris, which was not treated, to compare it with the white callus obtained after CP treatment. According to the results, it is evident that the metabolic profile of white plasma-derived calli was different from that of the white EC and yellow NEC. In addition, the EC was significantly richer in secondary metabolites than the yellow NEC and plasma-treated NEC. This difference may be related to the process of tissue differentiation during the development of somatic embryos that occurs in embryogenic tissues.

In the current study, we confirmed the synthesis of isoflavonoids in the callus of *I. rechenbachii*, which was previously reported only for the callus tissue of *I. pseudacorus* by Tarbeeva et al. [70]. Isoflavones and their glycosides are the main classes of polyphenolic compounds found in *Iris* species and were primarily recorded in the rhizome. These phytochemicals are very important as the use of iris rhizomes in both traditional and modern medicine relies on their presence in the extracts [37,38]. Many isoflavones have been named after the species of iris from which they were commonly found or were first isolated, such as tectorigenin from *Iris tectorum*, nigricin from *Iris nigricans* [125] or kashmirianin isolated from *Iris kashmiriana* [126]. The identification of isoflavones and their derivatives in our study, in the absence of standards, was based on the available literature on phytochemicals previously isolated or just identified in some iris species [94,95,127–131] and on the study of their MS fragmentation patterns (exact mass and MS^4 fragmentation).

According to our results, three flavonoids (narigenin, apigenin and acacetin) were only detected in plasma-treated callus. The synthesis of these phytochemicals was recorded in treated tissues one week and one year after CP treatment, whereas they were not present in the other two untreated calli samples (NEC and EC). While apigenin and acacetin were previously detected in irises [38,44], narigenin is characteristic of citrus fruits and was only recorded in the leaves of *I. sibirica* [63]. All these plasma-induced flavonoids have a great potential to inhibit the generation of ROS and reduce their levels as well as to fulfill antioxidant functions in tissues. We postulated that the synthesis of these compounds presents an enzymatic long-term defense response of the iris callus to ROS attack during plasma treatment. Apigenin [132], acacetin [133] and narigenin [134,135] induced in iris callus after CP treatment are excellent free radical scavengers with wide-ranging effects on human health, such as antioxidant, anti-inflammatory, anti-infection, anticancer and cardiovascular protective functions. In our study, we recorded the continuous synthesis of two isoflavonoid aglycones in *I. reichenbachii* callus, irisolidone and irilone, after CP treatment. These isoflavonoid aglycones were found in nature in rhizomes of *I. germanica* [136], endemic dwarf iris *I. adriatica* [137] and *I. pseudopumila* [138] as well as in the leaves of *I. sibirica* [63]. Irilone has strong cytotoxic and anti-inflammatory effects and also high antioxidant and estrogenic activity [139], while irisolidone possesses antioxidant, anti-inflammatory, antidiabetic and immunomodulatory activity [136,140]. This is the first report on the induction of secondary metabolites in the calli of any plant species after CP treatment without being the only stimulation of pre-existing synthesis, such as the synthesis of cannabinoids in *Cannabis sativa* callus [34] and atropine in *Datura inoxia* callus [10]. In addition, the synthesis of isoflavonoids in *I. domestica* callus is only possible after induction with $CuCl_2$ [71]. No polyphenolic phytochemicals were found in untreated callus, while induction started 7 days after $CuCl_2$ treatment and was maximally enriched 42 days after treatment and strongly decreased 49 days after elicitation treatment [71]. According to the available research results, isoflavone production in in vitro culture is influenced by various factors such as the type of culture medium but also by some physical factors (light intensity and temperature). Light, especially UV radiation, acts as a regulatory factor for the biosynthesis of isoflavonoids, as postulated for the calli culture of *Genista tinctoria* [141,142]. It is well documented that the effect of plasma on biological targets is mediated not only by chemically reactive species but also by visible light and UV radiation [7]. In this context, we postulated that the induction of the synthesis of isoflavonoids after plasma treatment in yellow iris callus was due to the physical effects of different types of light that were present in the plasma during treatment. In addition, two isoflavonoid aglycones—irigenin and iris-tectorigenin A/B—were synthetized in both the plasma-treated and plasma untreated iris callus (NEC and EC). This is the first report on the synthesis of these medicinally valuable compounds in plant callus tissue. The synthesis of the above-mentioned isoflavonoids was also recorded in the adventitious roots of *I. germanica* [62] and in the CuCl2-treated callus of *I. domestica* (previously referred as *Belamcanda chinensis*) [71].

So far all studies on the effects of plasma treatment of callus followed changes in biosynthesis up to two weeks after plasma treatment [10,34]. However, no one has yet addressed the question of long-term changes in the quality of specialized metabolites and antioxidants in plant callus due to CP treatment. In the current study, a significant decrease in isoflavonoid synthesis was observed in plasma-treated callus one year after treatment. Only constitutively present (irigenin and iristectorigenin A/B) and plasma-induced flavonoids (narigenin, apigenin and acacetin) and isoflavonoids (irisolidone and irilone) were found in the iris callus one year after plasma treatment. Polyphenols can be oxidized chemically or enzymatically by polyphenol oxidase or POX. However, new phenolic compounds can also be synthesized as a result of stress by increasing the activity of the enzyme phenylalanine ammonia lyase as a defense mechanism [143]. Therefore, follow-up studies and associated gene expression in phenylpropanoid metabolism are required to fully investigate the regulatory mechanism of CP treatment on the antioxidant capacity of yellow iris calli at the molecular level. Although the mechanism is still unclear, we hypothesize that the degradation of isoflavones results from the known ability of phenolic compounds to scavenge free radicals (antioxidant capacity). This opens up numerous possibilities for further work, including a comprehensive quantitative chemical analysis of iris calli and the changes in their antioxidant activity.

5. Conclusions

In this work, we have shown that the plasma treatment of calli acts as a stress factor that not only induces changes in morphology, physiology and growth but also contributes to increasing the content of biologically active constituents in plants. The morphological and physiological response of non-regenerative iris callus tissue induced by plasma treatment suggests that the plasma technique could be applied as an alternative and valuable approach for promoting regeneration and multiplication in plant meristematic tissue culture and as a potential elicitor for the production of various desirable secondary metabolites in plant cell and tissue culture. Here, we have shown that direct plasma treatment can significantly alter the production of secondary metabolites in *I. reichenbachii* calli, leading to qualitative changes in their phytochemical profile both in the short and long term. Therefore, our future work will involve the investigation of the molecular mechanism underlying the observed effects of CP treatment on calli.

Supplementary Materials: The following supporting information can be downloaded at https://www.mdpi.com/article/10.3390/horticulturae11070781/s1, Table S1: MS^{2-4} fragmentation data of specialized metabolites found in calli of *Iris reichenbachii*.

Author Contributions: Conceptualization, S.J., S.Ž., M.M. and N.P.; investigation, S.J., S.Ž., M.M., K.V., U.G., N.Š. and N.P.; writing—original draft preparation, S.J., S.Ž. and M.M.; writing—review and editing, S.J., S.Ž., M.M., K.V., U.G., N.Š. and N.P.; visualization, S.J., S.Ž., M.M. and N.P. All authors have read and agreed to the published version of the manuscript.

Funding: This research was funded by the Ministry of Science, Technological Development and Innovation of the Republic of Serbia, grant numbers 451-03-136/2025-03/200007, 451-03-136/2025-03/200024, 451-03-65/2024-03/200178 and 451-03-66/2024-03/200178. The results presented in this manuscript are in line with Sustainable Development Goal 2 (End Hunger) and Sustainable Development Goal 9 (Industry, Innovation and Infrastructure) of the United Nations 2030 Agenda.

Data Availability Statement: The original contributions presented in this study are included in this article. Further inquiries can be directed to the corresponding author.

Conflicts of Interest: The authors declare no conflicts of interest.

Abbreviations

The following abbreviations are used in this manuscript:

CP	Cold Plasma
RF	Radio Frequency
ROS	Reactive Oxygen Species
RONS	Reactive Oxygen and Nitrogen Species
SOD	Superoxide Dismutase
MnSOD	Manganese Superoxide Dismutase
CuZnSOD	Copper Zink Superoxide Dismutase
CAT	Catalase
POX	Peroxidase
NAA	Naphthylacetic Acid
2,4-D	2,4-Dichlorophenoxyacetic Acid
EC	Embryogenic Callus
NEC	Non-Embryogenic Callus
OC	Organogenic Callus
NBT	Nitro Blue Tetrazolium
DAB	3,3′-Diaminobenzidine
PCs	Parenchymatous Cells
MCs	Meristematic Centers

References

1. Veerana, M.; Mumtaz, S.; Rana, J.N.; Javed, R.; Panngom, K.; Ahmed, B.; Akter, K.; Choi, E.H. Recent Advances in Non-Thermal Plasma for Seed Germination, Plant Growth, and Secondary Metabolite Synthesis: A Promising Frontier for Sustainable Agriculture. *Plasma Chem. Plasma Process.* **2024**, *44*, 2263–2302. [CrossRef]
2. Bruggeman, P.J.; Iza, F.; Brandenburg, R. Foundations of atmospheric pressure non-equilibrium plasmas. *Plasma Sources Sci. Technol.* **2017**, *26*, 123002. [CrossRef]
3. Brandenburg, W.R.; Weltmann, K.-D. Atmospheric pressure plasma jets: An overview of devices and new directions. *Plasma Sources Sci. Technol.* **2015**, *24*, 064001. [CrossRef]
4. Park, G.Y.; Park, S.J.; Choi, M.Y.; Koo, I.G.; Byun, J.H.; Hong, J.W.; Sim, J.Y.; Collins, G.J.; Lee, J.K. Atmospheric-pressure plasma sources for biomedical applications. *Plasma Sources Sci. Technol.* **2012**, *21*, 043001. [CrossRef]
5. Fridman, A. Introduction to Theoretical and Applied Plasma Chemistry. In *Plasma Chemistry*; Cambridge University Press: Cambridge, UK, 2008; pp. 1–11. [CrossRef]
6. Bradu, C.; Kutasi, K.; Magureanu, M.; Puač, N.; Živković, S. Reactive nitrogen species in plasma-activated water: Generation, chemistry and application in agriculture. *J. Phys. D Appl. Phys.* **2020**, *53*, 223001. [CrossRef]
7. Bilea, F.; Garcia-Vaquero, M.; Magureanu, M.; Mihaila, I.; Mildažienė, V.; Mozetič, M.; Pawłat, J.; Primc, G.; Puač, N.; Robert, E.; et al. Non-Thermal Plasma as Environmentally-Friendly Technology for Agriculture: A Review and Roadmap. *Crit. Rev. Plant Sci.* **2024**, *43*, 428–486. [CrossRef]
8. Domonkos, M.; Tichá, P.; Trejbal, J.; Demo, P. Applications of Cold Atmospheric Pressure Plasma Technology in Medicine, Agriculture and Food Industry. *Appl. Sci.* **2021**, *11*, 4809. [CrossRef]
9. Puač, N.; Gherardi, M.; Shiratani, M. Plasma and agriculture. *Plasma Process Polym.* **2017**, e1700174. [CrossRef]
10. Tardast, Z.; Iranbakhsh, A.; Ebadi, M.; Oraghi Ardebili, Z. Corona discharge plasma stimulated production of atropine in callus of *Datura inoxia* by DNA hypomethylation and gene regulation: A novel technology for plant cell and tissue culture. *Protoplasma* **2023**, *260*, 1515–1525. [CrossRef]
11. Puač, N.; Škoro, N.; Spasić, K.; Živković, S.; Milutinović, M.; Malović, G.; Petrović, Z.L. Activity of catalase enzyme in *Paulownia tomentosa* seeds during the process of germination after treatments with low pressure plasma and plasma activated water. *Plasma Process Polym.* **2018**, *15*, e1700082. [CrossRef]
12. Bormashenko, E.; Grynyov, R.; Bormashenko, Y.; Drori, E. Cold radiofrequency plasma treatment modifies wettability and germination speed of plant seeds. *Sci. Rep.* **2012**, *2*, 741. [CrossRef] [PubMed]

13. Sivachandiran, L.; Khacef, A. Enhanced seed germination and plant growth by atmospheric pressure cold air plasma: Combined effect of seed and water treatment. *RSC Adv.* **2017**, *7*, 1822–1832. [CrossRef]
14. Lo Porto, C.; Ziuzina, D.; Los, A.; Boehm, D.; Palumbo, F.; Favia, P.; Tiwari, B.; Bourke, P.; Cullen, P.J. Plasma activated water and airborne ultrasound treatments for enhanced germination and growth of soybean. *Innov. Food Sci. Emerg.* **2018**, *49*, 13–19. [CrossRef]
15. Waskow, A.; Howling, A.; Furno, I. Mechanisms of Plasma-Seed Treatments as a Potential Seed Processing Technology. *Front. Phys.* **2021**, *9*, 617345. [CrossRef]
16. Ito, M.; Oh, J.-S.; Ohta, T.; Shiratani, M.; Hori, M. Current status and future prospects of agricultural applications using atmospheric-pressure plasma technologies. *Plasma Process Polym.* **2018**, *15*, 1700073. [CrossRef]
17. Gierczik, K.; Vukušić, T.; Kovács, L.; Székely, A.; Szalai, G.; Milošević, S.; Kocsy, G.; Kutasi, K.; Galiba, G. Plasma-activated water to improve the stress tolerance of barley. *Plasma Process Polym.* **2020**, *17*, 1900123. [CrossRef]
18. Mildaziene, V.; Ivankov, A.; Pauzaite, G.; Naucienė, Z.; Zukiene, R.; Degutyte-Fomins, L.; Pukalskas, A.; Rimantas Venskutonis, P.; Filatova, I.; Lyushkevichet, V. Seed treatment with cold plasma and electromagnetic field induces changes in red clover root growth dynamics, flavonoid exudation, and activates nodulation. *Plasma Process Polym.* **2021**, *18*, 2000160. [CrossRef]
19. Lee, H.; Kim, J.E.; Chung, M.-S.; Min, S.C. Cold plasma treatment for the microbiological safety of cabbage, lettuce, and dried figs. *Food Microbiol.* **2015**, *51*, 74–80. [CrossRef]
20. Filatova, I.I.; Lyushkevich, V.A.; Kalatskaja, J.N.; Goncharik, S.V.; Mildaziene, V.; Pauzaite, G. Influence of plasma and radiowave treatment of seeds on the accumulation of some secondary metabolites in plants. In *Proceedings of the 29 Summer School and International Symposium on the Physics of Ionized Gases; SPIG 2018, Contributed Papers & Abstracts of Invited Lectures, Topical Invited Lectures, Progress Reports and Workshop Lectures, Belgrade, Serbia, 28 August–1 September 2018*; Poparić, G., Obradović, B., Borka, D., Rajković, M., Eds.; Vinča Institute of Nuclear Sciences, University of Belgrade: Belgrade, Serbia, 2018; pp. 234–237.
21. Yodpitak, S.; Mahatheeranont, S.; Boonyawan, D.; Sookwong, P.; Roytrakul, S.; Norkaew, O. Cold plasma treatment to improve germination and enhance the bioactive phytochemical content of germinated brown rice. *Food Chem.* **2019**, *289*, 328–339. [CrossRef]
22. Judickaitė, A.; Venckus, J.; Koga, K.; Shiratani, M.; Mildažienė, V.; Žūkienė, R. Cold Plasma-Induced Changes in Stevia rebaudiana Morphometric and Biochemical Parameter Correlations. *Plants* **2023**, *12*, 1585. [CrossRef]
23. Singh, H.; Niharika; Lamichhane, P.; Gupta, R.; Kaushik, N.; Choi, E.H.; Kaushik, N.K. Enhancing crop health and sustainability: Exploring the potential of secondary metabolites and non-thermal plasma treatment as alternatives to pesticides. *Plant Biotechol. Rep.* **2023**, *17*, 803–820. [CrossRef]
24. Puač, N.; Petrović, Z.L.; Malović, G.; Đorđević, A.; Živković, S.; Giba, Z.; Grubišić, D. Measurements of voltage–current characteristics of a plasma needle and its effect on plant cells. *J. Phys. D Appl. Phys.* **2006**, *39*, 3514–3519. [CrossRef]
25. Puač, N.; Živković, S.; Selaković, N.; Milutinović, M.; Boljević, J.; Malović, G.; Petrović, Z.L. Long and short term effects of plasma treatment on meristematic plant cells. *Appl. Phys. Lett.* **2014**, *104*, 214106. [CrossRef]
26. Škoro, N.; Živković, S.; Jevremović, S.; Puač, N. Treatment of Chrysanthemum Synthetic Seeds by Air SDBD Plasma. *Plants* **2022**, *11*, 907. [CrossRef]
27. Tsuboyama, S.; Okumura, T.; Attri, P.; Koga, K.; Shiratani, M.; Kuchitsu, K. Growth control of *Marchantia polymorpha* gemmae using nonthermal plasma irradiation. *Sci. Rep.* **2024**, *14*, 3172. [CrossRef]
28. Yanagawa, Y.; Suenaga, Y.; Iijima, Y.; Endo, M.; Sanada, N.; Katoh, E.; Toki, S.; Okino, A.; Mitsuhara, I. Genome editing by introduction of Cas9/sgRNA into plant cells using temperature-controlled atmospheric pressure plasma. *PLoS ONE* **2023**, *18*, e0281767. [CrossRef]
29. Ozyigit, I.I.; Dogan, I.; Hocaoglu-Ozyigit, A.; Yalcin, B.; Erdogan, A.; Yalcin, I.E.; Cabi, E.; Kaya, Y. Production of secondary metabolites using tissue culture-based biotechnological applications. *Front. Plant Sci.* **2023**, *14*, 1132555. [CrossRef]
30. De Bellis, R.; Chiarantini, L.; Potenza, L.; Gorassini, A.; Verardo, G.; De Marco, R.; Benayada, L.; Stocchi, V.; Albertini, M.C.; Fraternale, D. High production of secondary metabolites and biological activities of *Cydonia oblonga* Mill. pulp fruit callus. *J. Funct. Foods* **2022**, *94*, 105133. [CrossRef]
31. Ozden, M. Secondary metabolite production in callus cultures of *Vitis vinifera*: Influence of genotype and sucrose concentration in the medium on antioxidant activity. *Acta Physiol. Plant* **2024**, *46*, 6. [CrossRef]
32. Lee, K.Y.; Nam, D.-H.; Jeon, Y.; Park, S.U.; Cho, J.; Gulandaz, M.A.; Chung, S.-O.; Lee, G.-J. Exploring the Production of Secondary Metabolites from a Halophyte *Tetragonia tetragonoides* through Callus Culture. *Horticulturae* **2024**, *10*, 244. [CrossRef]
33. Maqbool, M.; Ishtiaq, M.; Mazhar, M.W.; Casini, R.; Mahmoud, E.A.; Elansary, H.O. Enhancing Bioactive Metabolite Production in *Aerva sanguinolenta* Callus Cultures through Silver Nanoparticle and Salicylic Acid Elicitation. *Sustainability* **2023**, *15*, 10395. [CrossRef]
34. Darigh, F.; Iranbakhsh, A.; Oraghi Ardebili, Z.; Ebadi, M. 2022 Non-thermal plasma improved callogenesis performance and elicited the production of cannabinoids by modifying DNA methylome, expression of WRKY1 and ERF1B transcription factors, and expression of genes that contributed to the biosynthesis of cannabinoids. *Protoplasma* **2023**, *260*, 159–170. [CrossRef] [PubMed]

35. Kaššák, P. Secondary metabolites of the choosen genus Iris species. *Acta Univ. Agric. Silvic. Mendel. Brun.* **2012**, *60*, 269–280. [CrossRef]
36. Hoang, L.; Beneš, F.; Fenclová, M.; Kronusová, O.; Švarcová, V.; Řehořová, K.; Baldassarre Švecová, E.; Vosátka, M.; Hajšlová, J.; Kaštánek, P.; et al. Phytochemical Composition and In Vitro Biological Activity of *Iris* spp. (Iridaceae): A New Source of Bioactive Constituents for the Inhibition of Oral Bacterial Biofilms. *Antibiotics* **2020**, *9*, 403. [CrossRef]
37. Öztaş, F.; Türkmen, A.; Öztaş, H.; Türkmen, M. The Medical Properties of Iris and its Usage in Pharmaceutical, Perfumery and Cosmetic Industries. In *Medical Research and Its Applications*; Veeramani, V.P., Ed.; BP International: New Delhi, India, 2024; Volume 4, pp. 114–124. [CrossRef]
38. Khatib, S.; Faraloni, C.; Bouissane, L. Exploring the use of Iris species: Antioxidant properties, phytochemistry, medicinal andindustrial applications. *Antioxidants* **2022**, *11*, 526. [CrossRef]
39. Kukula-Koch, W.; Sieniawska, E.; Widelski, J.; Urjin, O.; Głowniak, P.; Skalicka Woźniak, K. Major secondary metabolites of *Iris* spp. *Phytochem. Rev.* **2015**, *14*, 51–80. [CrossRef]
40. Mizuno, T.; Okuyama, Y.; Iwashina, T. Phenolic compounds from *Iris rossii*, and their chemotaxonomic and systematic significance. *Biochem. System Ecol.* **2012**, *44*, 157–160. [CrossRef]
41. Kostić, A.Ž.; Gašić, U.M.; Pešić, M.B.; Stanojević, S.P.; Barać, M.B.; Mačukanović-Jocić, M.P.; Avramov, S.N.; Tešić, Ž.L. Phytochemical Analysis and Total Antioxidant Capacity of Rhizome, Above-Ground Vegetative Parts and Flower of Three *Iris* Species. *Chem. Biodivers.* **2019**, *16*, e1800565. [CrossRef]
42. Mykhailenko, O.; Gudžinskas, Z.; Kovalyov, V.; Desenko, V.; Ivanauskas, L.; Bezruk, I.; Georgiyants, V. Effect of ecological factors on the accumulation of phenolic compounds in Iris species from Latvia, Lithuania and Ukraine. *Phytochem. Anal.* **2020**, *31*, 545–563. [CrossRef]
43. Schütz, C.; Quitschau, M.; Hamburger, M.; Potterat, O. Profiling of isoflavonoids in *Iris germanica* rhizome extracts by microprobe NMR and HPLC-PDA-MS analysis. *Fitoterapia* **2011**, *82*, 1021–1026. [CrossRef]
44. Iwashina, T.; Mizuno, T. Flavonoids and Xanthones from the genus Iris: Phytochemistry, relationships with flower colors and taxonomy, and activities and function. *Nat. Prod. Commun.* **2020**, *15*. [CrossRef]
45. Xie, G.Y.; Zhu, Y.; Shu, P.; Qin, X.Y.; Wu, G.; Wang, Q.; Qin, M.J. Phenolic metabolite profiles and antioxidants assay of three Iridaceae medicinal plants for traditional Chinese medicine "She-gan" by on-line HPLC-DAD coupled with chemiluminescence (CL) and ESI-Q-TOF-MS/MS. *J. Pharm. Biomed. Anal.* **2014**, *98*, 40–51. [CrossRef] [PubMed]
46. Mykhailenko, O.; Korinek, M.; Ivanauskas, L.; Bezruk, I.; Myhal, A.; Petrikaitė, V.; El-Shazly, M.; Lin, G.-H.; Lin, C.-Y.; Yen, C.-H.; et al. Qualitative and Quantitative Analysis of Ukrainian Iris Species: A Fresh Look on Their Antioxidant Content and Biological Activities. *Molecules* **2020**, *25*, 4588. [CrossRef] [PubMed]
47. Reshi, Z.A.; Ahmad, W.; Lukatkin, A.S.; Javed, S.B. From Nature to Lab: A Review of Secondary Metabolite Biosynthetic Pathways, Environmental Influences, and In Vitro Approaches. *Metabolites* **2023**, *13*, 895. [CrossRef]
48. Ascough, G.D.; Erwin, J.E.; Van Staden, J. Micropropagation of iridace-A review. *Plant Cell Tissue Organ. Cult.* **2009**, *97*, 1–19. [CrossRef]
49. Al-Gabbiesh, A.; Hassawi, D.S.; Afifi, F.U. In vitro propagation of endagared Iris species. *J. Biol. Sci.* **2006**, *6*, 1035–1040. [CrossRef]
50. Bae, K.; Yoo, K.; Lee, H.; Yoon, E. Callus induction and plant regeneration of *Iris dichotoma* Pall. in endangered species. *J. Plant Biotechnol.* **2012**, *39*, 182–188. [CrossRef]
51. Safari, N.; Tehranifar, A.; Kharrazi, M.; Shoor, M. Micropropagation of endangered Iris Ferdowsii Joharchi & Memariani through callus induction. *Plant Cell Tissue Organ. Cult.* **2023**, *154*, 595–604. [CrossRef]
52. Wang, Y.; Jeknic, Z.; Ernst, R.C.; Chen, T.H.H. Efficient plant regeneration from suspension-cultured cells of tall bearded iris. *HortScience* **1999**, *34*, 730–735. [CrossRef]
53. Yabuya, T.; Ikeda, Y.; Adachi, T. In vitro propagation of Japanese garden Iris, *Iris ensata* Thunb. *Euphytica* **1991**, *57*, 77–81. [CrossRef]
54. Laublin, G.; Saini, H.S.; Cappadocia, M. In vitro plant regeneration via somatic embryogenesis from root culture of some rhizomatous irises. *Plant Cell Tissue Organ. Cult.* **1991**, *27*, 21–27. [CrossRef]
55. Liu, H.; Fan, L.; Song, H.; Tang, B.; Wang, L. Establishment of regeneration system of callus pathway *for Iris sanguinea* Donn ex Horn. *Vitr. Cell Dev. Biol. Plant* **2020**, *56*, 694–702. [CrossRef]
56. Gozu, Y.; Yokoyama, M.; Nakamura, M.; Namba, R.; Yomogida, K.; Yanagi, M.; Nakamura, S. In vitro propagation of *Iris pallida*. *Plant Cell Rep.* **1993**, *13*, 12–16. [CrossRef] [PubMed]
57. Jehan, H.; Courtois, D.; Ehret, C.; Lerch, K.; Petiard, V. Plant regeneration of *Iris pallida* Lam. and *Iris germanica* L. via somatic embryogenesis from leaves, apices and young flowers. *Plant Cell Rep.* **1994**, *13*, 671–675. [CrossRef]
58. Shimizu, K.; Nagaike, H.; Yabuya, T.; Adachi, T. Plant regeneration from suspension culture of *Iris germanica* L. *Plant Cell Tissue Organ. Cult.* **1997**, *50*, 27–31. [CrossRef]
59. Meucci, A.; Ghelardi, C.; Maggini, R.; Malorgio, F.; Chietera, G.; Mensuali, A. Micropropagation via somatic embryogenesis of *Iris pallida* Lam. Ecotypes. *Plant Cell, Tissue Organ Cult (PCTOC)* **2024**, *158*, 21. [CrossRef]

60. Boltenkov, E.V.; Rybin, V.G.; Zarembo, E.V. Cultivation of *Iris ensata* Thunb. callus tissue. *Prikl. Biokhimiia Mikrobiol.* **2004**, *40*, 244–251.
61. Boltenkov, E.V.; Zarembo, E.V. In vitro regeneration and callogenesis in tissue culture of floral organs of the genus *Iris* (Iridaceae). *Izv. Akad. Nauk. Ser. Biol.* **2005**, *2*, 174–179. [CrossRef]
62. Akashi, T.; Ishizaki, M.; Aoki, T.; Ayabe, S.-I. Isoflavonoid production by adventitious root cultures of *Iris germanica* (Iridaceae). *Plant Biotechnol.* **2005**, *22*, 207–215. [CrossRef]
63. Karpitskiy, D.A.; Bessonova, E.A.; Kartsova, L.A.; Tikhomirova, L.I. Development of approach for flavonoid profiling of biotechnological raw materials *Iris sibirica* L. by HPLC with high-resolution tandem mass spectrometry. *Phytochem. Anal.* **2022**, *33*, 869–878. [CrossRef]
64. Michalak, A.; Krauze-Baranowska, M.; Migas, P.; Kawiak, A.; Kokotkiewicz, A.; Królicka, A. *Iris pseudacorus* as an easily accessible source of antibacterial and cytotoxic compounds. *J. Pharm. Biomed. Anal.* **2021**, *195*, 113863. [CrossRef] [PubMed]
65. Para, G.; Baratti, J. Irones and precursors synthesized by *Iris sibirica* tissue culture. *Biosci. Biotechnol. Biochem.* **1992**, *56*, 1132–1133. [CrossRef] [PubMed]
66. Marner, F.J.; Simic, K.; Scholz, B.; Küster, B. Isolation and structure determination of new iridal from *Iris sibirica*. *J. Nat. Prod.* **1995**, *58*, 299–301. [CrossRef]
67. Ritzdorf, I.; Bartels, M.; Kerp, B.; Kasel, T.; Klonowski, S.; Marner, F.J. Identification of 10-desoxyiridal as an intermediate in the biosynthesis of iridals. *Phytochemistry* **1999**, *50*, 995–1003. [CrossRef]
68. Afifi, F.U.; Al-Gabbiesh, A.; Hassawi, D.S. Essential Oil Production from the Callus of Threatened Iris Species of Jordan. In *Floriculture, Ornamental, and Plant Biotechnology*; Global Science Books: London, UK, 2008; Volume V, pp. 227–233.
69. Jevremović, S.; Kristić-Milosević, D.; De Carlo, A.; Benelli, K.; Lambardi, M.; Antonić, D.; Trifunović Momchilov, M.; Lojić, M.; Subotić, A. Kultura tkiva i krioprezervacija kao metode za povećanje produkcije mangiferina kod balkanske perunike (*Iris reichenbachii* Heuffel). In Proceedings of the Zbornik radova: III Simpozijum biologa i ekologa Republike Srpske, Banja Luka, Republika Srpska, 12–14 November 2016; Golub, D., Ed.; Prirodno-matematički fakultet, Univerzitet u Banja Luci: Banja Luka, Republika Srpska, 2016; pp. 41–53.
70. Tarbeeva, D.V.; Fedoreyev, S.A.; Veselova, M.V.; Kalinovskiy, A.I.; Seletskaya, L.D.; Mazurok, T.I.; Bulgakov, V.P. Polyphenolic compounds from callus cultures of *Iris pseudacorus*. *Nat. Prod. Commun.* **2013**, *8*, 1419–1420. [CrossRef]
71. Zhu, Y.; Chen, Y.; Zhang, X.; Xie, G.Y.; Qin, M.J. Copper stress-induced changes in biomass accumulation, antioxidant activity and flavonoid contents in *Belamcanda chinensis* calli. *Plant Cell Tissue Organ. Cult.* **2020**, *142*, 299–311. [CrossRef]
72. Michalak, A.; Królicka, A. The influence of elicitors for total phenolic content in tissues of *Iris pseudacorus* plants cultured in vitro. In Proceedings of the XI Conference "In Vitro Cultures In Biotechnology and Plant Physiology", Kraków, Poland, 4–6 December 2019; Volume 61, Suppl 1. p. 57.
73. Jeknić, Z.; Jeknić, S.; Jevremović, S.; Subotić, A.; Chen, T.H.H. Alteration of flower color in *Iris germanica* L. 'Fire Bride' through ectopic expression of phytoene synthase gene (crtB) from *Pantoea agglomerans*. *Plant Cell Rep.* **2014**, *33*, 1307–1321. [CrossRef]
74. Aslanyants, L.K.; Marshavina, Z.V. Volatile oil synthesized by Iris sibirica tissue cultures. *Appl. Biochem. Microbiol.* **1979**, *15*, 581–585.
75. Mazko, O.N.; Tikhomirova, L.I.; Shcherbakova, L.V.; Bazarnova, N.G.; Karpitsky, D.A. Flavonoids of *Iris sibirica* L. grown in vitro culture. *Khimiya Rastit. Syr'ya* **2021**, *3*, 301–308. (In Russian) [CrossRef]
76. Tikhomirova, L.I.; Ilyicheva, T.N. Preparation of biotechnological raw materials of *Iris sibirica* L. with a given content of mangiferin and antiviral activity. *IOP Conf. Ser. Earth Environ. Sci.* **2020**, *421*, 022049. [CrossRef]
77. Jevremović, S.; Lojić, M.; Jeknić, Z.; Trifunović-Momčilov, M.; Antonić, D.; Petrić, M.; Subotić, A.; Radojević, L. In vitro propagation of *Iris reichenbachii* Heuff. and clonal fidelity of regenerated plants. *Bot. Serb.* **2015**, *39*, 129–136.
78. Williams, C.A.; Harborne, J.B.; Goldblatt, P. Correlations between phenolic patterns and tribal classification in the family iridaceae. *Phytochemistry* **1986**, *25*, 2135–2154. [CrossRef]
79. Williams, C.A.; Harborne, J.B.; Colasante, M. Flavonoid and xanthone patterns in bearded iris species and the pathway of chemical evolution in the genus. *Biochem. System. Ecol.* **1997**, *25*, 309–325. [CrossRef]
80. Šavikin-Fodulović, K.; Stojanović, D.; Menković, N. Fitohemijska i anatomska analiza vrste *Iris reichenbachii* Heuff. *Lek. Sirovine* **2000**, *20*, 21–26.
81. Šarić Medić, B.; Jerković-Mujkić, A.; Cubara, B.; Durmić Pašić, A.; Hanjalić Kurtović, J.; Bajrović, K.; Omeragić, E.; Dedić, M.; Bogunić, F.; Pojskic, L. Crude Extracts of Three *Iris* Species as Sources of MRSA Antimicrobial Compounds. *Eur. J. Biol.* **2024**, *83*, 182–188. [CrossRef]
82. Jevremović, S.; Subotić, A.; Radojević, L.J. In vitro morphogenesis of dwarf irises. In *Floriculture, Ornamental and Plant Biotechnology: Advances and Topical Issues*; Teixeira da Silva, J., Ed.; Global Science Books: London, UK, 2006; Volume II, pp. 551–557.
83. Jevremović, S.; Jeknić, Z.; Subotić, A. Micropropagation of irises. In *Protocols for Micropropagation of Selected Economically-Important Horticultural Plants. Methods in Molecular Biology*; Lambardi, M., Ozudogru, E.A., Jain, S.M., Eds.; Humana Press-Springer: Totowa, NJ, USA, 2012; Volume 994, pp. 291–304. [CrossRef]

84. Živković, S.; Jevremović, S.; Gašić, U.; Milutinović, M.; Petrović, Z.L.; Škoro, N.; Puač, N. Long term effects in dwarf bearded iris (*Iris reichenbachii* Heuff.) calli metabolism induced by plasma treatment. In *Proceedings of the Book of Abstracts: 9th Central European Symposium on Plasma Chemistry (CESPC-9) Joint with COST Action CA19110 Plasma Applications for Smart and Sustainable Agriculture (PlAgri), Vysoké Tatry, Slovakia, 4–9 September 2022*; Hensel, K., Cimerman, R., Lavrikova, A., Janda, M., Machala, Z., Eds.; FMFI UK: Bratislava, Slovakia, 2022; p. 137.
85. Radojević, L.; Subotić, A. Plant regeneration of *Iris setosa* Pall. through somatic embryogenesis and organogenesis. *J. Plant Physiol.* **1992**, *139*, 690–696. [CrossRef]
86. Kim, T.D.; Ahn, C.H.; Bae, K.H.; Choi, Y.E. The embryogenic competency and morphological changes during somatic embryogenesis in *Iris pseudacorus*. *Plant Biotechnol. Rep.* **2009**, *3*, 251–257. [CrossRef]
87. Murashige, T.; Skoog, F. A revised medium for rapid growth and bioassay with tobacco tissue cultures. *Physiol. Plant.* **1962**, *15*, 473–497. [CrossRef]
88. Puač, N.; Miletić, M.; Mojović, M.; Popović-Bijelić, A.; Vuković, D.; Miličić, B.; Maletić, D.; Lazović, S.; Malović, G.; Petrović, Z. Sterilization of bacteria suspensions and identification of radicals deposited during plasma treatment. *Open Chem.* **2015**, *13*, 000010151520150041. [CrossRef]
89. Betekhtin, A.; Milewska-Hendel, A.; Chajec, L.; Rojek, M.; Nowak, K.; Kwasniewska, J.; Wolny, E.; Kurczynska, E.; Hasterok, R. 5-Azacitidine Induces Cell Death in a Tissue Culture of *Brachypodium distachyon*. *Int. J. Mol. Sci.* **2018**, *19*, 1806. [CrossRef]
90. Jambunathan, N. Determination and Detection of Reactive Oxygen Species (ROS), Lipid Peroxidation, and Electrolyte Leakage in Plants. In *Plant Stress Tolerance. Methods in Molecular Biology*; Sunkar, R., Ed.; Humana Press: Totowa, NJ, USA, 2010; Volume 639, pp. 292–298. [CrossRef]
91. Dmitrović, S.; Perišić, M.; Stojić, A.; Živković, S.; Boljević, J.; Živković, J.N.; Aničić, N.; Ristić, M.; Mišić, D. Essential oils of two *Nepeta* species inhibit growth and induce oxidative stress in ragweed (*Ambrosia artemisiifolia* L.) shoots in vitro. *Acta physiol. Plant* **2015**, *37*, 64. [CrossRef]
92. Božunović, J.; Živković, S.; Gašić, U.; Glamočlija, J.; Ćirić, A.; Matekalo, D.; Šiler, B.; Soković, M.; Tešić, Ž.; Mišić, D. In vitro and in vivo transformations of *Centaurium erythraea* secoiridoid glucosides alternate their antioxidant and antimicrobial capacity. *Ind. Crops Prod.* **2018**, *111*, 705–721. [CrossRef]
93. Banjanac, T.; Dragićević, M.; Šiler, B.; Gašić, U.; Bohanec, B.; Nestorović Živković, J.; Trifunović, S.; Mišić, D. Chemodiversity of two closely related tetraploid *Centaurium* species and their hexaploid hybrid: Metabolomic search for high-resolution taxonomic classifiers. *Phytochemistry* **2017**, *140*, 27–44. [CrossRef] [PubMed]
94. Zhang, Y.Y.; Wang, Q.; Qi, L.W.; Qin, X.Y.; Qin, M.J. Characterization and determination of the major constituents in *Belamcandae* Rhizoma by HPLC–DAD–ESI-MSn. *J. Pharm. Biomed. Anal.* **2011**, *56*, 304–314. [CrossRef]
95. Chen, D.; Meng, Y.; Zhu, Y.; Wu, G.; Yuan, J.; Qin, M.; Xie, G. Qualitative and quantitative analysis of C-glycosyl-flavones of iris lactea leaves by liquid chromatography/tandem mass spectrometry. *Molecules* **2018**, *23*, 3359. [CrossRef]
96. Ibrahim, S.; Al-Ahdal, A.; Khedr, A.; Mohamed, G. Antioxidant α-amylase inhibitors flavonoids from *Iris germanica* rhizomes. *Rev. Bras. De Farmacogn.* **2017**, *27*, 170–174. [CrossRef]
97. Yu, X.; Li, M.; O'Doherty, G.A. De novo symmetric approach to the disaccharide portion of SCH-47554. *Heterocycles* **2011**, *82*, 1577–1584. [CrossRef]
98. Malović, G.; Puač, N.; Lazović, S.; Petrović, Z. Mass analysis of an atmospheric pressure plasma needle discharge. *Plasma Sources Sci. Technol.* **2010**, *19*, 034014. [CrossRef]
99. Živković, S.; Jevremović, S.; Puač, N.; Škoro, N.; Petrović, Z.L. Morphological and physiological aspects of plasma treatment-induced changes in plant cells and tissues. In Proceedings of the XXIInd International Conference on Gas Discharges and their Applications, Novi Sad, Serbia, 2–7 September 2018; Petrović, Z.L., Puač, N., Dujko, S., Škoro, N., Eds.; Serbian Academy of Sciences and Arts: Belgrade, Serbia, 2018; pp. 646–649.
100. Radojevic, L.; Landré, P. Cytological features of developing embryogenic callus and somatic embryos of *Iris pumila* L. and *Iris setosa* Pall. *G. Bot. Ital. Off. J. Soc. Bot. Ital.* **1995**, *129*, 1213–1224. [CrossRef]
101. Stanišić, M.; Raspor, M.; Ninković, S.; Milošević, S.; Ćalić, D.; Bohanec, B.; Trifunović, M.; Petrić, M.; Subotić, A.; Jevremović, S. Clonal fidelity of *Iris sibirica* plants regenerated by somatic embryogenesis and organogenesis in leaf-base culture—RAPD and flow cytometer analyses. *S. Afr. J. Bot.* **2015**, *96*, 42–52. [CrossRef]
102. Li, D.; Ji, C.; Shao, L.; Xu, T.; Chen, X.; Wang, X.; Zhang, R.; Cui, Y.; Zhang, J.; Xia, Y. An Efficient and Environmentally Friendly Strategy for Embryogenic Callus Induction and Plant Regeneration of *Iris pseudacorus* L. *J. Plant Growth Regul.* **2025**. [CrossRef]
103. Živković, S.; Jevremović, S.; Milutinović, M.; Puač, N.; Petrović, Z.L.; Škoro, N. Alteration in plant tissue morphology and physiology induced by plasma treatment. In *Program Abstracts of the Twenty-First International Summer School on Vacuum, Electron and Ion Technologies (VEIT 2019)*; Heron Press: Birmingham, UK, 2019; pp. 63–64.
104. Kurczynska, E.; Godel-Jedrychowska, K. Apoplastic and symplasmic markers of somatic embryogenesis. *Plants* **2023**, *12*, 1951. [CrossRef] [PubMed]

105. Karimi, J.; Bansal, S.A.; Kumar, V.; Pasalari, H.; Badr, A.A.; Nejad, Z.J. Effect of cold plasma on plant physiological and biochemical processes: A review. *Biologia* **2024**, *79*, 3475–3487. [CrossRef]
106. Adhikari, B.; Adhikari, M.; Park, G. The effects of plasma on plant growth, development, and sustainability. *Appl. Sci.* **2020**, *10*, 6045. [CrossRef]
107. Priatama, R.A.; Pervitasari, A.N.; Park, S.; Park, S.J.; Lee, Y.K. Current Advancements in the Molecular Mechanism of Plasma Treatment for Seed Germination and Plant Growth. *Int. J. Mol. Sci.* **2022**, *23*, 4609. [CrossRef]
108. Li, K.; Zhong, C.; Shi, Q.; Bi, H.; Gong, B. Cold plasma seed treatment improves chilling resistance of tomato plants through hydrogen peroxide and abscisic acid signaling pathway. *Free Radic. Biol. Med.* **2021**, *172*, 286–297. [CrossRef]
109. Waszczak, C.; Carmody, M.; Kangasjärvi, J. Reactive oxygen species in plant signaling. *Annu. Rev. Plant Biol.* **2018**, *69*, 209–236. [CrossRef]
110. Wang, H.; Zhang, Y.; Jiang, H.; Cao, J.; Jiang, W. A comprehensive review of effects of electrolyzed water and plasma-activated water on growth, chemical compositions, microbiological safety and postharvest quality of sprouts. *Trends Food Sci. Technol.* **2022**, *129*, 449–462. [CrossRef]
111. Zahoranová, A.; Henselová, M.; Hudecová, D.; Kaliňáková, B.; Kováčik, D.; Medvecká, V.; Černák, M. Effect of cold atmospheric pressure plasma on the wheat seedlings vigor and on the inactivation of microorganisms on the seeds surface. *Plasma Chem. Plasma Proc.* **2016**, *36*, 397–414. [CrossRef]
112. Hosseini, S.I.; Mohsenimehr, S.; Hadian, J.; Ghorbanpour, M.; Shokri, B. Physico-chemical induced modification of seed germination and early development in artichoke (*Cynara scolymus* L.) using low energy plasma technology. *Phys. Plasmas* **2018**, *25*, 013525. [CrossRef]
113. Kostoláni, D.; Ndiffo Yemeli, G.B.; Švubová, R.; Kyzek, S.; Machala, Z. Physiological responses of young pea and barley seedlings to plasma-activated water. *Plants* **2021**, *10*, 1750. [CrossRef] [PubMed]
114. Rahman, M.M.; Sajib, S.A.; Rahi, M.S.; Tahura, S.; Roy, N.C.; Parvez, S.; Reza, M.A.; Talukder, M.R.; Kabir, A.H. Mechanisms and signaling associated with LPDBD plasma mediated growth improvement in wheat. *Sci. Rep.* **2018**, *8*, 10498. [CrossRef] [PubMed]
115. Cui, D.; Yin, Y.; Sun, H.; Wang, X.; Zhuang, J.; Wang, L.; Ma, R.; Jiao, Z. Regulation of cellular redox homeostasis in Arabidopsis thaliana seedling by atmospheric pressure cold plasma-generated reactive oxygen/nitrogen species. *Ecotoxicol. Environ. Saf.* **2022**, *240*, 113703. [CrossRef] [PubMed]
116. Bailly, C.; Leymarie, J.; Lehner, A.; Rousseau, S.; Côme, D.; Corbineau, F. Catalase activity and expression in developing sunflower seeds as related to drying. *J. Exp. Bot.* **2004**, *55*, 475–483. [CrossRef] [PubMed]
117. Kokkinidis, M.; Glykos, N.M.; Fadouloglou, V.E. Catalytic activity regulation through post-translational modification: The expanding universe of protein diversity. *Adv. Protein Chem. Struct. Biol.* **2020**, *122*, 97–125. [CrossRef]
118. Soares, C.; Carvalho, M.E.; Azevedo, R.A.; Fidalgo, F. Plants facing oxidative challenges—A little help from the antioxidant networks. *Environ. Exp. Bot.* **2019**, *161*, 4–25. [CrossRef]
119. Ivankov, A.; Nauciene, Z.; Zukiene, R.; Degutyte-Fomins, L.; Malakauskiene, A.; Kraujalis, P.; Venskutonis, P.R.; Filatova, I.; Lyushkevich, V.; Mildaziene, V. Changes in growth and production of non-psychotropic cannabinoids induced by pre-sowing treatment of hemp seeds with cold plasma, vacuum and electromagnetic field. *Appl. Sci.* **2020**, *10*, 8519. [CrossRef]
120. Mildaziene, V.; Pauzaite, G.; Nauciene, Z.; Zukiene, R.; Malakauskiene, A.; Norkeviciene, E.; Stukonis, V.; Slepetiene, A.; Olsauskaite, V.; Padarauskas, A.; et al. Effect of seed treatment with cold plasma and electromagnetic field on red clover germination, growth and content of major isoflavones. *J. Phys. D Appl. Phys.* **2020**, *53*, 264001. [CrossRef]
121. Ivankov, A.; Zukiene, R.; Nauciene, Z.; Degutyte-Fomins, L.; Filatova, I.; Lyushkevich, V.; Mildaziene, V. The effects of red clover seed treatment with cold plasma and electromagnetic field on germination and seedling growth are dependent on seed color. *Appl. Sci.* **2021**, *11*, 4676. [CrossRef]
122. Judickaitė, A.; Lyushkevich, V.; Filatova, I.; Mildažienė, V.; Žūkienė, R. The potential of cold plasma and electromagnetic field as stimulators of natural sweeteners biosynthesis in *Stevia rebaudiana* Bertoni. *Plants* **2022**, *11*, 611. [CrossRef]
123. Pérez-Alonso, N.; Jiménez, E. Producción de metabolitos secundarios de plantas mediante el cultivo in vitro. *Biotechnol. Veg* **2011**, *11*, 195–211.
124. McDonald, K.A.; Jackman, A.P.; Thorup, J.E.; Dandekar, A.M. Plant callus as a source of biochemicals. *Appl. Biochem. Biotechnol.* **1995**, *54*, 93–108. [CrossRef]
125. Al-Khalil, S.; Tosa, H.; Iinuma, M. A xanthone C-glycoside from *Iris nigricans*. *Phytochemistry* **1995**, *38*, 729–731. [CrossRef]
126. Razdan, T.K.; Kachroo, P.K.; Qadri, B.; Kalla, A.K.; Taneja, S.C.; Koul, S.K.; Dhar, K.L. Isocladrastin and kashmigenin—Two isoflavones from *Iris kashmiriana*. *Phytochemistry* **1996**, *41*, 947–949. [CrossRef]
127. Zhang, X.; Qiao, G.-X.; Zhao, G.-F.; Zhao, S.-F. Characterization of the metabolites of irisflorentin by using ultra-high performance liquid chromatography combined with quadrupole/orbitrap tandem mass spectrometry. *J. Pharm. Biomed. Anal.* **2021**, *203*, 114222. [CrossRef]
128. Fang, R.; Veitch, N.C.; Kite, G.C.; Howes, M.J.R.; Porter, E.A.; Simmonds, M.S. Glycosylated Constituents of *Iris fulva* and *Iris brevicaulis*. *Chem. Pharm. Bull.* **2011**, *59*, 124–128. [CrossRef]

129. Yehia, S.M.; Ayoub, I.M.; Watanabe, M.; Devkota H., P.; Singab, A.N.B. Metabolic profiling, antioxidant, and enzyme inhibition potential of Iris pseudacorus L. from Egypt and Japan: A comparative study. *Sci. Rep.* **2023**, *13*, 5233. [CrossRef]
130. Song, Y.Y.; Liu, Y.; Yan, Y.M.; Lu, X.F.; Cheng, Y.X. Phenolic compounds from *Belamcanda chinensis* seeds. *Molecules* **2018**, *23*, 580. [CrossRef]
131. Roger, B.; Jeannot, V.; Fernandez, X.; Cerantola, S.; Chahboun, J. Characterisation and quantification of flavonoids in *Iris germanica* L. and *Iris pallida* Lam. resinoids from Morocco. *Phytochem. Anal.* **2012**, *23*, 450–455. [CrossRef]
132. Salehi, B.; Venditti, A.; Sharifi-Rad, M.; Kregiel, D.; Sharifi-Rad, J.; Durazzo, A.; Lucarini, M.; Santini, A.; Souto, E.B.; Novellino, E.; et al. The therapeutic potential of apigenin. *Int. J. Mol. Sci.* **2019**, *20*, 1305. [CrossRef]
133. Chen, K.; Gao, Z. Acacetin, a Natural Flavone with Potential in Improving Liver Disease Based on Its Anti-Inflammation, Anti-Cancer, Anti-Infection and Other Effects. *Molecules* **2024**, *29*, 4872. [CrossRef] [PubMed]
134. Pashaei, R. Features of Apigenin, Luteolin, Hesperetin and Naringenin in Crop and Body. *Int. J. Food Sci. Nutr. Diet.* **2016**, *5*, 300–304. [CrossRef]
135. Moghaddam, H.R.; Samimi, Z.; Zachariah Moradi, S.; Little, P.J.; Xu, S.; Farzaei, M.H. Naringenin and naringin in cardiovascular disease prevention: A preclinical review. *Eur. J. Pharmacol.* **2020**, *887*, 173535. [CrossRef] [PubMed]
136. Amin, H.I.M.; Hussain, F.H.S.; Najmaldin, S.K.; Thu, Z.M.; Ibrahim, M.F.; Gilardoni, G.; Vidari, G. Phytochemistry and Biological Activities of Iris Species Growing in Iraqi Kurdistan and Phenolic Constituents of the Traditional Plant *Iris postii*. *Molecules* **2021**, *26*, 264. [CrossRef]
137. Alperth, F.; Mitić, B.; Mayer, S.; Maleš, Ž.; Kunert, O.; Hruševar, D.; Bucar, F. Metabolic profiling of rhizomes of native populations of the strictly endemic Croatian species Iris adriatica. *Plant Biosyst.* **2019**, *153*, 317–324. [CrossRef]
138. Conforti, F.; Menichini, F.; Rigano, D.; Senatore, F. Antiproliferative activity on human cancer cell lines after treatment with polyphenolic compounds isolated from *Iris pseudopumila* flowers and rhizomes. *Z. Naturforsch C.* **2009**, *64*, 490–494. [CrossRef]
139. Lutter, S.; Schmalbach, K.; Esch, H.L.; Lehmann, L. The isoflavone irilone contributes to the estrogenic potential of dietary supplements containing red clover. *Arch. Toxicol.* **2014**, *88*, 309–321. [CrossRef]
140. Kang, K.A.; Zhang, R.; Piao, M.J.; Ko, D.O.; Wang, Z.H.; Kim, B.J.; Park, J.W.; Kim, H.S.; Kim, D.H.; Hyun, J.W. Protective effect of irisolidone, a metabolite of kakkalide, against hydrogen peroxide induced cell damage via antioxidant effect. *Bioorg. Med.Chem.* **2008**, *16*, 1133–1141. [CrossRef]
141. Tůmová, L.; Tůma, J. The efect of UV light on isofavonoid production in *Genista tinctoria* culture in vitro. *Acta Physiol. Plant* **2011**, *33*, 635–640. [CrossRef]
142. Vidya, N.; Saravanan, K.; Halka, J.; Kowsalya, K.; Preetha, J.S.Y.; Gurusaravanan, P.; Radhakrishnan, R.; Nanthin, A.U.R.; Arun, M. An insight into in vitro strategies for bioproduction of isoflavones. *Plant Biotechnol. Rep.* **2021**, *15*, 717–740. [CrossRef]
143. Heredia, J.B.; Cisneros-Zevallos, L. The effect of exogenous ethylene and methyl jasmonate on pal activity, phenolic profiles and antioxidant capacity of carrots (*Daucus carota*) under different wounding intensities. *Postharvest Biol. Technol.* **2009**, *51*, 242–249. [CrossRef]

horticulturae

MDPI

Article

Exploring the Production of Secondary Metabolites from a Halophyte *Tetragonia tetragonoides* through Callus Culture

Ka Youn Lee [1,†], Do-Hyeon Nam [1,2], Yongsam Jeon [1], Sang Un Park [3], Jongki Cho [4], Md Ashrafuzzaman Gulandaz [2,5], Sun-Ok Chung [5] and Geung-Joo Lee [1,2,*]

1 Department of Horticulture, Chungnam National University, 99 Daehak-ro, Yuseong-gu, Daejeon 34134, Republic of Korea; kayoun200@korea.kr (K.Y.L.); namdohyeon@o.cnu.ac.kr (D.-H.N.); plantotaku@naver.com (Y.J.)

2 Department of Smart Agriculture Systems, Chungnam National University, Daejeon 34134, Republic of Korea; gulandazbari@o.cnu.ac.kr

3 Department of Crop Science, Graduate School, Chungnam National University, Daejeon 34134, Republic of Korea; supark@cnu.ac.kr

4 College of Veterinary Medicine, Chungnam National University, Daejeon 34134, Republic of Korea; cjki@cnu.ac.kr

5 Department of Agricultural Machinery Engineering, Graduate School, Chungnam National University, Daejeon 34134, Republic of Korea; sochung@cnu.ac.kr

* Correspondence: gjlee@cnu.ac.kr

† Current address: Floriculture Research Division, National Institute of Horticultural and Herbal Science, Rural Development Administration, Wanju 55365, Republic of Korea.

Abstract: Considering Korea's gradual shift toward an aging society, consumer interest in compounds with physiological benefits, including antioxidant and anticancer effects, has surged. This study explored the potential of *Tetragonia tetragonoides* (Pall.) Kunze, commonly known as New Zealand Spinach (NZS), a halophyte with reported health benefits, including efficacy in treating gastrointestinal diseases, high blood pressure, diabetes, and obesity. This study also introduced a novel callus culture system for NZS, allowing for the rapid in vitro production of secondary metabolites. Optimal callus induction (100%) and biomass production (0.416 g) were achieved by adding 2.0 mg·L^{-1} 6-BA (6-Benzylaminopurine) and 0.5 mg·L^{-1} 2,4-D (2,4-Dichlorophenoxyacetic acid) among five auxin and cytokinin combinations. Two distinct callus types, TGC [TDZ (Thidiazuron)-supplemented Green Callus] and TNYC [TDZ + NAA (Naphthalene acetic acid)-supplemented Yellow Callus], were identified, each with unique characteristics. The calli showed total phenolic and flavonoid contents comparable to those of NZS leaves grown in the greenhouse. An expression analysis of six genes (*CHS, CHI, F3H, F3′H, FLS*, and *DFR*) involved in the kaempferol biosynthesis revealed an enhanced flavonoid biosynthesis-related gene expression in TGC, emphasizing its potential for compound production. GC-MS analysis identified distinct compound profiles in TGC and TNYC, with 2,3-butanediol and succinic acid being the predominant compounds among the nine and forty-four components, respectively. These calli offer a stable supply of functional compounds and present an environmentally sustainable solution. The derived callus culture system is anticipated to contribute to the development of healthy functional foods or pharmaceuticals from halophyte NZS.

Keywords: callus induction; gas chromatography–mass spectrometry; gene expression; kaempferol biosynthesis; New Zealand spinach; secondary metabolite

1. Introduction

Recently, as Korea has gradually become an aging society, consumers' interest in compounds with physiological activities, such as antioxidant and anticancer effects, is increasing. Plants produce and accumulate a variety of secondary metabolites to respond and defend themselves against biological stress as well as abiotic stress such as environmental stress [1–4]. These metabolites, renowned for their antioxidant properties, possess diverse anticancer and antibacterial properties. Not only have they found industrial applications in chemicals, spices, food additives, and medicines, but they also serve as vital resources [5,6].

However, the production of secondary metabolites in plants may be limited because of slow growth and endangered plant species [7]. Recently, secondary metabolite production through cell culture has attracted attention because it can be performed in large quantities quickly without being affected by external weather [8–10]. Additionally, calli grow faster than seedlings and have shorter biosynthetic cycles. Therefore, they can produce secondary metabolites in a shorter time than plants [11]. Anjum et al. [12] reported that 1.0 mg·L^{-1} of NAA was treated to induce the callus of *Linum usitatissimum* L. followed by lignans and neo-lignans production. And Adil et al. [13] reported that high TPC and TFC were identified in the callus of *Cnidium officalale* Makino and that the content of phthalides was higher than that of seeding plants. In addition, 45 aromatic compounds, including rosemaryic acid, were identified in a callus derived from the leaves of *Rosmarinus openis* L., and calli treated with 200 µM of melatonin doubled the rosemaryic acid content [14].

Tetragonia tetragonoides (Pall.) Kunze is a conditional salt plant, commonly known as New Zealand Spinach (NZS). This plant grows naturally on sandy soil or coastal cliffs in New Zealand, Australia, Japan, and the Republic of Korea [15]. NZS has leaves and has a soft taste, with a texture similar to that of regular spinach. NZS is used to treat stomach cancer, gastritis, gastric ulcers, and hyper acidosis and to improve food intake [16]. In addition, anti-inflammatory, anti-blood pressure, and antioxidant agents have been reported [17–20]. In addition, extracts from NZS have been shown to enhance hippocampal insulin signaling, reduce insulin resistance, and enhance gut microbiota and memory function in amyloid-β-infused mice in an Alzheimer's disease animal model [21]. These findings confirm that NZS extract can provide a variety of health benefits.

In this study, we explored methods to produce beneficial secondary metabolites from NZS using callus cultures. Hormonal conditions were established to effectively induce callus growth on NZS leaves. Furthermore, we assessed the potential of the selected callus as an extractant raw material for functional compounds using the total phenolic compound (TPC) and total flavonoid (TFC) measurements and GC-MS screening. The results of this study contribute to the development of this industry with a raw material for healthy functional foods, pharmaceuticals, and cosmetics.

2. Materials and Methods

2.1. Combinations of Plant Growth Regulators Used to Induce Callus from NZS Leaves

To investigate the effects of plant growth regulators (PGRs) (Duchefa, Haarlem, Netherlands) on callus induction, explants were cultured in MS medium containing different levels of auxins (IAA, NAA, and 2,4-D) and cytokinins (zeatin and 6-BA) (Figure S1). The explants were prepared from the apical site to the first and second nodes of in vitro-grown NZS, with leaves sized approximately 0.5 cm × 0.5 cm. Ten explants were placed in the medium for each treatment, and the procedure was repeated three times. We placed the prepared implants in the medium and measured the induction rate and fresh weight of the callus after four weeks. Additionally, to obtain calli of various colors and shapes, calli derived from the optimal media were transferred to MS medium supplemented with TDZ alone or a combination of TDZ and NAA. The concentrations of TDZ and NAA were set based on previous reports indicating a high induction rate of calli or sprouts in the Aizoaceae family [22]. Ultimately, we selected a vigorously growing green hard tissue callus in a medium containing 2.5 mg·L^{-1} TDZ, named TGC. Simultaneously, we chose a yellow hard tissue callus labeled TNYC, grown on a medium with 2.5 mg·L^{-1} TDZ and 0.5 mg·L^{-1} NAA.

2.2. Explant Material and In Vitro Culture Conditions

The mature seeds, collected from the seashore of Jeonnam Goheung-Gun, Republic of Korea, were stored at 4 °C until used in the experiment. The seeds were sterilized through dipping in 70% (*v*/*v*) ethanol for 1 min and in a 2% sodium hypochlorite solution for 20 min, and then washed with sterile distilled water three times. Germination was performed on 1/2 MS (Duchefa, Haarlem, Netherlands) medium supplemented with 30 $g \cdot L^{-1}$ sucrose (Duchefa, Haarlem, Netherlands) and 7 $g \cdot L^{-1}$ plant agar (Duchefa, Haarlem, The Netherlands). Plantlets established from 8-week-old seedlings were used as explant sources (Figure 1). Before the medium was autoclaved at 120 °C for 20 min, the pH was adjusted to 5.7. All the cultures were incubated in a growth room at 25 ± 1 °C under a 16 h light/8 h dark photoperiod, using an irradiance of 20 $\mu mol \cdot m^{-2} \cdot s^{-1}$ from white fluorescent lamps (Osram GmbH, Munich, Germany).

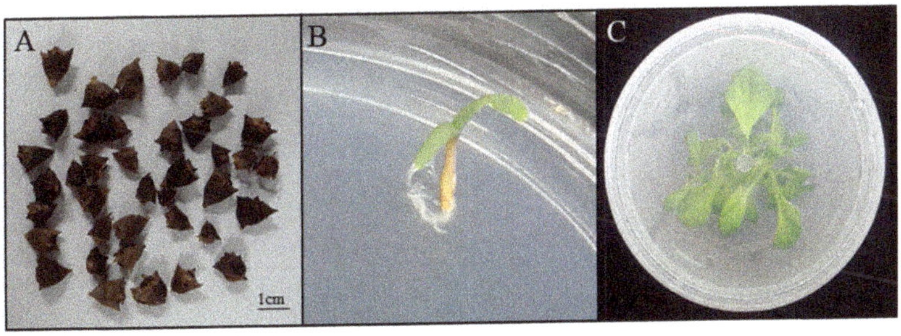

Figure 1. Tissue culture seedlings of NZS used for the induction of callus in this study. (**A**) Seeds of NZS collected in native habitat. (**B**) Young seedling germinated from seed. (**C**) Seedlings grown in vitro for eight weeks.

2.3. Measuring Total Phenol Content (TPC) and Total Flavonoid Content (TFC) of Selected Callus

To obtain the extracts of the callus grown in vitro and leaves grown in greenhouses, they were extracted with 80% ethanol for three hours at 85 °C and filtered through a 0.45 μm PTFE syringe filter (Hyundai micro, Seoul, Republic of Korea).

The TPC was determined using the Folin–Ciocaltzeu method [23]. A standard solution was prepared with concentrations of 13.25, 62.5, 125, 250, 500, and 1000 $mg \cdot L^{-1}$ from gallic acid (Sigma Chemical, St. Louis, MO, USA). Subsequently, 0.05 mL of each solution was mixed with 0.25 mL of Folin–Ciocalteu solvent (Sigma Chemical, St. Louis, MO, USA), followed by the addition of 2.5% Na_2CO_3 (0.5 mL). The mixture was stored in the dark for 40 min, and the extract was measured three times.

The TFC was analyzed using the aluminum chloride colorimetric method. Standard solutions were prepared at concentrations of 13.25, 125, 250, and 500 $mg \cdot L^{-1}$ using Quercetin (Sigma Chemical Co., St. Louis, MO, USA). In the measurement, 0.02 mL of 10% aluminum nitrate and 0.02 mL of 1 M potassium acetate were added to 0.02 mL of the extract, and then 80% ethanol was added to make a total volume of 1 mL. After incubating the mixture at room temperature for 40 min, absorbance was measured at 415 nm.

2.4. Analysis of Gene Expression Involved in the Biosynthesis of Flavonoids

Three major compounds of the kaempferol backbone were detected in NZS, as reported previously [20]. Thus, we investigated the expression levels of genes, including *chalcone synthase* (*CHS*), *chalcone isomerase* (*CHI*), *flavanone 3-hydroxylase* (*F3H*), *flavonoid 3′-hydroxylase* (*F3′H*), *flavonol synthase* (*FLS*), and *dihydroflavonol 4-reductase* (*DFR*), involved in the biosynthesis of kaempferol. The expression patterns of the six genes from the selected calli were compared with those of seedlings grown in the greenhouse.

The total RNA was extracted using an RNeasy mini kit (Qiagen, Hilden, Germany) after powdering each portion (leaf, stem, and root) of the NZS and callus using liquid nitrogen. Subsequently, approximately 1 μg of cDNA was synthesized from the extracted RNA using the PrimeScript 1st Strand cDNA Synthesis Kit (Takara Bio, Shiga, Japan).

Kaempferol biosynthesis-related genes were identified by searching the Kyoto Encyclopedia of Genes and Genomes pathway database. Gene-specific primers were designed to amplify products ranging from 200 to 550 bp (Table S1). The amplification conditions included 35 cycles of denaturation for 30 s at 94 °C, annealing for 30 s at 60 °C, extension for 30 s at 72 °C, and a final extension for 10 min at 72 °C. PCR was performed using a TaKaRa PCR Thermal Cycler Dice® gradient (Takara Bio, Shiga, Japan). The RT-PCR products were digested on a 1.2% agarose gel, and the electrophoretic bands were visualized under UV fluorescence using gel documentation.

2.5. Phytochemical Analysis of Selected Callus via GC-MS

To detect phytochemicals in the two selected calli, we prepared extracts and profiled them using gas chromatography–mass spectrometry (GC-MS) according to the manufacture's manual. Callus extracts were prepared through soaking in 70% ethanol for 24 h at room temperature, followed by centrifugation at 13,000 rpm for 30 min to collect the supernatants. The freeze-dried powder obtained from the supernatant was dissolved in 100% ethanol for further analysis.

For GC-MS analysis, an HP-1MS capillary column (30 m × 320 μm × 1 μm film thickness, Agilent, Santa Clara, CA, USA) was utilized, with an MS (5975 inert Mass Selective Detector, Agilent) serving as the detector. The GC oven temperature was initially set at 70 °C for 1 min, and then increased from 70 °C to 180 °C at 10 °C/min, followed by an increase from 180 °C to 220 °C at 5 °C/min, maintaining at 220 °C for 1 min. Subsequently, the temperature was further raised from 220 °C to 300 °C at 25 °C/min, and a post time was held at 320 °C for 1 min. The total run time for the GC-MS analysis was 24.20 min. Helium was used as the carrier gas at a flow rate of 1 ml/min. The sample inlet was operated in splitless mode, and 1 μL of the sample was injected using an Autosampler (7683B Series injector, Agilent, Santa Clara, CA, USA). The syringe used for the autosampler had a volume of 10 μL. To minimize contamination-related errors, the syringe was washed thrice with acetone before injection, the sample was pumped thrice for collection, and washed thrice with acetone after injection. To reduce errors due to residual amounts in the column during sample analysis, the sample was analyzed once and subsequently cleaned once with acetone.

2.6. Statistical Analysis

The statistical analysis was performed using one-way analysis of variance (ANOVA) and Duncan's multiple range test for the comparison of mean values using the SPSS statistical software Version 20.0 (IBM, New York, NY, USA).

3. Results

3.1. Effect of PGR on Callus Induction

In this study, leaf explants were placed in MS medium containing cytokinin and auxin to determine the effects of the PGRs on callus induction through NZS. When zeatin was used, the callus induction speed and fresh weight decreased in the order of 2,4-D > NAA > IAA (Table 1). In particular, calli induced in the MS medium with NAA showed root development (Figure S1). When 6-BA was used, the callus induction rate and fresh weight were higher in the 2,4-D treatment than in the NAA treatment (Table 2).

Table 1. Induction rates and fresh weights of calli in MS medium with various concentrations of zeatin and auxin (IAA, NAA, 2,4-D).

Plant Growth Regulators ($mg{\cdot}L^{-1}$)				Callus Induction (%) [Z]	Fresh Weight (g) [Z]
Zeatin	IAA	NAA	2,4-D		
0.5	0.5			0.0 ± 0.0 b	0.046 ± 0.002 ij
0.5	1.0			0.0 ± 0.0 b	0.064 ± 0.002 g–j
0.5	1.5			0.0 ± 0.0 b	0.065 ± 0.004 g–j
0.5	2.0			60.0 ± 0.0 ab	0.041 ± 0.003 ij
1.0	0.5			75.0 ± 2.5 ab	0.053 ± 0.001 h–j
1.0	1.0			0.0 ± 0.0 b	0.054 ± 0.003 h–j
1.0	1.5			0.0 ± 0.0 b	0.058 ± 0.001 h–j
1.0	2.0			0.0 ± 0.0 b	0.069 ± 0.005 g–j
1.5	0.5			65.0 ± 2.5 ab	0.045 ± 0.004 ij
1.5	1.0			25.0 ± 2.5 b	0.055 ± 0.004 h–j
1.5	1.5			25.0 ± 2.5 b	0.058 ± 0.005 h–j
1.5	2.0			0.0 ± 0.0 b	0.055 ± 0.003 h–j
2.0	0.5			0.0 ± 0.0 b	0.044 ± 0.001 ij
2.0	1.0			0.0 ± 0.0 b	0.037 ± 0.000 j
2.0	1.5			0.0 ± 0.0 b	0.039 ± 0.001 ij
2.0	2.0			0.0 ± 0.0 b	0.055 ± 0.004 h–j
0.5		0.5		75.0 ± 7.5 ab	0.070 ± 0.004 g–j
0.5		1.0		80.0 ± 10.0 ab	0.117 ± 0.010 c–j
0.5		1.5		95.0 ± 2.5 ab	0.081 ± 0.006 e–j
0.5		2.0		90.0 ± 0.0 ab	0.207 ± 0.034 a–c
1.0		0.5		65.0 ± 2.5 ab	0.095 ± 0.013 d–j
1.0		1.0		70.0 ± 5.0 ab	0.057 ± 0.003 h–j
1.0		1.5		100.0 ± 0.0 ab	0.081 ± 0.006 e–j
1.0		2.0		90.0 ± 0.0 ab	0.075 ± 0.004 f–j
1.5		0.5		65.0 ± 7.5 ab	0.090 ± 0.005 d–j
1.5		1.0		65.0 ± 7.5 ab	0.124 ± 0.007 c–j
1.5		1.5		70.0 ± 15.0 ab	0.093 ± 0.011 d–j
1.5		2.0		45.0 ± 17.5 ab	0.035 ± 0.003 j
2.0		0.5		45.0 ± 17.5 ab	0.087 ± 0.010 d–j
2.0		1.0		70.0 ± 15.0 ab	0.076 ± 0.001 f–j
2.0		1.5		65.0 ± 12.5 ab	0.057 ± 0.001 h–j
2.0		2.0		95.0 ± 2.5 ab	0.148 ± 0.007 a–g
0.5			0.5	60.0 ± 20.0 ab	0.176 ± 0.014 a–e
0.5			1.0	100.0 ± 0.0 ab	0.081 ± 0.004 e–j
0.5			1.5	75.0 ± 2.5 ab	0.088 ± 0.009 d–j
0.5			2.0	85.0 ± 2.5 ab	0.044 ± 0.002 ij
1.0			0.5	65.0 ± 17.5 ab	0.172 ± 0.002 a–f
1.0			1.0	100.0 ± 0.0 a	0.180 ± 0.008 a–d
1.0			1.5	100.0 ± 0.0 a	0.234 ± 0.009 a
1.0			2.0	100.0 ± 0.0 a	0.133 ± 0.010 c–j
1.5			0.5	100.0 ± 0.0 a	0.228 ± 0.027 ab
1.5			1.0	100.0 ± 0.0 a	0.139 ± 0.009 c–j
1.5			1.5	100.0 ± 0.0 a	0.076 ± 0.001 f–j
1.5			2.0	100.0 ± 0.0 a	0.060 ± 0.004 h–j
2.0			0.5	100.0 ± 0.0 a	0.161 ± 0.013 a–g
2.0			1.0	100.0 ± 0.0 a	0.112 ± 0.002 d–j
2.0			1.5	100.0 ± 0.0 a	0.237 ± 0.004 a
2.0			2.0	100.0 ± 0.0 a	0.072 ± 0.002 g–j

[Z] Means followed by the same letter are not significantly different from each other ($p < 0.001$), as determined using Duncan's multiple range test (DMRT).

Table 2. Induction rates and fresh weights of calli in MS medium with various concentrations of 6-BA auxin (NAA, 2,4-D).

Plant Growth Regulators ($mg \cdot L^{-1}$)			Callus Induction (%) [Z]	Fresh Weight (g) [Z]
6-BA	NAA	2,4-D		
0.0	0.0		0.0 ± 0.0 d	0.014 ± 0.001 d
0.0	0.5		0.0 ± 0.0 d	0.016 ± 0.001 d
0.0	1.0		0.0 ± 0.0 d	0.017 ± 0.001 d
0.0	1.5		0.0 ± 0.0 d	0.013 ± 0.001 d
1.0	0.0		40.0 ± 2.9 bc	0.215 ± 0.005 d
1.0	0.5		16.7 ± 1.7 b–d	0.026 ± 0.003 d
1.0	1.0		0.0 ± 0.0 b–d	0.013 ± 0.001 d
1.0	1.5		0.0 ± 0.0 cd	0.015 ± 0.001 d
1.5	0.0		26.7 ± 1.7 b–d	0.211 ± 0.004 d
1.5	0.5		10.0 ± 2.9 d	0.023 ± 0.002 d
1.5	1.0		10.0 ± 5.0 b–d	0.026 ± 0.003 d
1.5	1.5		26.7 ± 7.3 cd	0.035 ± 0.001 d
2.0	0.0		16.7 ± 3.3 b–d	0.139 ± 0.001 d
2.0	0.5		3.3 ± 1.7 d	0.019 ± 0.001 d
2.0	1.0		3.3 ± 1.7 b–d	0.020 ± 0.004 d
2.0	1.5		46.7 ± 13.0 b	0.047 ± 0.007 d
0.0		0.0	0.0 ± 0.0 d	0.013 ± 0.001 d
0.0		0.5	86.7 ± 3.3 b	0.060 ± 0.003 cd
0.0		1.0	100.0 ± 0.0 b–d	0.026 ± 0.015 d
0.0		1.5	100.0 ± 0.0 b–d	0.021 ± 0.004 d
1.0		0.0	46.7 ± 3.3 a	0.124 ± 0.005 bc
1.0		0.5	100.0 ± 0.0 a	0.221 ± 0.001 b
1.0		1.0	100.0 ± 0.0 a	0.211 ± 0.003 b
1.0		1.5	100.0 ± 0.0 a	0.207 ± 0.006 b
1.5		0.0	33.3 ± 4.4 a	0.220 ± 0.002 b
1.5		0.5	100.0 ± 0.0 a	0.344 ± 0.011 a
1.5		1.0	100.0 ± 0.0 a	0.337 ± 0.003 a
1.5		1.5	100.0 ± 0.0 a	0.392 ± 0.034 a
2.0		0.0	20.0 ± 7.6 a	0.210 ± 0.004 b
2.0		0.5	100.0 ± 0.0 a	0.416 ± 0.019 a
2.0		1.0	100.0 ± 0.0 a	0.380 ± 0.027 a
2.0		1.5	100.0 ± 0.0 a	0.375 ± 0.021 a

[Z] Means followed by the same letter are not significantly different from each other ($p < 0.001$), as determined using Duncan's multiple range test (DMRT).

In media containing 6-BA and 2,4-D, the callus induction (0.851) and fresh weight (0.770) were significantly positively correlated with 2,4-D ($p < 0.01$). Ultimately, in media with 2.0 $mg \cdot L^{-1}$ 6-BA and 0.5 $mg \cdot L^{-1}$ 2,4-D, the maximum callus induction and fresh weight reached 100% and 0.416 ± 0.019 g, respectively. These results highlight the importance of specific PGR combinations in optimizing callus induction and growth in NZS.

3.2. Total Phenolic and Total Flavonoid Concentration of Selected Callus

Following callus induction in the optimized medium, the cultured calli were subsequently cultured in a medium containing TDZ alone or a combination of TDZ and NAA (Figure 2). The two selected callus species were labeled TGC (induced in the medium with TDZ) and TNYC (induced in the medium with TDZ and NAA). Selected calli were used to measure the TPC and TFC to determine the secondary metabolite content. The TPC of TGC was measured at 62.9 g GAE/g DW, showing no significant difference when compared with the leaves of NZS grown in the greenhouse (Figure 3). In contrast, the TNYC calli displayed approximately six-times-lower content than that of the NZS leaves, indicating a significant difference ($p < 0.01$). The total flavonoid content did not differ significantly between the treatment groups. The measured values were 19.9, 10.9, and 18.5 g QE/g DW for the NZS leaves grown in the greenhouse, TGC, and TNYC, respectively.

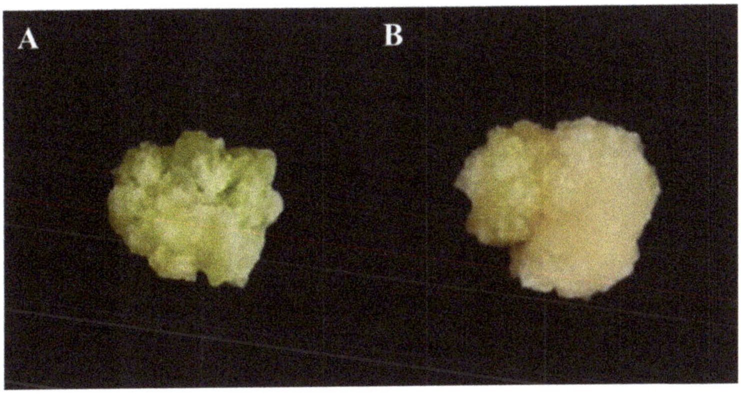

Figure 2. Two types of calli selected after induction from the leaves of NZS. (**A**) Hard tissue and green callus induced in MS medium with 2.5 mg·L^{-1} TDZ (TGC). (**B**) Hard tissue and yellow callus induced in MS medium with 2.5 mg·L^{-1} TDZ and 0.5 mg·L^{-1} NAA (TNYC).

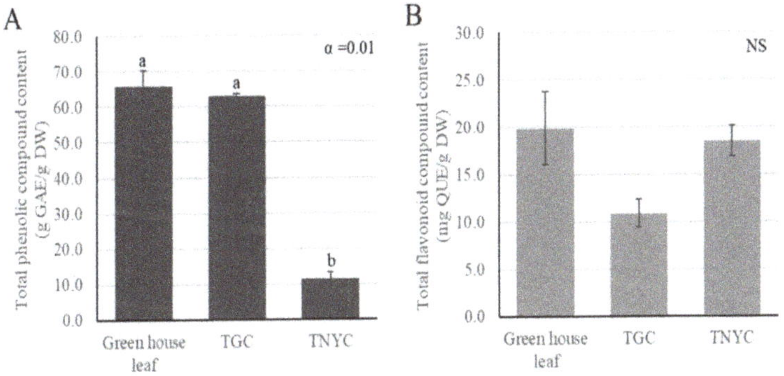

Figure 3. The contents of total phenolics (**A**) and flavonoids (**B**) in the selected two types of calli. Groups sharing the same letter do not exhibit statistically significant differences, as determined by Duncan's test. NS: not significant.

3.3. Comparison of Gene Expressions Related to Flavonoid Biosynthesis

Lee et al. [24] identified the major compounds in NZS as 6-methoxykaempferol-3-O-β-d-glucosyl (1‴→2″)-β-d-glucopyranoside (1), 6-methoxykaempferol-3-O -β-d-glucosyl (1‴→2″)-β-d-glucopyranosyl-(6⁗-caffeoyl)-7-O-β-d-glucopyranoside (2), and 6,4 ′-dime thoxykaempferol-3-O-β-d-glucosyl (1‴→2″)-β-d-glucopyranosyl-(6⁗-caffeoyl)-7-O-β-d-glucopyranoside (3). We examined the expressions of the genes involved in kaempferol biosynthesis to confirm the production of these major compounds in the selected calli. The genes targeted *chalcone synthase* (*CHS*), *chalcone isomerase* (*CHI*), *flavanone 3-hydroxylase* (*F3H*), *flavonoid 3′-hydroxylase* (*F3′H*), *flavonol synthase* (*FLS*), and *dihydroflavonol 4-reductase* (*DFR*), and their expressions were compared. The RT-PCR results demonstrate the elevated production of kaempferol, indicated by low expression levels of *F3′H* and high expression levels of *F3H* and *FLS* in the leaves of NZS (Figure 4). In contrast, the roots exhibited minimal *FLS* expression, implying the predominant production and accumulation of flavonoids. In TGC, all six flavonoid biosynthesis-related genes showed elevated expression levels. TNYC displayed notably low expression levels of *F3′H* and *FLS*. In particular, high DFR expression levels of *DFR* were confirmed in both callus types, suggesting the production of a more diverse array of flavonoids in calli than in plants.

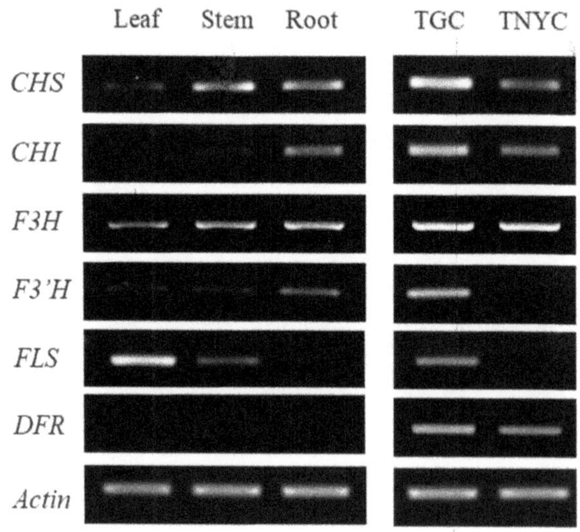

Figure 4. Comparison of the relative expression levels of genes involved in flavonoid synthesis between the seedlings of NZS and selected calli via semi-quantitative RT-PCR analysis. *CHS*: *chalcone synthase*, *CHI*: *chalcone isomerase*, *F3H*: *flavanone 3-hydroxylase*, *F3′H*: *flavonoid 3′-hydroxylase*, *FLS*: *flavonol synthase*, *DFR*: *dihydroflavonol 4-reductase*.

3.4. Analysis of the Phytochemicals of TGC and TNYC

We extracted callus components using 70% ethanol and analyzed the compounds in each extract via GC-MS. Nine components were detected in TGC, whereas TNYC exhibited a higher number of 44 detected components (Figures S2 and S3). There was a substantial difference in the composition of volatile compounds between the two types of calli. In TGC, the predominant components were 2,3-Butanediol (68.6%), cytidine (20.3%), and dimethylamine (5.6%). Conversely (Table 3), TNYC showed a higher abundance of butanedioic acid (23.1%); 3-piperidinol (14.1%); ethylamine, N-ethyl-nitroso (8.2%); and 1,3-propanediol, 2-ethyl-2-(hydroxymethyl)- (6.4%) (Table 4).

Table 3. Classification and content (%) of phytochemicals identified via GC-MS analysis in TGC.

Peak No.	R.T (min) ^{z}	Identified Metabolites	Area (%) ^{y}
1	7.5825	2,3-Butanediol	68.55
2	9.3225	Ethanol	5.544
3	9.3779	Cyclopentasiloxane, decamethyl-	-
4	11.8558	Dimethylamine	5.647
5	17.3404	Cytidine	20.258
6	23.0586	5,6,8,9-TETRAMETHOXY-2-METHYLPEPERO (3,4,5-JK)-9,10-DIHYDROPHENANTHRACENE	-
7	23.3906	Silane, 1,4-phenylenebis[trimethyl-	-
8	23.6858	Silicone grease, Siliconfett	-
9	23.7903	Silicone grease, Siliconfett	-

^{z,y} means the time each substance stays in the column (R.T) and the calculated area of each peak (Area), respectively.

Table 4. Classification and content (%) of phytochemicals identified via GC-MS analysis in TNYC.

Peak No.	R.T (min) [z]	Identified Metabolites	Area (%) [y]
1	4.5327	Butanedioic acid	23.053
2	4.7971	Ethanol	1.855
3	5.4305	Ethanol	-
4	6.156	Ethyl 4-(N-(2-nitro)benzylidene)amino-benzoate	3.528
5	6.2052	Cyclotetrasiloxane, octamethyl-	3.519
6	7.2135	5,6,8,9-TETRAMETHOXY-2-METHYLPEPERO (3,4,5-JK)-9,10-DIHYDROPHENANTHRACENE	1.605
7	7.4165	Propanoic acid, 3-(acetylthio)-2-methyl-	3.473
8	7.6316	Thioacetonitrile	4.078
9	8.7384	N-(Methoxycarbonylmethyl)-N-ethylnitrosamine	4.666
10	8.9659	1-PROPANOL-O-D	2.314
11	9.2549	6-Aza-5,7,12,14-tetrathiapentacene	-
12	9.7714	Cyclopentasiloxane, decamethyl-	1.243
13	10.374	Ammonia	1.726
14	10.4785	6-Aza-5,7,12,14-tetrathiapentacene	1.243
15	10.706	(S)-N-(Ethoxycarbonyl)valine	-
16	11.4869	2,4,5-Trioxoimidazolidine	1.588
17	11.5668	2,3,4-Tridesoxy-5,6:8,9-di-O-isopropyliden-3-methyl-4-nitro-.beta.-D-manno-4-nonulo-4,7-furanosononitrile	5.062
18	11.7635	1,3,5-Triazine, hexahydro-1,3,5-trimethyl-	2.855
19	12.0402	1H-Imidazole-4-carboxamide, 5-amino-	-
20	12.7104	Ethanamine, N-ethyl-N-nitroso-	8.237
21	12.858	L-Alanine, N-glycyl-	2.35
22	13.2515	2,3-Dihydro-3,5-dihydroxy-6-methyl-4H-pyran-4-one	-
23	14.1554	2-Furancarboxaldehyde, 5-(hydroxymethyl)-	2.104
24	15.0654	1-(2-Adamantylidene)semicarbazide	-
25	15.9877	2-Methoxy-4-vinylphenol	1.529
26	16.7193	3-Piperidinol	14.137
27	16.7501	3-Piperidinol	-
28	17.3588	1,3-Propanediol, 2-ethyl-2-(hydroxymethyl)-	6.378
29	17.697	1-Propanol, 3-amino-	-
30	19.6031	E-1-phenylbutetene	-
31	19.7383	E-1-phenylbutetene	-
32	19.7875	3,27-Dioxa-2,28-disilanonacosane, 2,2,4,28,28-pentamethyl-	0.854
33	19.9597	Phenol, 2,2′-[(1-methyl-1,2-ethanediyl)bis(nitrilomethylidyne)]bis-	0.854
34	21.1095	Urea	-
35	21.4784	4-HYDROXY-3-NITROCOUMARIN	0.938
36	21.6628	4-HYDROXY-3-NITROCOUMARIN	-
37	21.8166	Phenol, 2,2′-[(1-methyl-1,2-ethanediyl)bis(nitrilomethylidyne)]bis-	-
38	22.911	5,6,8,9-TETRAMETHOXY-2-METHYLPEPERO(3,4,5-JK)-9,10-DIHYDROPHENANTHRACENE	-
39	23.1324	5,6,8,9-TETRAMETHOXY-2-METHYLPEPERO(3,4,5-JK)-9,10-DIHYDROPHENANTHRACENE	-

Table 4. *Cont.*

Peak No.	R.T (min) [z]	Identified Metabolites	Area (%) [y]
40	23.2246	5,6,8,9-TETRAMETHOXY-2-METHYLPEPERO(3,4,5-JK)-9,10-DIHYDROPHENANTHRACENE	-
41	23.4275	5,6,8,9-TETRAMETHOXY-2-METHYLPEPERO(3,4,5-JK)-9,10-DIHYDROPHENANTHRACENE	-
42	23.5751	1,1,1,3,5,5,5-Heptamethyltrisiloxane	-
43	23.6058	1,3-Bis(trimethylsilyl)benzene	0.854
44	23.7657	Silicone grease, Siliconfett	0.854

[z,y] means the time each substance stays in the column (R.T) and the calculated area of each peak (Area), respectively.

4. Discussion

In this study, we introduced callus culture techniques to identify the functional compounds in NZS. To obtain abundant compounds from calli within a short period, it is essential to establish effective callus induction conditions and select fast-growing calli. PGRs play crucial roles in plant cell growth under laboratory conditions. Therefore, in this study, we investigated the effect of PGRs on callus induction from NZS leaves and their subsequent effects on callus growth by examining induction rates and biomasses.

The PGR used in in vitro cultures generally depends on the type of explant and is strongly correlated with the amount of endogenous PGRs [25]. Auxins containing NAA and 2,4-D increase the vacuole size, thereby promoting the formation of smooth and friability calli [26,27]. In particular, 2,4-D is effective at low concentrations and exhibits high stability even at high temperatures, making it widely applicable for callus induction in various plant species [28,29]. Rad et al. [30] reported that the medium containing 0.5 mg·L^{-1} 2,4-D and 2 mg·L^{-1} 6-BA exhibited the most effective induction of calli in *Digitalis purpurea* L. These results suggest that the callus induction rate tends to increase with increasing concentrations of 2,4-D in the medium. We found that NZS callus induction increased effectively when 2,4-D was added compared to that in controls without 2,4-D. We observed a significant increase in the NZS callus induction when 2,4-D was added compared to the control conditions without 2,4-D. Additionally, a maximal callus induction was observed in the medium supplemented with 2.0 mg·L^{-1} 6-BA and 0.5 mg·L^{-1} 2,4-D. Consequently, our results demonstrate a trend similar to that observed in previous studies.

Furthermore, we selected calli with good growth by culturing them on media with TDZ alone or a combination of TDZ and NAA. TDZ induces callus formation and facilitates somatic embryo development. It plays a crucial role in the initiation and proliferation of calli, shoots, and roots [31–33]. However, it is noteworthy that increasing the concentration of TDZ may promote the growth of calli and plants but may result in browning. A careful selection of the TDZ concentration is essential because it can stimulate endogenous ethylene production [34,35]. Singh et al. [36] achieved optimal callus growth with 1.2 mg·L^{-1} TDZ alone in *Santalum album* L. Srivastava et al. [37] induced up to 85.56% callus using 9.1 μM of TDZ and 0.5 μM of NAA. In our study, we successfully obtained calli with 2.5 mg·L^{-1} TDZ, which were non- browning and had excellent growth.

Ultimately, we selected two distinct callus types based on the differences in color and shape. In general, a friable callus is required for cell differentiation cultures. Friable calli are easy to separate and facilitate cell proliferation and are a necessary factor for mass production. We did not select friable calli in the present study, but we additionally compared the growth of calli under different medium conditions. For TNYC, placing 1 g of calli in 50 mL liquid medium resulted in approximately eight- and two-fold increases in fresh and dry weights after five weeks, respectively. In addition, when 0, 1, 3, and 5% sucrose was added to the liquid medium, the fresh weight increased eight folds without a change in the browning in the 5% sucrose-added group (data not shown). Through the results of these further studies, we were able to assess the applicability of TNYC to liquid cultures and obtain supplemental information about growth conditions.

The TPC and TFC of the callus extract were measured to assess the functional potential of the calli. Numerous studies have reported on the antioxidant activity and effectiveness of callus extracts. However, the contents and activities of functional compounds differ depending on the culture conditions and methods used for plant callus culture. In this study, the TPC of the TGC was measured to be 62.9 g GAE/g DW. The TNYC showed 11.3 g GAE/g DW, which was lower than that of the TDZ treatment group. Upon comparing TPC and TFC after callus induction with various PGR combinations in *Lallemantia Iberica*, the TDZ single treatment group exhibited the highest total phenol (3.93 ± 0.018 mgPYRO/g) and flavonoid (51.27 ± 0.036 mgQUE/g) contents. Mok et al. [38] reported that TDZ inhibits nitrate transport through the stimulation of endogenous cytokinin biosynthesis or through its involvement in metabolism. Nitrate inhibition stimulates secondary metabolites [39]. Considering previous studies, we expected that the biosynthesis of secondary metabolites would be promoted in calli induced using a single TDZ treatment. Moreover, auxins and cytokinins may interact to exhibit synergistic and antagonistic effects, and thus have lower content in mixed treatments compared to single treatments [40]. Hence, variations in TPC and TFC among the selected calli may be attributed to this phenomenon.

In our previous experiments, we observed no statistically significant differences in TPC and TFC between leaves and calli. Consequently, we investigated whether the main compounds in the kaempferol backbone are also biosynthesized in calli. To achieve this, we compared and examined the expressions of six genes involved in the kaempferol and quercetin biosynthesis in callus and plant tissues. The Kaempferol and quercetin productions require dihydrokaempferol and dihydroquercetin, respectively. Furthermore, dihydrokaempferol acts as a substrate for *F3′H* to produce dihydroquercetin. Finally, *FLS* converts dihydrokaempferol into kaempferol, and dihydroquercetin into quercetin [41]. Therefore, the production of kaempferol or quercetin was expected to depend on the expression ratios of *F3′H* and *FLS*. According to the experimental results, there was a clear difference in the gene expression patterns between the two callus types. TGC exhibited high expressions for all six genes, whereas TNYC had low expressions for *F3′H* and *FLS*. Consequently, the two callus types were expected to produce different metabolites.

Plant color results from the accumulation of metabolites (pigments) [42]. In particular, it has been reported that yellow color occurs because of differences in flavonoid or carotenoid content [43]. Therefore, we expected TNYC (yellow) to have a higher flavonoid content than TGC (green) during callus screening. However, the total flavonoid content did not differ significantly between the groups (Figure 3). Therefore, we compared the expressions of flavonoid biosynthesis-related genes. The gene expression patterns of TNYC were high in the expressions of *F3H* and *DFR* and low in the expressions of *F3′H* and *FLS* (Figure 4). Therefore, the accumulation of dihydroflavonol was expected to be higher in the TNYC than in the other groups. *F3H* is also an important enzyme involved in anthocyanin and dihydroflavonol [44]. Thus, the high expression of *DFR*, which regulates anthocyanin accumulation, together with the high expression of *F3H*, indicates the potential for anthocyanin synthesis in TNYC. Therefore, further studies on the anthocyanin content are required. For TGC, the strong expression of *CHS* and the high expressions of the five genes indicate the rapid progression of the synthesis and conversion of various flavonoids. As there was no difference in the flavonoid content of greenhouse-grown NZS and TGC, it is expected that various flavonoids will be synthesized in TGC.

Finally, we performed a GC-MS analysis of the secondary metabolite profiles of the two calli. As a result, 9 and 44 components were detected in TGC and TNYC, respectively. None of these components matched the main compounds in NZS. Differences between the secondary metabolite profiles of in vitro and native plants have been revealed in previous reports. In addition, when cultured in vitro, the types and contents of the compounds differed depending on the growth conditions. Locally cultivated *Mentha arvensis* L. produces essential oils composed of 30 or more components, whereas in vitro-cultivated plant compounds differ depending on culture conditions. Pulegone and isopulegone have been isolated from calli [45]. This pattern has also been observed in *Lallemantia iverica*. When explants were treated with six different concentrations of TDZ, 55 components were detected and distributed variably among each treatment group [35]. In *Justicia gendarussa* [46], PGRs exert a significant influence on the production and accumulation of secondary metabolites as well as the growth and development of plants [47]. In the present study, we identified a novel compound, NZS, in the two types of calli selected under different growth conditions.

Among the compounds detected, 2,3-Butanediol (2,3-BD), which is the most abundant in TGC and is used in various chemical synthesis processes, has substantial industrial value in the pharmaceutical, cosmetic, perfume, wetting agent, and food industries [48–50]. Additionally, 2,3-BD promotes plant growth and enhances resistance to disease and drought stress by strengthening cell walls [51,52]. Traditionally, 2,3-BD has been produced through chemical processes using petroleum; however, biological processes using microorganisms have recently attracted attention because of their low production costs and minimal waste. In the TGC selected for this study, a high 2,3-Butanediol (2,3-BD) content of 68.6% was detected, suggesting the potential utility of this callus as a new production material. The second most abundant compound in the TGC was cytidine, constituting 20.3% of the composition. Forced swimming experiments, conducted to evaluate the antidepressant effects of cytidine, demonstrated a strong antidepressant effect [53]. Additionally, cytidine stimulates the synthesis of cytidine 5′-diphosphocholine(citicoline). Citicoline has been shown to enhance the production of membrane phospholipids in both cellular and animal models, and it has demonstrated a capacity to reduce cocaine withdrawal symptoms in mice [54,55].

The most prominent compound detected in the TNYC was butanedioic acid (23.1%), also known as succinic acid. Succinic acid is widely used in various industries, including food additives, detergents, spices, cement additives, and antimicrobials. In particular, the various pharmacological effects of succinic acid, such as anti-inflammatory and antibacterial effects, have been reported in many studies [56–58]. The demand for succinic acid is steadily increasing and is projected to reach 768 million metric tons by 2025 [59]. Traditionally, succinic acid has been produced through chemical pathways; however, in recent years, there has been an emphasis on biological production methods because of concerns about nonrenewable resources and environmental issues stemming from greenhouse gas emissions. Consequently, the TNYC selected in this study has potential as material capable of producing succinic acid, a crucial industrial chemical.

5. Conclusions

In this study, we utilized callus culture technology to detect and analyze the functional compounds in NZS. The initial optimization of induction conditions with PGRs, particularly the addition of 2.0 mg·L^{-1} 6-BA and 0.5 mg·L^{-1} 2,4-D, resulted in a maximum callus induction rate (100%) and biomass (0.416 $\pm$ 0.019 g). Subsequently, two distinct callus types, TGC and TNYC, were selected based on their color and shape variations. The TPC and TFC contents of the selected calli showed no statistically significant differences when compared to leaves grown in greenhouses. Furthermore, the gene involved in the biosynthesis of kaempferol, the backbone of NZS's main compound of NZS, displayed different expression patterns in the two callus types, suggesting the potential synthesis of flavonoids distinct from those in NZS seedlings.

A GC-MS analysis revealed that 2,3-butanediol (2,3-BD) and succinic acid were the most abundant compounds in the TGC and TNYC, respectively. The current global chemical industry is pursuing biologically based production processes to address the impact of climate change. The calli selected in this study not only provide an effective method for the stable supply of functional compounds but also offer a sustainable solution without causing environmental issues. Moreover, calli containing diverse compounds are expected to serve as high-value raw materials in the industrial sector.

Supplementary Materials: The following supporting information can be downloaded from https://www.mdpi.com/article/10.3390/horticulturae10030244/s1: Figure S1: Appearance of calli induced from NZS leaves after four weeks in the medium; Figure S2: Phytochemical profile of TGC extracts via gas chromatography–mass spectrometry; Figure S3: Phytochemical profile of TNYC extracts via gas chromatography–mass spectrometry; Table S1: Primers used in this study.

Author Contributions: Conceptualization, G.-J.L.; investigation, K.Y.L., D.-H.N., Y.J. and M.A.G.; data curation, K.Y.L., D.-H.N., Y.J. and M.A.G.; formal analysis, K.Y.L., D.-H.N. and Y.J.; funding acquisition, S.U.P., J.C., S.-O.C. and G.-J.L.; methodology, K.Y.L. and G.-J.L.; project administration, G.-J.L.; writing—original draft, K.Y.L., D.-H.N., Y.J. and M.A.G.; writing—review and editing, S.U.P., J.C., S.-O.C. and G.-J.L. All authors have read and agreed to the published version of the manuscript.

Funding: This work was supported by the National Research Foundation of Korea Grant 2021, funded by the Korean Government (MOE).

Data Availability Statement: Data are contained within the article or Supplementary Material.

Conflicts of Interest: The authors declare no conflicts of interest regarding the publication of this article.

References

1. Karakas, F.P.; Turker, A.U. Improvement of shoot proliferation and comparison of secondary metabolites in shoot and callus cultures of *Phlomis armeniaca* by LC-ESI-MS/MS analysis. *In Vitro Cell. Dev. Biol.-Plant.* **2016**, *52*, 608–618. [CrossRef]
2. Bennett, R.N.; Wallsgrove, R.M. Secondary metabolites in plant defense mechanisms. *New Phytol.* **1994**, *127*, 617–633. [CrossRef] [PubMed]
3. Wink, M. Plant secondary metabolism: Diversity, function and its evolution. *Nat. Prod. Commun.* **2008**, *3*, 1934578X0800300801. [CrossRef]
4. Van Loon, L.C.; Rep, M.; Pieterse, C.M. Significance of inducible defense-related proteins in infected plants. *Annu. Rev. Phytopathol.* **2006**, *44*, 135–162. [CrossRef] [PubMed]
5. Balandrin, M.F.; Klocke, J.A.; Wurtele, E.S.; Bollinger, W.H. Natural plant chemicals: Sources of industrial and medicinal materials. *Science* **1985**, *228*, 1154–1160. [CrossRef] [PubMed]
6. Ashraf, M.A.; Iqbal, M.; Rasheed, R.; Hussain, I.; Riaz, M.; Arif, M.S. Environmental stress and secondary metabolites in plants: An overview. In *Plant Metabolites and Regulation under Environmental Stress*; Academic Press: Cambridge, MA, USA, 2018; pp. 153–167.
7. Yue, W.; Ming, Q.L.; Lin, B.; Rahman, K.; Zheng, C.J.; Han, T.; Qin, L.P. Medicinal plant cell suspension cultures: Pharmaceutical applications and high-yielding strategies for the desired secondary metabolites. *Crit. Rev. Biotechnol.* **2016**, *36*, 215–232. [CrossRef]
8. Steward, F.C.; Mapes, M.O.; Mears, K. Growth and organized development of cultured cells. II. Organization in cultures grown from freely suspended cells. *Am. J. Bot.* **1958**, *45*, 705–708. [CrossRef]
9. Plasson, C.; Michel, R.; Lienard, D.; Saint-Jore-Dupas, C.; Sourrouille, C.; March, G.G.D.; Gomord, V. Production of recombinant proteins in suspension–cultured plant cells. In *Recombinant Proteins from Plants: Methods and Protocols*; Humana: New York, NY, USA, 2009; pp. 145–161.
10. Lee, K.Y.; Shin, J.Y.; Ahn, M.S.; Kim, S.J.; An, H.R.; Kim, Y.J.; Kwon, O.H.; Lee, S.Y. Callus Derived from Petals of the *Rosa hybrida* Breeding Line 15R-12-2 as New Material Useful for Fragrance Production. *Plants* **2023**, *12*, 2986. [CrossRef]
11. Dörnenburg, H.; Knorr, D. Strategies for the improvement of secondary metabolite production in plant cell cultures. *Enzyme Microb. Technol.* **1995**, *17*, 674–684. [CrossRef]
12. Anjum, S.; Abbasi, B.H.; Hano, C. Trends in accumulation of pharmacologically important antioxidant-secondary metabolites in callus cultures of *Linum usitatissimum* L. *Plant Cell Tissue Organ Cult.* **2017**, *129*, 73–87. [CrossRef]
13. Adil, M.; Ren, X.; Kang, D.I.; Thi, L.T.; Jeong, B.R. Effect of explant type and plant growth regulators on callus induction, growth and secondary metabolites production in *Cnidium officinale* Makino. *Mol. Biol. Rep.* **2018**, *45*, 1919–1927. [CrossRef]
14. Coskun, Y.; Duran, R.E.; Kılıc, S. Striking effects of melatonin on secondary metabolites produced by callus culture of rosemary (*Rosmarinus officinalis* L.). *Plant Cell Tissue Organ Cult.* **2019**, *138*, 89–95. [CrossRef]
15. Wilson, C.; Lesch, S.M.; Grieve, C.M. Growth stage modulates salinity tolerance of New Zealand Spinach (*Tetragonia tetragonioides*, Pall) and Red Orach (*Atriplex hortensis* L.). *Ann. Bot.* **2000**, *85*, 501–509. [CrossRef]
16. Kato, M.; Takeda, T.; Ogihara, Y.; Shimu, M.; Nomura, T.; Tomita, Y. Studies on the structure of polysaccharide from *Tetragonia tetragonioides*. *Chem. Pharm. Bull.* **1985**, *33*, 3675–3680. [CrossRef]
17. Choi, H.J.; Kang, J.S.; Choi, Y.W.; Jeong, Y.K.; Joo, W.H. Inhibitory activity on the diabetes related enzymes of *Tetragonia tetragonioides*. *KSBB J.* **2008**, *23*, 419–424.
18. Lee, M.A.; Choi, H.J.; Kang, J.S.; Choi, Y.W.; Joo, W.H. Antioxidant activities of the solvent extracts from *Tetragonia tetragonioides*. *J. Life Sci.* **2008**, *18*, 220–227. [CrossRef]
19. Choi, H.J.; Yee, S.T.; Kwon, G.S.; Joo, W.H. Anti-inflammatory and anti-tumor effects of *Tetragonia tetragonoides* extracts. *Microbiol. Biotechnol. Lett.* **2015**, *43*, 391–395. [CrossRef]

20. Choi, H.J.; Yee, S.T.; Joo, W.H. Antidiabetic activity of polysaccharide extract from *Tetragonia tetragonoides* in streptozotocin-induced diabetic mice. *J. Life Sci.* **2017**, *27*, 579–583. [CrossRef]
21. Kim, D.S.; Ko, B.S.; Ryuk, J.A.; Park, S. *Tetragonia tetragonioides* protected against memory dysfunction by elevating hippocampal amyloid-β deposition through potentiating insulin signaling and altering gut microbiome composition. *Int. J. Mol. Sci.* **2020**, *21*, 2900. [CrossRef]
22. Sunagawa, H.; Agarie, S.; Umemoto, M.; Makishi, Y.; Nose, A. Effect of urea-type cytokinins on the adventitious shoots plant, *Mesembryanthemum crystallinum*. *Plant Prod. Sci.* **2007**, *10*, 47–56. [CrossRef]
23. Kumazawa, S.; Hamasaka, T.; Nakayama, T. Antioxidant activity of propolis of various geographic origins. *Food Chem.* **2004**, *84*, 329–339. [CrossRef]
24. Lee, Y.S.; Kim, S.H.; Yuk, H.J.; Lee, G.J.; Kim, D.S. *Tetragonia tetragonoides* (Pall.) Kuntze (New Zealand Spinach) prevents obesity and hyperuricemia in high-fat diet-induced obese mice. *Nutrients* **2018**, *10*, 1087. [CrossRef] [PubMed]
25. Centeno, M.L.; Rodríguez, A.; Feito, I.; Fernández, B. Relationship between endogenous auxin and cytokinin levels and morphogenic responses in *Actinidia deliciosa* tissue cultures. *Plant Cell Rep.* **1996**, *16*, 58–62. [CrossRef] [PubMed]
26. Gueye, B.; Morcillo, F.; Collin, M.; Gargani, D.; Overvoorde, P.; Aberlenc-Bertossi, F.; Tranbarger, T.; Sane, D.; Tregear, J.W.; Borgel, A.; et al. Acquisition of callogenic capacity in date palm leaf tissues in response to 2,4-D treatment. *Plant Cell Tissue Organ Cult.* **2009**, *99*, 35–45. [CrossRef]
27. Berejsza-Wysecki, W.; Hrazdin, G. Establishment of callus and cell suspension cultures of raspberry (*Rubus idaeus* cv. Royalty). *Plant Cell Tissue Organ Cult.* **1994**, *37*, 213–216. [CrossRef]
28. Gaj, M.D. Factors influencing somatic embryogenesis induction and plant regeneration with particular reference to *Arabidopsis thaliana* (L.) Heynh. *Plant Growth Regul.* **2004**, *43*, 27–47. [CrossRef]
29. George, E.F.; Hall, M.A.; De Klerk, G.J. *Plant Propagation by Tissue Culture*, 3rd ed.; Springer: Dordrecht, The Netherlands, 2008; 501p.
30. Rad, M.M.; Abdossi, V.; Moradi, P.; Rakhshandehroo, F.; Mehrafarin, A. Phytochemical changes of *Digitalis purpurea* L. in response to polyamines and methyl jasmonate application in callus culture. *J. Plant Biochem. Biotechnol.* **2022**, *31*, 310–319. [CrossRef]
31. Linden, L.; Riikonen, A. Effects of 6-benzylaminopurine, thidiazuron and type of explant on in vitro shoot development of *Acer platanoides* L. *Propag. Ornam. Plants* **2006**, *6*, 201–204.
32. Jones, M.P.; Yi, Z.; Murch, S.J.; Saxena, P.K. Thidiazuron-induced regeneration of *Echinacea purpurea* L.: Micropropagation in solid and liquid culture systems. *Plant Cell Rep.* **2007**, *26*, 13–19. [CrossRef]
33. Wang, J.; Bao, M.Z. Plant regeneration of pansy (*Viola wittrockiana*) 'Caidie' via petiole-derived callus. *Sci. Hortic.* **2007**, *111*, 266–270. [CrossRef]
34. Suttle, J.C. Involvement of ethylene in the action of the cotton defoliant thidiazuron. *Plant Physiol.* **1985**, *78*, 272–276. [CrossRef]
35. Pourebad, N.; Motafakkerazad, R.; Kosari-Nasab, M.; Farsad Akhtar, N.; Movafeghi, A. The influence of TDZ concentrations on in vitro growth and production of secondary metabolites by the shoot and callus culture of *Lallemantia iberica*. *Plant Cell Tissue Organ Cult.* **2015**, *122*, 331–339. [CrossRef]
36. Singh, C.K.; Raj, S.R.; Jaiswal, P.S.; Patil, V.R.; Punwar, B.S.; Chavda, J.C.; Subhash, N. Effect of plant growth regulators on in vitro plant regeneration of sandalwood (*Santalum album* L.) via organogenesis. *Agrofor. Syst.* **2016**, *90*, 281–288. [CrossRef]
37. Srivastava, S.; Krishna, R.; Sinha, R.P.; Singh, M. TDZ-induced plant regeneration in *Brassica oleracea* L. var. *botrytis*: Effect of antioxidative enzyme activity and genetic stability in regenerated plantlets. *In Vitro Cell. Dev. Biol.-Plant.* **2017**, *53*, 598–605.
38. Mok, M.C.; Martin, R.C.; Mok, D.W.S. Cytokinins: Biosynthesis metabolism and perception. *In Vitro Cell. Dev. Biol.-Plant.* **2000**, *36*, 102–107. [CrossRef]
39. Sakakibara, H. Cytokinins: Activity, biosynthesis, and translocation. *Annu. Rev. Plant Biol.* **2006**, *57*, 431–449. [CrossRef] [PubMed]
40. Coenen, C.; Lomax, T.L. Auxin—Cytokinin interactions in higher plants: Old problems and new tools. *Trends Plant Sci.* **1997**, *2*, 351–356. [CrossRef] [PubMed]
41. Crozier, A.; Jaganath, I.B.; Clifford, M.N. Dietary phenolics: Chemistry, bioavailability and effects on health. *Nat. Prod. Rep.* **2009**, *26*, 1001–1043. [CrossRef] [PubMed]
42. Mol, J.; Grotewold, E.; Koes, R. How genes paint flowers and seeds. *Trends Plant Sci.* **1998**, *3*, 212–217. [CrossRef]
43. Deng, J.; Chen, S.; Yin, X.; Wang, K.; Liu, Y.; Li, S.; Yang, P. Systematic qualitative and quantitative assessment of anthocyanins, flavones and flavonols in the petals of 108 lotus (*Nelumbo nucifera*) cultivars. *Food Chem.* **2013**, *139*, 307–312. [CrossRef] [PubMed]
44. Cheynier, V.; Comtem, G.; Davies, K.M.; Lattanzio, V.; Martens, S. Plant phenolics: Recent advances on their biosynthesis, genetics, and ecophysiology. *Plant Physiol. Biochem.* **2013**, *72*, 1–20. [CrossRef]
45. Phatak, S.V.; Heble, M.R. Organogenesis and terpenoid synthesis in *Mentha arvensis*. *Fitoterapia* **2002**, *73*, 32–39. [CrossRef]
46. Wahyuni, D.K.; Huda, A.; Faizah, S.; Purnobasuki, H.; Wardojo, B.P.E. Effects of light, sucrose concentration and repetitive subculture on callus growth and medically important production in *Justicia gendarussa* Burm. f. *Plant Biotechnol. Rep.* **2020**, *27*, e00473.
47. Richard, C.; Lescot, M.; Inzé, D.; De Veylder, L. Effect of auxin, cytokinin, and sucrose on cell cycle gene expression in *Arabidopsis thaliana* cell suspension cultures. *Plant Cell Tissue Organ Cult.* **2002**, *69*, 167–176. [CrossRef]
48. Garg, S.K.; Jain, A. Fermentative production of 2,3-butanediol: A review. *Bioresour. Technol.* **1995**, *51*, 103–109. [CrossRef]
49. Celińska, E.; Grajek, W. Biotechnological production of 2,3-butanediol—Current state and prospects. *Biotechnol. Adv.* **2009**, *27*, 715–725. [CrossRef] [PubMed]

50. Syu, M.J. Biological production of 2,3-butanediol. *Appl. Microbiol. Biotechnol.* **2001**, *55*, 10–18. [CrossRef] [PubMed]
51. Liu, X.J.; Ma, Y.; Shi, Y.; Ma, H.L. 2,3-Butanediol induces brown blotch resistance in creeping bentgrass by strengthening cell wall structure and promoting lignin synthesis of precursor phenolic acid. *Acta Physiol. Plant.* **2023**, *45*, 40. [CrossRef]
52. Yasmin, H.; Rashid, U.; Hassan, M.N.; Nosheen, A.; Naz, R.; Ilyas, N.; Sajjad, M.; Azmat, A.; Alyemeni, M.N. Volatile organic compounds produced by *Pseudomonas pseudoalcaligenes* alleviated drought stress by modulating defense system in maize (*Zea mays* L.). *Physiol. Plant.* **2021**, *172*, 896–911. [CrossRef] [PubMed]
53. Carlezon, W.A., Jr.; Pliakas, A.M.; Parow, A.M.; Detke, M.J.; Cohen, B.M.; Renshaw, P.F. Antidepressant-like effects of cytidine in the forced swim test in rats. *Biol. Psychiatry* **2002**, *51*, 882–889. [CrossRef] [PubMed]
54. Knapp, S.; Wurtman, R.J. Enhancement of free fatty acid incorporation into phospholipids by choline plus cytidine. *Brain Res.* **1999**, *822*, 52–59. [CrossRef] [PubMed]
55. Markou, A.; Hauger, R.L.; Koob, G.F. Desmethylimipramine attenuates cocaine withdrawal in rats. *Psychopharmacology* **1992**, *109*, 305–314. [CrossRef] [PubMed]
56. Radkowski, M.; Zdrodowska, B.; Gomółka-Pawlicka, M. Effect of succinic acid on elimination of *Salmonella* in chicken meat. *J. Food Prot.* **2018**, *81*, 1491–1495. [CrossRef] [PubMed]
57. Nissen, M.D.; Lau, E.T.; Cabot, P.J.; Steadman, K.J. Baltic amber teething necklaces: Could succinic acid leaching from beads provide anti-inflammatory effects? *BMC Complement. Altern. Med.* **2019**, *19*, 162. [CrossRef]
58. Harber, K.J.; de Goede, K.E.; Verberk, S.G.; Meinster, E.; de Vries, H.E.; van Weeghel, M.; de Winther, M.P.; Van den Bossche, J. Succinate is an inflammation-induced immunoregulatory metabolite in macrophages. *Metabolites* **2020**, *10*, 372. [CrossRef]
59. Global Succinic Acid Market Analysis & Trends [2013–2017]—Industry Forecast to 2025: $1.76 Billion Growth Opportunities/Investment Opportunities—Research and Markets. Available online: https://ceo.ca/@newswire/global-succinic-acid-market-analysis-trends-2013-2017 (accessed on 6 March 2017).

horticulturae

MDPI

Article

In Vitro Propagation and Genetic Stability Assessment Using the ISSR Markers of *Stachys byzantina* K. Koch, a Promising Ornamental Species

Stefanos Hatzilazarou [1], Chara Kantere [1], Aikaterini-Angeliki Kotoula [1], Athanasios Economou [1], Konstantinos Bertsouklis [2], Anastasios Darras [3] and Stefanos Kostas [1,*]

[1] Laboratory of Floriculture, School of Agriculture, Aristotle University, 54124 Thessaloniki, Greece; hatzilaz@agro.auth.gr (S.H.); charakantere@outlook.com (C.K.); aikakoto@agro.auth.gr (A.-A.K.); aseconom@agro.auth.gr (A.E.)

[2] Laboratory of Floriculture and Landscape Architecture, School of Plant Sciences, Agricultural University of Athens, 11855 Athens, Greece; kber@aua.gr

[3] Laboratory of Floriculture and Landscape Architecture, Department of Agriculture, University of Peloponnese, 24100 Kalamata, Greece; a.darras@uop.gr

* Correspondence: skostas@agro.auth.gr

Abstract: In this study, a reliable and efficient micropropagation protocol was developed for *Stachys byzantina*, a valuable and promising ornamental species. For the initial in vitro cultures on the Murashige and Skoog (MS) medium, shoot tips were used as explants. The addition of 5 μM of kinetin (KIN) resulted in the production of multiple (6.0 shoots/explant) and elongated (3.6 cm) shoots. The MS medium supplemented with 10 μM of a-Naphthaleneacetic acid (NAA) proved efficient for the in vitro rooting (73.3%) of the microshoots. For the ex vitro rooting of the microshoots, the treatment with 0.5 g L^{-1} of Indole-3-butyric acid potassium salt (K-IBA), before planting in 1:1 (v/v) peat and perlite substrate and placed in a fog system, led to 86.7% rooting. The acclimatization stage was successful, and 96.7% survival was recorded for the ex vitro-rooted plantlets. Inter Simple Sequence Repeat (ISSR) markers were employed to examine the genetic uniformity of the in vitro-derived plantlets with the mother *S. byzantina* plants. The monomorphic banding pattern in the micropropagated plants and the mother plant confirmed the genetic uniformity of the in vitro-derived plantlets and revealed the reliability of the proposed in vitro protocol for *S. byzantina*. As far as we know, this is the first study on a combined micropropagation and genetic uniformity assessment of the species, the findings of which could be further used to apply new in vitro cultivation techniques or to produce elite genotypes of *S. byzantina*.

Keywords: acclimatization; genetic uniformity; IBA; NAA; new ornamental crop

1. Introduction

Modern ornamental horticulture supports the need to explore and introduce new plant species, adapted to climate change. However, the selection of suitable species is a challenge due to the diversity of plants and their varied morphological traits. Plants with specific ornamental and morphological characteristics, such as colorful foliage or naturally compact growth, can be effectively used in the landscape architecture and floriculture industry, either as potted or ground cover plants [1].

Stachys byzantina K. Koch (syn. *S. lanata*) is an evergreen, herbaceous perennial species of the cosmopolitan genus *Stachys* (f. Lamiaceae). The plant is one of approximately

54 species and subspecies of *Stachys*, which can be found in Greece [2]. It is also called lamb's ear, because of the curved shape of its leaves and its white, soft, fur-like hair coating. *S. byzantina* is native to Armenia, Iran, and Mediterranean regions, and is found in semi-arid grasslands and phrygana ecosystems [3]. The plant blooms in late spring to early summer, forming spike-like stems, 50–60 cm in height, bearing small, pink–purple/lavender flowers [4]. Various *Stachys* species are considered of great ornamental value, due to their grey-colored foliage, and are recommended for use in Botanical Gardens [5]. The plant can be used as a unique species in urban landscapes as well, due to its extraordinary foliage, which makes it suitable for green roof establishment, contributing to storm water management [6]. Additionally, according to the literature, *S. byzantina* has anti-inflammatory, antitumor, and anti-cancer properties [3,7,8]. It contains valuable chemical constituents, such as fatty acids, alkaloids, and other bioactive compounds [9].

S. byzantina is characterized by bushy forms, with numerous shoots, and its closely spaced internodes, which make rooting and propagation challenging [10]. In vitro propagation methods could reinforce the development of optimized in vitro protocols for the exploitation of new biotypes, in large quantities [11–15]. The main advantages of micropropagation are that it enables the mass production of better new plants, which are free from pests and pathogens, in less time and space, and all year round [16]. It is estimated that the global annual in vitro production of ornamental plants has increased from EUR 800 million to EUR 2 billion over the past decade [17]. A recent study states that the micropropagation market size will reach EUR 2.5 billion by 2030 [15]. Europe plays a significant role in the global micropropagation market, with approximately 140 commercial in vitro production laboratories [17].

There is a number of studies on the in vitro propagation of some *Stachys* species [18–21]. However, to the best of our knowledge, there is no specific protocol for the in vitro propagation of *S. byzantina*. Legkobit and Khadeeva reported preliminary data on the in vitro culture of *S. byzantina*, using B5 Gamborg's nutrient medium, with 1 to 20 mg/L of Benzylaminopurine, but limited data for the explant growth was provided and no information on the acclimatization performance was presented [10].

Stressful conditions during in vitro culture may lead to somaclonal variation in new plants with or without ornamental value [22,23]. Thus, it is crucial to assess the genetic uniformity of in vitro-derived individuals and the mother plants and evaluate the cultural potential of these new ornamentals for commercial utilization [24]. Various molecular markers have been used to assess genetic variation, such as the ISSR (Inter Simple Sequence Repeat). The technique is cost efficient, requiring small amounts of DNA [25]. Moreover, ISSRs have been used successfully in regard to a number of species, such as *Ardisia crenata, Ficus religiosa, Lycium chinense, Nepeta asterotricha, Prunus salicina, Sideritis gulendamii, Solanum khasianum, and Gardenia jasminoides* [26–33].

The present research aims to investigate the potential for in vitro establishment, multiple shoot production, and in vitro and ex vitro rooting of *S. byzantina* microshoots, as well as the acclimatization of rooted plantlets. Additionally, the genetic uniformity of the produced plantlets was assessed in comparison to the mother plants. Therefore, the goal of this research was to develop an efficient and reliable protocol for the quick and large-scale production of healthy, standardized propagating material. This approach could support the broader use of *S. byzantina* in ornamental horticulture and in landscape architecture.

2. Materials and Methods

2.1. Plant Material and Explant Disinfestation

Shoot tips (1–1.5 cm long, 3–4 nodes each), excised from one-year-old *S. byzantina* plants, were used as explants. The mother plants were grown in 5 L pots, containing a substrate of 3:1 (v/v) peat (TS2, Klasmann-Deilmann, Geeste, Germany) and perlite (Isocon, Athens, Greece), and maintained in a glass greenhouse as part of the Floriculture Laboratory at Aristotle University, Thessaloniki. For explant disinfestation, they were first washed under running tap water for 20 min. Then, the explants were transferred to a thread-flow bank and dipped in a dilute solution of 0.08% $HgCl_2$ for 5 min, where the immersion and simultaneous agitation of the explants was performed. This step was followed by two rinses with sterilized–distilled water, 4 min each. Finally, a 1% NaOCl solution was used, followed by three rinses with sterilized–distilled water, 4 min each. The above process was carried out in spring and summer and 200 explants were used during each period.

2.2. Establishment of In Vitro Cultures

The initial culture of the explants was carried out in MS nutrient medium [34], supplemented with 2% (w/v) sucrose and 0.8% (w/v) agar (Technobiochem, Athens, Greece). Glass test tubes (10 × 3 cm, 10 mL medium each) were used for the cultures. The pH of the nutrient media was adjusted to 5.8, prior to agar inclusion. The sterilization of the media involved autoclaving for 20 min at 121 °C and 122 kPa, prior to explant placement. All the in vitro cultures were maintained in a plant growth chamber, with a photoperiod of 16 h, provided by cool white fluorescent lamps, with a photosynthetic photon flux density (PPFD) of 46 μmol m^{-2} s^{-1} at the glass tube level and a temperature of 22 ± 0.1 °C. After a 3-week period, the explants' survival rate was assessed. For each treatment, four replications of twenty-five explants each were used.

2.3. Shoot Multiplication

2.3.1. Effect of Growth Regulators

Following the initial culture, a subculture was conducted on solid MS media supplemented with four different concentrations (0, 5, 10, 20 µM) of 6-Benzylaminopurine (BAP; Sigma-Aldrich, St. Louis, MI, USA) and kinetin (KIN; Sigma-Aldrich, St. Louis, MI, USA). The in vitro cultures were maintained in a plant growth chamber with the same environmental conditions as described previously. In each treatment, thirty explants (three replications of ten explants each) were used. At the end of the 4-week culture period, shoot formation (%), as well as the number and the length (cm) of the new shoots, were measured.

2.3.2. Effect of Nutrient Medium

To find out the most efficient nutrient medium for the production of multiple shoots on the explants, two media, namely WPM [35] and MS, with different strengths (1:1, 1:2, 1:4), were tested. All the nutrient media contained 5 µM of KIN. The explants were derived from in vitro cultures from the previous stage and all the cultures were maintained in a growth chamber, as described previously. For each treatment, three replications of ten explants each were used. Measurements were taken 4 weeks after establishment and concerned the frequency of shoot formation, as well as the number and the length (cm) of the new shoots.

2.4. Rooting of Microshoots

2.4.1. In Vitro Rooting

Axillary shoots (1.5–2 cm in length, with 2–4 leaves), produced from cultures from the previous multiplication stage, were excised and transferred for rooting on a solid MS medium. The medium was supplemented with 0, 1, 5, or 10 µM of Indole-3-butyric acid (IBA; Sigma-Aldrich, St. Louis, MI, USA) or α-Naphthaleneacetic acid (NAA; Sigma-Aldrich, St. Louis, MI, USA). The cultures were maintained in a growth chamber, according to the environmental conditions mentioned previously. Ten shoots in three replications were used for each auxin and concentration. After 5 weeks, the rooting rate and the number and length of the roots (cm) were calculated.

2.4.2. Ex Vitro Rooting

Axillary shoots (1.5–2 cm in length, with 2–4 leaves), derived from in vitro cultures, were first treated as microcuttings, with or without Indole-3-butyric acid potassium salt (K-IBA; Sigma-Aldrich, St. Louis, MI, USA) and, then, were transferred for rooting in multi-cell propagation trays (cell volume 34 cc), filled with peat or peat–perlite (1:1, *v*/*v*). When K-IBA was used, the basal portion of the shoots was dipped, for 10 s, into a water solution containing 0.5 g L^{-1} of K-IBA. The trays were then placed into a fog rooting system. The temperature at the root zone was adjusted to 19 ± 1 °C, while the ambient temperature in the fog system was 21 ± 1 °C and the relative humidity (RH) was 96 ± 2%. A total of 40% shading was provided over the trays by a cloth net (photosynthetic photon flux density approximately 175 µmol m^{-2} s^{-1}). After 5 weeks, the shoot rooting frequency (%) and the number and length (cm) of the roots were measured. Thirty shoots in three replications were used for each substrate and K-IBA treatment.

2.5. Acclimatization of Young Plantlets

The rooted plantlets from both the in vitro and ex vitro rooting conditions were transplanted into nursery trays (8.5 × 8.5 × 9 cm), filled with a 3 peat/1 perlite (*v*/*v*) mixture, and placed in a greenhouse bed, with a natural photoperiod of 12–14 h and an ambient temperature of 24 ± 2 °C, under a polyethylene cover for shading, equipped with a fogging system. The initial RH of 95% under the polyethylene cover decreased gradually to 75%, while the light irradiance gradually increased, with the aim of approaching the conditions of the natural environment in a 3-week period. Three replications of ten plantlets each were used for each rooting condition. After three weeks, the plantlet survival rate was calculated.

2.6. Genetic Stability Assessment Using ISSR Markers

For the genetic assessment, genomic DNA (gDNA) was extracted from the leaves of 12 randomly selected *S. byzantina* plantlets, 5 weeks after acclimatization, and from the mother *S. byzantina* control plant, using the ammonium bromide (CTAB) method, as described by Tsaktsira et al. [36]. The quality and quantity of gDNA was verified using NanoDrop 2000 (Thermo Electron Corporation, Waltham, MA, USA) and the samples were diluted with TE buffer to 20 ng/µL (10 mM Tris: 0.1 mM EDTA; pH 8.0) and stored at −18 °C. Nine ISSR primers were used with the codes ISSR UBC 807, 809, 810, 811, 812, 815, 816, 818, and 821 (University of British Columbia, Vancouver, BC, Canada), with the most intense and sharpest zones, selected following tests conducted on a total of 20 primers.

The ISSR amplifications were carried out in solution with a volume of 20 µL, containing 40 ng total DNA, 0.16 µL Kapa Taq DNA Polymerase, 20 ng dNTPs, 2 µL Kapa PCR buffer A, and 20 ng from each primer, and ultra-pure water. The amplifications were performed using a chain of processes: an initial denaturation step at 94 °C for 5 min, 35 cycles comprising of a denaturation cycle at 94 °C for 1 min, 1 min of annealing at a specific temperature for each primer, and annealing at 72 °C for 2 min. The final extension step was carried out at 72 °C for 10 min. For the PCR amplifications, the SimpliAmp™ Thermal Cycler (Life Technologies, Thermo Fischer Scientific, Waltham, MA, USA) was used. The DNA amplification products were separated on 1.5% agarose gel, using 1 × TAE buffer (Tris, acetic acid and EDTA), and stained with ethidium bromide. The gels were photographed under UV light and the size of the amplification products (zones) was estimated using a 1 kb DNA ladder. Two independent PCR amplifications were performed for each sample.

The ISSR zones were compared with each other for all the DNA samples to ascertain the genetic similarity of the plant material.

2.7. Statistical Analysis

The micropropagation experiments were conducted using complete randomized designs. The statistical analysis of the data was based on an analysis of variance (ANOVA). Prior to the statistical analysis, the data expressed as percentages were subjected to arcsine transformation in terms of their proportions and were then transformed back to percentages in order to be presented in tables. Mean separation was performed using Duncan's multiple range test, at $p \leq 0.05$. All the statistical analysis results were generated using the SPSS 27 software (IMB, Statistical Package for the Social Sciences, Chicago, IL, USA).

3. Results and Discussion

3.1. Establishment of In Vitro Cultures

The disinfestation treatment, with 1% NaOCl for 3 min, followed by 0.08% $HgCl_2$ for 5 min, proved effective, and the highest percentage of healthy *S. byzantina* explants of 70% was observed during the summer period (Table 1). On the other hand, the explants collected in the spring showed a lower survival rate of 61%. In a previous study, 0.1% mercury bichloride was employed for the disinfestation of explants grown on B5 Gamborg's nutrient medium, with no reported success rates [10]. The variable response of the explants to disinfestation could be due to the seasonal effect of explant collection, which has been reported for many woody species [37–39]. The application of a fungicide solution in conjunction with sodium hypochlorite could improve the overall efficacy of the disinfestation process [40,41].

Table 1. Survival rates (%) following disinfestation of *S. byzantina* explants, observed during the spring and summer period (mean ± standard deviation).

Explant Collection Period	Survival Rate (%)
Spring	61.0 ± 3.3 b *
Summer	70.0 ± 3.3 a

* The means with different letters are significantly different, according to Duncan's multiple range test, at $p \leq 0.05$.

3.2. Shoot Multiplication

3.2.1. Effect of Growth Regulators

The *S. byzantina* explants fully responded to (100%) and produced new shoots during the multiplication stage. The highest shoot number was observed on the MS nutrient medium containing 5 μM of KIN, resulting in an average of six shoots/explant (Figure 1B). The lowest number of new shoots was observed on the media supplemented with BAP (1.7–2.9 shoots/explant) (Table 2, Figure 1A). In contrast, the use of 1–20 mg L^{-1} of BAP (~4.6–90 μM) in other *Stachys* species, like *S. sieboldii* and *S. ocymastrum*, led to the production of multiple shoots [10]. The same trend was reported by Mantovska et al. [42] in a study carried out on *Stachys leucoglossa* (a Balkan endemic species), which revealed that the most effective medium for in vitro multiplication was the MS medium containing 0.5 mg L^{-1} (~2.3 μM) of BAP. The application of KIN resulted in more than a two-fold increase in the number of new shoots compared to BAP (Figure 1A,B). The shoot elongation was greater (5.7 cm), as expected, on the control media, while on the KIN-containing media, the shoot length was greater than the BAP-containing media (3.6–4.3 cm and 2.4–2.4 cm, respectively) (Table 2).

Figure 1. In vitro response of *S. byzantina* explants during multiplication stage in terms of shoot formation, after 5 weeks of culture on MS media, containing: (**A**) 5 μM BAP; and (**B**) 5 μM KIN. The black bar in the images corresponds to a size of 1 cm.

Table 2. Effect of different concentrations (0.5, 10, 20 μM) of BAP and KIN on shoot number and shoot length of *S. byzantina* explants in vitro (mean ± standard deviation).

Concentration (μM)	Number of Shoots		Length of Shoots (cm)	
	BAP	KIN	BAP	KIN
0 (control)	3.7 ± 0.25 c *	3.7 ± 0.25 c	5.7 ± 0.35 a *	5.7 ± 0.35 a
5	2.9 ± 0.28 d	6.0 ± 0.50 a	2.7 ± 0.28 d	3.6 ± 0.50 c
10	1.9 ± 0.28 e	4.9 ± 0.28 b	2.9 ± 0.38 d	4.3 ± 0.30 b
20	1.7 ± 0.28 e	3.0 ± 0.50 d	2.4 ± 0.15 d	3.7 ± 0.20 c

* The means, among the eight treatments, with different letters are significantly different, according to Duncan's multiple range test, at $p \leq 0.05$.

The proliferation rate observed in the present study was more than twice that reported by Legkobit and Khadeeva [10] on the B5 medium, which yielded 3.1 shoots per explant, with an average shoot length of 2.7 cm. Notably, even in the absence of growth regulators (control), both the shoot formation and proliferation rates were satisfactory, aligning with the findings by Panayotova et al. [43], who reported an excellent regeneration rate for *Stachys maritima*, with vigorous growth on a hormone-free MS medium. Nevertheless, an improvement in the shoot formation, and a sufficient length as well, was noticed when 5 μM of KIN was added to the nutrient medium. Hence, KIN in a low concentration (5 μM) was found to be the most efficient cytokinin for *S. byzantina* explant in vitro multiplication.

3.2.2. Effect of Nutrient Medium

All the explants cultured in either the MS media (Figure 2A) or WPM (Figure 2B) fully responded and formed new shoots. However, shoots formed in the MS media were more vigorous, while those cultured in WPM showed chlorosis after five weeks. The shoot proliferation rate was higher in the MS media compared to WPM, with the best results observed at a strength of 1:1 and 1:4 (2.9 shoots/explant) (Table 3). On the other hand, in WPM at a strength of 1:4, only 1.8 shoots/explant were formed. Legkobit and Khadeeva reported that the B5 medium was effective in regard to the in vitro establishment and multiplication of *S. byzantina* explants [10]. The present study did not evaluate the B5 medium due to the fact that the MS medium was shown to be effective in regard to in vitro shoot multiplication. Furthermore, the MS medium has also been reported as the most suitable medium for both the initial and multiplication stages for other *Stachys* species, i.e., *S. leucoglossa* [42] and *S. maritima* [43]. The strength of the nutrient medium was not a crucial factor for shoot multiplication in vitro. However, the MS medium proved to be superior to WPM in terms of shoot production (Table 3). The measurements of the shoot length indicated that MS strengths of 1:1 and 1:2 resulted in longer shoots (2.9–3.2 cm) compared to those of WPM (2.2–2.5 cm). At a strength of 1:4, the shoot lengths were not statistically different between the MS and WPM media (2.9 cm and 2.8 cm, respectively) (Table 3).

Figure 2. Multiple shoot production by *S. byzantina* explants in: (**A**) the full-strength MS medium; and (**B**) the WPM medium. The yellow bar in the images corresponds to a size of 1 cm.

Table 3. Effect of different strengths (1:1, 1:2, 1:4) of WPM or MS nutrient media on the number and length of shoots of *S. byzantina* explants produced in vitro, after five weeks of culture (mean ± standard deviation).

Nutrient Medium Strength	Number of Shoots		Length of Shoots (cm)	
	WPM	MS	WPM	MS
1:1	2.2 ± 0.20 cd *	2.9 ± 0.10 a	2.2 ± 0.11 c *	3.2 ± 0.15 a
1:2	2.1 ± 0.23 cd	2.5 ± 0.10 bc	2.5 ± 0.10 bc	3.0 ± 0.11 a
1:4	1.8 ± 0.20 d	2.9 ± 0.50 a	2.8 ± 0.47 ab	2.9 ± 0.10 ab

* The means, among the six treatments, with different letters are significantly different, according to Duncan's multiple range test, at $p \leq 0.05$.

The effect of the nutrient medium strength may vary among different species and depends on the type and the physiological condition of the explants. Thus, the in vitro multiplication and growth of *Typhonium flagelliforme* was improved when the MS strength was decreased [44]. The same trend was also noticed in *Delphinium cardinale*, in which the best multiplication rate was assessed when the strength of the MS nutrient medium was reduced to 1:3 [45], as well as in *Swertia chirata* Buch, wherein the reduction in the strength of the medium resulted in a decline in the number and length of the shoots [46]. On the contrary, in *Pogostemon cablin*, more shoots were produced in full-strength MS compared to ½ strength MS [47]. In this study, the influence of various nutrient media strengths on the shoot multiplication of *S. byzantina* explants was rather limited in the MS medium, while a ¼ strength of WPM produced longer shoots than ½ or full-strength WPM.

3.3. Rooting of Microshoots

3.3.1. In Vitro Rooting

The rooting of shoots achieved high rates (73.3%), in a 5-week period, with the application of 10 μM of NAA, while no rooting was observed either in the control medium or in the medium containing 1 μM of IBA (Table 4). The rooting rate in the other two IBA treatments did not exceed 17%. On the other hand, the addition of NAA was crucial for the in vitro rooting of the shoots. As the concentration of NAA increased, higher rooting rates were achieved. Thus, a lower rooting rate was noticed in the treatment with 1 μM of NAA (10%); the application of 5 μM of NAA led to a rooting rate of 46.7%, while when 10 μM of NAA was applied, a rooting rate of 73.3% was recorded (Table 4, Figure 3B). The same trend was also noticed in regard to the number and length of the roots formed. Hence, when 1 μM of NAA was applied, two roots, 0.7 cm in length, were formed, while the addition of 10 μM of NAA, led to the development of more and longer roots (2.9 roots, 1.6 cm in length) (Table 4). Overall, the application of 5 or 10 μM of NAA led to the production of nearly twice the number and length of the roots compared to those formed with IBA (Figure 3A,C).

Table 4. In vitro rooting (%), number and length of roots of *S. byzantina* shoots, in MS media, with the addition of different concentrations (0, 1, 5, or 10 μM) of IBA and NAA (mean ± standard deviation).

Concentration (μM)	Rooting (%)		Number of Roots		Length of Roots (cm)	
	IBA	NAA	IBA	NAA	IBA	NAA
0 (control)	0	0	0	0	0	0
1	0	10.0 ± 2.70 c *	0	2.0 ± 1.0 c *	0	0.7 ± 0.2 cd *
5	13.3 ± 8.82 c	46.7 ± 3.35 b	1.3 ± 0.57 c	2.6 ± 0.2 b	0.5 ± 0.1 d	1.1 ± 0.1 b
10	16.7 ± 3.35 c	73.3 ± 6.65 a	1.7 ± 0.57 c	2.9 ± 0.2 a	0.8 ± 0.2 c	1.6 ± 0.2 a

* The means, among the eight treatments, of each rooting regulator with different letters are significantly different, according to Duncan's multiple range test, at $p \leq 0.05$.

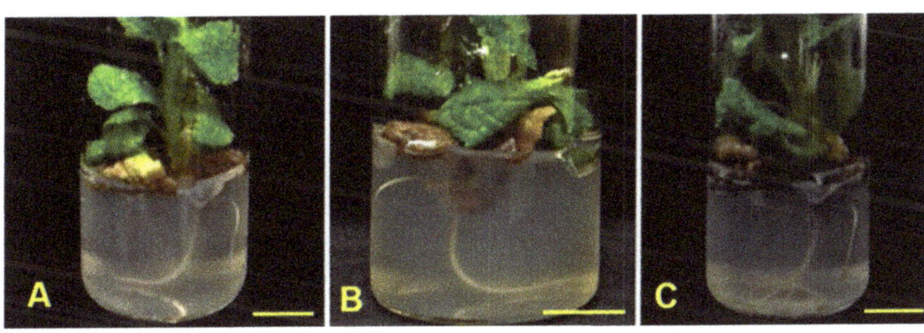

Figure 3. In vitro rooting of *S. byzantina* shoots in MS nutrient medium, supplemented with: (**A**) 5 μM IBA; (**B**) 5 μM NAA; and (**C**) 10 μM IBA. The yellow bar in the images corresponds to a size of 1 cm.

The shoots of *S. byzantina* formed roots promptly in high concentrations of NAA, exhibiting an adequate rooting ability and a well-formed rooting system. In contrast, Legkobit and Khadeeva [10] reported that bushy forms, with numerous shoots, whose internodes were closer to each other, were usually difficult to root or propagate via cuttings. Thus, they proposed the division of the bush into two or three parts and the supply of the nutrient medium with 0.5–1.0 mg L^{-1} of IBA and Gibberellin (~2.5–5 μM IBA and 1.5–3 μM Gibberellin). The bushy form of the plants is usually not desired by nurseries, which prefer single-stemmed plants, so that they can shape them in any way they wish. Thus, in our study, from each in vitro culture, 2–3 rooted plantlets were produced, ready for transplanting.

The lack of rooting regulators in the nutrient medium resulted in the inability of *S. byzantina* shoots to form roots, which is in line with the findings by Kevers et al. [48], who stated that during the rooting induction phase, the exogenous application of auxins, even at low concentrations, promotes rooting. The formation of adventitious roots is generally a complex genetic trait, regulated by the interaction of environmental and endogenous factors, and auxin is a key regulator in this process [49]. It is evidence that some woody species, like *Viburnum dentatum* [50], root rather easily in vitro without auxin application, while other species, like *Pyrus elaeagrifolia* 'Pallas', require various auxins for in vitro rooting induction [51]. Mantovska et al. [42] reported that the rooting rate of *Stachys leucoglossa* in WPM media supplemented with 2.5 μM of IBA and 3 μM of NAA was 60% and 40%, respectively. As Legkobit and Khadeeva [10] stated, in vitro cultures of different *Stachys* species, growing under the same conditions, differed in regard to their behavior. Thus, the media and the concentrations of rooting regulators should be selected individually, for each case. In the present study, NAA enhanced the rooting of the shoots in terms of the root formation rate (%), root number, and root length. However, a high auxin concentration can be toxic when inducing ethylene production, which, in turn, can result in growth inhibition [52].

3.3.2. Ex Vitro Rooting

The ex vitro shoot rooting was affected by both the K-IBA application and the substrate used (Table 5). The application of 0.5 g L^{-1} of K-IBA was crucial for the rooting success and led to a 4–6-fold increase in the shoot rooting rates compared to untreated shoots. Thus, when K-IBA was applied in peat substrate, the rooting rate of the shoots increased from 16.7 (Figure 4C) to 73.3% (Figure 4A). Similarly, in the mixture of peat and perlite, the application of K-IBA led to an increase in the rooting rate from 13.3 to 86.7%. Furthermore, the mixture of peat/perlite (1:1) was more effective in regard to the rooting of *S. byzantina* shoots treated with K-IBA (Figure 4B) than those in peat (Figure 4A), whereas no statistically significant differences were found in the rooting of untreated shoots between the two substrates (Table 5). The number of roots was 4–5 times higher in substrates with K-IBA (Figure 4A,B) compared to substrates without K-IBA (Figure 4C,D). In the peat–perlite mixture, 6.2 roots/shoot were formed following the application of K-IBA (Table 5). The same trend was also noticed in regard to the length of the roots. Thus, longer roots were recorded in the peat–perlite mixture with the use of K-IBA (5.1 cm long), followed by peat with K-IBA (4.2 cm long) (Table 5).

Table 5. Ex vitro rooting rate (%), number, and length of roots of *S. byzantina* shoots in different substrates, with or without the use of K-IBA (0.5 g L^{-1}) (mean ± standard deviation).

Substrate	Rooting (%)		Number of Roots		Length of Roots (cm)	
	+K-IBA	−K-IBA	+K-IBA	−K-IBA	+K-IBA	−K-IBA
Peat	73.3 ± 3.3 b *	16.7 ± 3.35 c	4.8 ± 0.2 b *	1.0 ± 0.1 d	4.2 ± 0.26 b *	1.4 ± 0.3 c
Peat–Perlite (1:1)	86.7 ± 6.65 a	13.3 ± 3.35 c	6.2 ± 0.26 a	1.3 ± 0.2 c	5.1 ± 0.26 a	0.9 ± 0.2 d

* The means, among the four treatments, with different letters are significantly different, according to Duncan's multiple range test, at $p \leq 0.05$.

Figure 4. Ex vitro rooting of *S. byzantina* shoots following: (**A**) immersion of their base in 0.5 g L^{-1} K-IBA and planting in peat; (**B**) immersion of their base in 0.5 g L^{-1} K-IBA and planting in the peat–perlite mixture (1:1); (**C**) planting without K-IBA in peat; and (**D**) planting without K-IBA in the peat–perlite (1:1) mixture. The yellow bar in the images corresponds to a size of 1 cm.

A successful micropropagation system presumes that the formation of healthy roots has occurred for the subsequent establishment of the plantlets in the soil and the production of well-growing plants. Many studies report higher rooting and survival rates, as well as greater plant uniformity, as reasons for proposing the use of the in vitro rooting method for plant production, involving species such as *Spiraea betulifolia* [53], *Pseudostellaria heterophylla* [54], *Siraitia grosvenorii* [55], and *Acer rubrum* L. 'Red Sunset', *Betula nigra* L., and *Malus* × *domestica* Borkh 'McIntosh' [56]. This study highlighted the high rooting rate of microshoots treated ex vitro with K-IBA, due to their ability to adapt from a sterile and controlled in vitro environment (temperature, humidity, light intensity) to the natural environment of the greenhouse. The same trends were also reported for other plant species like *Ceropegia bulbosa* Roxb [57] and *Siraitia grosvenorii* [55].

The mixture of peat–perlite (1:1 v/v) was more efficient than peat in regard to promoting the rooting rate of *S. byzantina* shoots. Peat has low porosity and retains more moisture than perlite. The addition of perlite to a substrate aims to increase the porosity of the substrate, as well as to efficiently drain water, and, at the same time, allows oxygen to penetrate the soil [58].

The water-holding capacity and air in the substrate determine the amount of water to be added. A desirable substrate is one that is moistened immediately after irrigation and retains at least 10% or more of the volume of air afterwards [59]. An aerated substrate facilitates rooting and the development of a better root system. Conversely, peat retains a higher water potential than the peat–perlite mixture and has a lower aeration potential, with the possibility of retardation of rhizogenesis in the shoots. Furthermore, the fog system is used to maintain a high relative humidity in the atmosphere, which leads to low evaporation rates and water losses from the shoots through reduced transpiration, thus avoiding substantial tissue water deficits [60]. Thus, under these substrate and environmental conditions, the stress in the shoots, especially during the first weeks of rooting, is much less, exhibiting rapid recovery and, in the following weeks, a functional root system is formed [61]. The formation of more and longer roots in a mixture of peat–perlite (1:1), as compared with peat, bearing root hairs, has also been reported for *Nerium oleander* [62].

3.4. Acclimatization of Young Plantlets

The acclimatization of the ex vitro-rooted plantlets was successful, with a high survival rate of 96.7%, compared to in vitro-rooted plantlets with a rate of 73.3% (Table 6; Figure 5A,B). The lower survival rate achieved by in vitro-rooted plantlets could be due to the structural characteristics of the rooting system. Kataoka [63] reported that the roots of *Carica papaya* growing in an agar medium showed structural abnormalities and lacked root hairs, which had adverse effects on their growth in field soil. In addition, McClelland et al. [56] stated that the availability of nutrients, water, and oxygen in an in vitro culture could negatively affect the development of a secondary phloem and xylem. However, long-term cultivation in the soil substrate eliminates these differences and the anatomy of the regenerated roots corresponds to that of normal plants.

Table 6. Survival rate (%) of plantlets of *S. byzantina*, rooted in vitro or ex vitro, five weeks after transplanting (mean ± standard deviation).

Rooting	Survival (%)
In vitro	73.3 ± 3.35 b *
Ex vitro	96.7 ± 3.86 a

* The means with different letters are significantly different, according to Duncan's multiple range test, at $p \leq 0.05$.

Figure 5. Plantlets of *S. byzantina*, which have been rooted ex vitro or in vitro (**A**,**B**), six weeks after transplanting into a mixture of three parts peat and one part perlite (v/v). The yellow bar in the image corresponds to a size of 10 cm.

The acclimatization of in vitro-produced plantlets requires their gradual adaptation to external environmental conditions and the transition of the root system from a non-functional structure to a functional one [61]. In the present study, the ex vitro-rooted plantlets formed a greater number of longer roots, with root hairs. On the other hand, the in vitro-rooted plantlets did not bear root hairs. This could be a crucial factor, influencing the survival rate of the plantlets at the acclimatization stage.

The physiology of the leaves of the plantlets derived from in vitro or ex vitro culture plays an important role in their successful acclimatization. In *Pseudostellaria heterophylla* plantlets produced in vitro, the stomata on the leaves were contiguous and open, the stomatal density was high and the stomatal function was abnormal, whereas plantlets derived from ex vitro cultures developed leaves with closed stomata and a low stomatal density [54]. The second case was attributed to the development of more functional leaves during rooting in inert substrates, facilitating the acclimatization of the plantlets to the external environment. Usually, a gradual reduction of the ambient relative humidity, coupled with a gradual increase in the light irradiance, greatly contributes to plantlet hardening towards successful acclimatization [61].

In the present study, the plantlets of *S. byzantina* that were rooted ex vitro easily acclimatized to the new environment, while those that were rooted in vitro were much more stressed during the first days of their adaptation to natural environmental conditions, resulting in a reduction in the survival rate by more than 26%. Nevertheless, all surviving plantlets, regardless of their rooting in vitro or ex vitro, subsequently had healthy and robust growth, with normal morphological characteristics (Figure 5A,B).

3.5. Genetic Stability

One of the main goals in the mass propagation of plants through tissue culture is the production of uniform plantlets of assured genetic fidelity. Hence, it is important to ensure that there is no genetic variation between the mother plant (the donor of the explants) and the produced plants. The genetic stability of plantlets produced by in vitro propagation should be examined prior to any commercial exploitation [64,65].

In this study, the screening involving nine ISSR primers (Table 7) generated 69 scorable and reproducible zones and a total of 897 zones (number of zones X number of gDNA samples), ranging in size from 200 to 1700 bp. The number of scorable zones for each primer varied from five (UBC 821) to nine (UBC 807 and 815) (Figure 6), with an average of 7.67 zones per primer. The ISSR-based zones were monomorphic and no genetic variation was detected among the 12 randomly selected plantlets, which were produced from the in vitro cultures and the *S. byzantina* mother plant. The results indicated that *S.*

byzantina plantlets derived from in vitro cultures exhibited genetic stability after five weeks of acclimatization and resembled the mother plant based on their ISSR profiles.

Table 7. Sequence, annealing temperature, and number of zones of nine ISSR primers, used to determine the genetic stability of *S. byzantina* plantlets produced from in vitro cultures.

Primer	Primer Sequence (5′-3′)	Annealing Temperature (°C)	Number of Bands
ISSR UBC 807	AGA GAG AGA GAG AGA GT	56	9
ISSR UBC 809	AGA GAG AGA GAG AGA GG	58	8
ISSR UBC 810	GAG AGA GAG AGA GAG AT	52	7
ISSR UBC 811	GAG AGA GAG AGA GAG AC	54	7
ISSR UBC 812	GAG AGA GAG AGA GAG AA	54	8
ISSR UBC 815	CTC TCT CTC TCT CTC TG	50	9
ISSR UBC 816	CAC ACA CAC ACA CAC AT	54	7
ISSR UBC 818	CAC ACA CAC ACA CAC AG	56	9
ISSR UBC 821	GTG TGT GTG TGT GTG TT	54	5

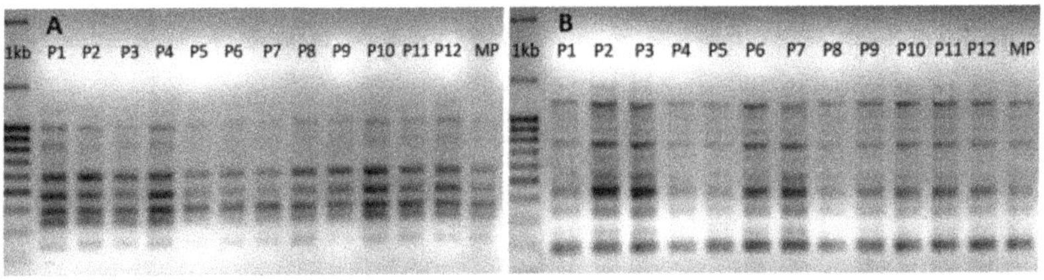

Figure 6. Amplification profiles of eight plantlets (P1–P8), derived from in vitro cultures and mother plant (MP) of *S. byzantina*, obtained using primers ISSR UBC 810 (**A**) and ISSR UBC 815 (**B**). The size of the amplified bands was calculated using a 1 kb DNA ladder (200–3000 bp). The image has been converted to a negative and to black and white.

These results confirm previous reports, wherein genetic stability using molecular markers has been identified in many other micropropagated ornamental plants, such as cold-stored artificial seeds of the woody species, *Rauvolfia tetraphylla* [66] and *Viburnum dentatum* [67]. ISSR markers, due to their simplicity and cost effectiveness, have been successfully employed to assess the genetic stability of many woody species [68–72].

The present study demonstrated that the in vitro-produced plantlets showed genetic uniformity both among themselves and compared to the mother plant (the donor of the explants), without any morphological differences being observed, underlining the reliability of the in vitro production protocol used.

4. Conclusions

The results of this study showed that in vitro cultures of *S. byzantina* can be successfully established in summer, with a protocol involving explant disinfestation, using a combination of NaOCl (1%, 3 min) and $HgCl_2$ (0.08%, 5 min). For shoot multiplication, the addition of 5 μM of KIN to an MS medium results in a high proliferation rate (six shoots per explant). Furthermore, the ex vitro rooting of microshoots, using K-IBA (0.5 g L^{-1}), in a mixture of peat–perlite (1:1), ensures a very high rooting rate (86.7%), which is also followed by a high survival rate (96.7%) of the plantlets after the acclimatization stage. The acclimatized plantlets showed genetic uniformity with the mother plant, as well as with each other, and, therefore, this protocol would be valuable in terms of its exploitation by the floriculture industry.

Author Contributions: Conceptualization, S.H. and A.E.; methodology, S.H., A.E., A.-A.K., K.B. and A.D.; software, S.K., C.K. and A.-A.K.; validation, S.H., C.K. and A.-A.K.; formal analysis, A.E. and C.K.; investigation, S.H., C.K., and, A.-A.K.; resources, S.K. and C.K.; data curation, S.H., C.K. and A.-A.K.; writing—original draft preparation, S.H., S.K., A.-A.K., A.E., K.B. and A.D.; writing—review and editing, S.H., S.K., A.E., K.B. and A.D.; visualization, S.K., A.E. and A.-A.K.; supervision, S.H., S.K. and A.E.; project administration, S.H. and S.K.; funding acquisition, S.H. and S.K. All authors have read and agreed to the published version of the manuscript.

Funding: This research received no external funding.

Data Availability Statement: All the data are presented in the article. Further inquiries can be directed to the corresponding author.

Acknowledgments: We would like to thank the technical staff at the Floriculture Laboratory as part of Aristotle University for their contribution to this publication.

Conflicts of Interest: The authors declare that there are no conflicts of interest.

References

1. Seaton, K.; Bettin, A.; Grüneberg, H. New ornamental plants for horticulture. In *Horticulture: Plants for People and Places, Volume 1*; Dixon, G.R., Aldous, D.E., Eds.; Springer: Dordrecht, The Netherlands, 2014; pp. 435–463, ISBN 978-94-017-8577-8.
2. Vascular Flora of Greece Web (Vascular Plants of Greece-An Annotated Checklist). Available online: https://portal.cybertaxonomy.org/flora-greece/intro (accessed on 1 August 2023).
3. Asnaashari, S.; Delazar, A.; Alipour, S.S.; Nahar, L.; Williams, A.S.; Pasdaran, A.; Mojarab, M.; Azad, F.F.; Sarker, S. Chemical composition, free-radical-scavenging and insecticidal activities of the aerial parts of *Stachys byzantina*. *Arch. Biol. Sci.* **2010**, *62*, 653–662. [CrossRef]
4. North Carolina Plant Toolbox, Stachys byzantina—Plant Information. Available online: https://plants.ces.ncsu.edu/plants/stachys-byzantina/?utm_source=chatgpt.com (accessed on 5 April 2025).
5. Devecchi, M. The use of *Labiatae* of ornamental interest in the design of parks and gardens. *Acta Hortic.* **2006**, *723*, 51–58. [CrossRef]
6. Kemp, S.; Hadley, P.; Blanuša, T. The Influence of plant type on green roof rainfall retention. *Urban Ecosyst.* **2019**, *22*, 355–366. [CrossRef]
7. Naghibi, F.; Mosadegh, M.; Mohammadi, M.S.; Ghorbani, A. Labiatae family in folk medicine in Iran: From ethnobotany to pharmacology. *Iran. J. Pharm. Res.* **2005**, *2*, 2.
8. Aminfar, P.; Abtahi, M.; Parastar, H. Gas chromatographic fingerprint analysis of secondary metabolites of *Stachys lanata* (*Stachys byzantine* C. Koch) Combined with Antioxidant Activity Modelling Using Multivariate Chemometric Methods. *J. Chromat. A* **2019**, *1602*, 432–440. [CrossRef]
9. Tomou, E.-M.; Barda, C.; Skaltsa, H. Genus *Stachys*: A review of traditional uses, phytochemistry and bioactivity. *Medicines* **2020**, *7*, 63. [CrossRef]
10. Legkobit, M.P.; Khadeeva, N.V. Variation and morphogenetic characteristics of different *Stachys* species during microclonal propagation. *Russ. J. Genet.* **2004**, *40*, 743–750. [CrossRef]
11. Grigoriadou, K.; Krigas, N.; Sarropoulou, V.; Papanastasi, K.; Tsoktouridis, G.; Maloupa, E. In vitro propagation of medicinal and aromatic plants: The case of selected Greek species with conservation priority. *Vitro Cell. Dev. Biol.-Plant* **2019**, *55*, 635–646. [CrossRef]
12. Katsanou, C.E.; Kostas, S.; Bantis, F.; Bertsouklis, K.; Hatzilazarou, S. Optimizing in vitro propagation of "Ladania" (*Cistus creticus* L.) through interaction of light spectra and plant growth regulators. *Agronomy* **2025**, *15*, 774. [CrossRef]
13. Rai, A.C.; Kumar, A.; Modi, A.; Singh, M. *Advances in Plant Tissue Culture: Current Developments and Future Trends*; Academic Press: Cambridge, MA, USA, 2022; ISBN 0-323-99819-4.
14. Ahmad, N.; Faisal, M.; Ahmad, A.; Alatar, A.A.; Qahtan, A.A.; Alok, A. Thidiazuron induced in vitro clonal propagation of *Lagerstroemia speciosa* (L.) Pers.—An important avenue Tree. *Horticulturae* **2022**, *8*, 359. [CrossRef]
15. Yasemin, S.; Beruto, M. A Review on flower bulb micropropagation: Challenges and opportunities. *Horticulturae* **2024**, *10*, 284. [CrossRef]
16. Bhojwani, S.S.; Dantu, P.K. Micropropagation. In *Plant Tissue Culture: An Introductory Text*; Springer: New Delhi, India, 2013; pp. 245–274, ISBN 978-81-322-1025-2.
17. Podwyszyńska, M.; Orlikowska, T.; Trojak-Goluch, A.; Wojtania, A. Application and improvement of in vitro culture systems for commercial production of ornamental, fruit, and industrial plants in Poland. *Acta Soc. Bot. Pol.* **2022**, *91*, 914. [CrossRef]

18. Bairu, M.W.; Fennell, C.W.; van Staden, J. The effect of plant growth regulators on somaclonal variation in *Cavendish banana* (Musa AAA Cv.'Zelig'). *Sci. Hortic.* **2006**, *108*, 347–351. [CrossRef]
19. Haisel, D.; Hofman, P.; Vágner, M.; Lipavska, H.; Tichá, I.; Schäfer, C.A.; Čapková, V. Ex vitro phenotype stability is affected by in vitro cultivation. *Biol. Plant.* **2001**, *44*, 321–324. [CrossRef]
20. Zietkiewicz, E.; Rafalski, A.; Labuda, D. Genome fingerprinting by Simple Sequence Repeat (SSR)-Anchored polymerase chain reaction amplification. *Genomics* **1994**, *20*, 176–183. [CrossRef]
21. Ai, X.; Wen, Y.; Wang, B. Assessment of genetic stability on in vitro propagation of *Ardisia crenata* var. bicolor using ISSR markers. *Int. J. Plant Biol.* **2023**, *14*, 218–227. [CrossRef]
22. Kader, A.; Sinha, S.N.; Ghosh, P. A standardized protocol for genetically stable artificial seed production of *Ficus religiosa* L. using ISSR fingerprinting. *J. Appl. Genet.* **2023**, *64*, 275–287. [CrossRef]
23. Jung, W.-S.; Chung, I.-M.; Kim, S.-H.; Chi, H.-Y.; Yu, C.Y.; Ghimire, B.K. Direct shoot organogenesis from *Lycium chinense* Miller leaf explants and assessment of genetic stability using ISSR markers. *Agronomy* **2021**, *11*, 503. [CrossRef]
24. Zamani, M.; Sonboli, A.; Goldansaz, M.; Mirjalili, M.H. In vitro micropropagation and conservation of endangered medicinal plant *Nepeta asterotricha* Rech. f. (Lamiaceae): Genetic fidelity, phytochemical and biological assessment. *Physiol. Mol. Biol. Plants* **2024**, *30*, 67–80. [CrossRef]
25. Thakur, M.; Sharma, V.; Luharch, R. Propagation of plum (*Prunus salicina* L.) cultivar frontier in vitro through control of shoot tip necrosis (stn) and validation of genetic integrity using ISSR markers. *Plant Physiol. Rep.* **2021**, *26*, 238–246. [CrossRef]
26. Yıldırım, M.; Tuğ, G.N.; Yaprak, A.E. Analyses of genetic diversity and population structure of endemic and endangered species *Sideritis gulendamii* (Lamiaceae) and implications for its conservation. *Genet. Resour. Crop Evol.* **2024**, *71*, 4331–4345. [CrossRef]
27. Chirumamilla, P.; Gopu, C.; Jogam, P.; Taduri, S. Highly efficient rapid micropropagation and assessment of genetic fidelity of regenerants by ISSR and SCoT markers of *Solanum khasianum* Clarke. *Plant Cell Tiss. Org. Cult.* **2021**, *144*, 397–407. [CrossRef]
28. Hatzilazarou, S.; Kostas, S.; Nendou, T.; Economou, A. Conservation, regeneration and genetic stability of regenerants from alginate-encapsulated shoot explants of *Gardenia jasminoides* Ellis. *Polymers* **2021**, *13*, 1666. [CrossRef]
29. Akin, B.; Bingöl, N.A.; Bulman, G.C. A first approach for the micropropagation of threatened endemic subspecies of *Stachys cretica* subsp. *kutahyensis*. *Biologia* **2024**, *79*, 1653–1661. [CrossRef]
30. Cüce, M.; Bekircan, T.; Laghari, A.H.; Sökmen, M.; Sökmen, A.; Uçar, E.Ö.; Kılıç, A.O. Antioxidant phenolic constituents, antimicrobial and cytotoxic properties of *Stachys annua* L. from both natural resources and micropropagated plantlets. *Indian J. Tradit. Knowl.* **2017**, *16*, 407–416.
31. Hadi, S.; Ahmadabadi, M.; Valizadeh Kamran, R. Efficient in vitro callus induction, regeneration and shoot multiplication protocols for *Stachys inflata* subsp. *caucasica* (Stschegl.); A rare medicinal plant. *J. Med. Plants -Prod.* **2024**, *13*, 570–576.
32. Mantovska, D.I.; Zhiponova, M.K.; Petrova, D.; Alipieva, K.; Bonchev, G.; Boycheva, I.; Evstatieva, Y.; Nikolova, D.; Tsacheva, I.; Simova, S.; et al. Exploring the phytochemical composition and biological potential of Balkan endemic species *Stachys scardica* Griseb. *Plants* **2023**, *13*, 30. [CrossRef]
33. Gamborg, O.; Murashige, T.; Thorpe, T.; Vasil, I. Plant tissue culture media. *In Vitro* **1976**, *12*, 473–478. [CrossRef]
34. Murashige, T.; Skoog, F. A revised medium for rapid growth and bio assays with tobacco tissue cultures. *Physiol. Plant.* **1962**, *15*, 473–497. [CrossRef]
35. McCown, B. Woody plant medium (WPM)-a mineral nutrient formulation for microculture for woody plant species. *Hort. Sci.* **1981**, *16*, 453.
36. Tsaktsira, M.; Chavale, E.; Kostas, S.; Pipinis, E.; Tsoulpha, P.; Hatzilazarou, S.; Ziogou, F.-T.; Nianiou-Obeidat, I.; Iliev, I.; Economou, A. Vegetative propagation and ISSR-based genetic identification of genotypes of *Ilex aquifolium* 'Agrifoglio commune'. *Sustainability* **2021**, *13*, 10345. [CrossRef]
37. Kartsonas, E.; Papafotiou, M. Mother plant age and seasonal influence on in vitro propagation of *Quercus euboica* Pap., an endemic, rare and endangered oak species of Greece. *Plant Cell Tiss. Org. Cult.* **2007**, *90*, 111–116. [CrossRef]
38. Wang, Y.; Yao, R. Plantlet regeneration of adult *Pinus massoniana* Lamb. Trees using explants collected in March and thidiazuron in culture medium. *J. For. Res.* **2017**, *28*, 1169–1175. [CrossRef]
39. Bertsouklis, K.; Paraskevopoulou, A.T.; Petraki, E. In vitro regeneration from adult node explants of *Juniperus oxycedrus*. *Not. Bot. Horti Agrobot.* **2023**, *51*, 13062. [CrossRef]
40. Loureiro, J.; Capelo, A.; Brito, G.; Rodriguez, E.; Silva, S.; Pinto, G.; Santos, C. Micropropagation of *Juniperus phoenicea* from adult plant explants and analysis of ploidy stability using flow cytometry. *Biol. Plant.* **2007**, *51*, 7–14. [CrossRef]
41. Babu, G.A.; Mosa Christas, K.; Kowsalya, E.; Ramesh, M.; Sohn, S.-I.; Pandian, S. Improved sterilization techniques for successful in vitro micropropagation. In *Commercial Scale Tissue Culture for Horticulture and Plantation Crops*; Gupta, S., Chaturvedi, P., Eds.; Springer Nature: Singapore, 2022; pp. 1–21, ISBN 978-981-19-0054-9.
42. Mantovska, D.I.; Kapchina, V.M.; Yordanova, Z.P. In vitro propagation of the Balkan endemic species *Stachys leucoglossa* Griseb. *Bulg. J. Agric. Sci.* **2019**, *25*, 1211–1215.

43. Panayotova, L.G.; Ivanova, T.A.; Bogdanova, Y.Y.; Gussev, C.V.; Stanilova, M.I.; Bosseva, Y.Z.; Stoeva, T.D. In vitro cultivation of plant species from sandy dunes along the Bulgarian Black Sea coast. *Phytol. Balc.* **2008**, *14*, 119–123.
44. Rezali, N.I.; Sidik, N.J.; Saleh, A.; Osman, N.I.; Adam, N.A.M. The effects of different strength of MS media in solid and liquid media on in vitro growth of *Typhonium flagelliforme*. *Asian Pac. J. Trop. Biomed.* **2017**, *7*, 151–156. [CrossRef]
45. Ohki, S.; Sawaki, S. The effects of inorganic salts and growth regulators on in vitro shoot proliferation and leaf chlorophyll content of *Delphinium cardinale*. *Sci. Hortic.* **1999**, *81*, 149–158. [CrossRef]
46. Pant, M.; Bisht, P.; Gusain, M.P. Effect of sucrose and media strength on in vitro multiplication in *Swertia shirata* Buch.-Ham ex wall: An endangered medicinal herb. *Int. J. Pharm. Pharm. Sci.* **2018**, *10*, 40–42. [CrossRef]
47. Hida, W.N.; Norrizah, J.; Aminah, S.S.; Ruzaina, S.S.; Faezah, P. Effect of medium strength and hormones concentration on regeneration of *Pogostemon cablin* using nodes explant. *Asian J. Biotechnol.* **2012**, *4*, 46–52. [CrossRef]
48. Kevers, C.; Hausman, J.-F.; Faivre-Rampant, O.; Evers, D.; Gaspar, T. Hormonal control of adventitious rooting: Progress and questions. *Angew. Bot.* **1997**, *71*, 71–79.
49. Pacurar, D.I.; Perrone, I.; Bellini, C. Auxin is a central player in the hormone cross-talks that control adventitious rooting. *Physiol. Plant.* **2014**, *151*, 83–96. [CrossRef] [PubMed]
50. Hatzilazarou, S.; Rifaki, N.; Patsou, M.; Kostas, S.; Economou, A.S. In vitro propagation of *Viburnum dentatum* L. 'Lucidum Aiton'. *Prop. Ornam. Plants* **2009**, *9*, 39–42.
51. Aygun, A.; Dumanoglu, H. In vitro shoot proliferation and in vitro and ex vitro root formation of *Pyrus elaeagrifolia* Pallas. *Front. Plant Sci.* **2015**, *6*, 225. [CrossRef]
52. Gaba, V.P. Plant growth regulators in plant tissue culture and development. In *Plant Development and Biotechnology*; CRC Press: Boca Raton, FL, USA, 2005; pp. 87–99.
53. Muraseva, D.; Kostikova, V. In vitro and ex vitro rooting of *Spiraea betulifolia* subsp. *aemiliana* (Rosaceae), an ornamental shrub. *EDP Sci.* **2020**, *24*, 00058. [CrossRef]
54. Wang, F.; Xin, X.; Wei, H.; Qiu, X.; Liu, B. In vitro regeneration, ex vitro rooting and foliar stoma studies of *Pseudostellaria heterophylla* (Miq.) Pax. *Agronomy* **2020**, *10*, 949. [CrossRef]
55. Yan, H.; Liang, C.; Yang, L.; Li, Y. In vitro and ex vitro rooting of *Siratia grosvenorii*, a traditional medicinal plant. *Acta Physiol. Plant.* **2010**, *32*, 115–120. [CrossRef]
56. McClelland, M.; Smith, M.; Carothers, Z. The effects of in vitro and ex vitro root initiation on subsequent microcutting root quality in three woody plants. *Plant Cell Tiss. Org. Cult.* **1990**, *23*, 115–123. [CrossRef]
57. Phulwaria, M.; Shekhawat, N.; Rathore, J.; Singh, R. An efficient in vitro regeneration and ex vitro rooting of *Ceropegia bulbosa* Roxb.—A threatened and pharmaceutical important plant of Indian Thar Desert. *Ind. Crops Prod.* **2013**, *42*, 25–29. [CrossRef]
58. Kormanek, M.; Małek, S.; Banach, J.; Durło, G.; Jagiełło-Leńczuk, K.; Dudek, K. Seasonal changes of perlite–peat substrate properties in seedlings grown in different sized container trays. *New For.* **2021**, *52*, 271–283. [CrossRef]
59. Cabrera, P.I.; Johnson, J.R. *Fundamentals of Container Media Management: Part 1. Greenhouse and Nursery Crops Fact Sheets & Bulletins*; Rutgers Fact Sheet FS 812; The State University of New Jersey: New Brunswick, NJ, USA, 2014; p. 3.
60. Mateja, Š.; Dominik, V.; Franci, Š.; Gregor, O. The effects of a fogging system on the physiological status and rooting capacity of leafy cuttings of woody species. *Trees* **2007**, *21*, 491–496. [CrossRef]
61. Economou, A.S. From microcutting rooting to microplant establishment: Key points to consider for maximum success in woody plants. *Acta Hortic.* **2013**, *988*, 43–56. [CrossRef]
62. Hatzilazarou, S.; Kostas, S.; Economou, A.; Scaltsoyiannes, A. Efficient propagation of *Nerium oleander* L. through tissue culture. *Prop. Ornam. Plants* **2017**, *17*, 64–74.
63. Kataoka, I. Influence of rooting substrates on the morphology of papaya root formed in vitro. *Jap. J. Trop. Agric.* **1994**, *38*, 251–257.
64. Biswas, P.; Kumar, N. Application of molecular markers for the assessment of genetic fidelity of in vitro raised plants: Current status and future prospects. In *Molecular Marker Techniques*; Kumar, N., Ed.; Springer Nature: Singapore, 2023; pp. 233–256, ISBN 978-981-99-1611-5.
65. Martínez, O. Selection of molecular markers for the estimation of somaclonal variation. In *Plant Cell Culture Protocols*; Loyola-Vargas, V.M., Ochoa-Alejo, N., Eds.; Methods in Molecular Biology; Springer: New York, NY, USA, 2018; Volume 1815, pp. 103–129, ISBN 978-1-4939-8593-7.
66. Alatar, A.; Faisal, M. Encapsulation of *Rauvolfia tetraphylla* microshoots as artificial seeds and evaluation of genetic fidelity using RAPD and ISSR markers. *J. Med. Plants Res.* **2012**, *6*, 1367–1374. [CrossRef]
67. Hatzilazarou, S.; Kostas, S.; Joachim, M.; Economou, A. Regeneration of *Viburnum dentatum* L. from alginate-encapsulated shoot explants after short-term cold storage and assessment of genetic stability using ISSR analysis. *Agronomy* **2020**, *10*, 1660. [CrossRef]
68. Bramhanapalli, M.; Thogatabalija, L.; Gudipalli, P. Efficient in vitro plant regeneration from seedling-derived explants and genetic stability analysis of regenerated plants of *Simarouba glauca* DC. by RAPD and ISSR markers. *Vitro Cell. Dev. Biol.-Plant* **2017**, *53*, 50–63. [CrossRef]

69. Ilczuk, A.; Jacygrad, E. In vitro propagation and assessment of genetic stability of acclimated plantlets of *Cornus alba* L. using RAPD and ISSR markers. *Vitro Cell. Dev. Biol.-Plant* **2016**, *52*, 379–390. [CrossRef]
70. Soares, D.M.M.; Sattler, M.C.; Ferreira, M.F.d.S.; Praça-Fontes, M.M. Assessment of genetic stability in three generations of in vitro propagated *Jatropha curcas* L. plantlets using ISSR markers. *Trop. Plant Biol.* **2016**, *9*, 229–238. [CrossRef]
71. Amin, S.; Wani, T.A.; Kaloo, Z.A.; Singh, S.; John, R.; Majeed, U.; Shapoo, G.A. Genetic stability using RAPD and ISSR markers in efficiently in vitro regenerated plants of *Inula royleana* DC. *Meta Gene* **2018**, *18*, 100–106. [CrossRef]
72. Ali, H.; Musa, I.F.; Abu Bakar, N.A.; Karsani, S.A.; Yaacob, J.S. In vitro regeneration and ISSR-based genetic fidelity analysis of *Orthosiphon stamineus* Benth. *Agronomy* **2019**, *9*, 778. [CrossRef]

horticulturae

MDPI

Communication

Histochemical Localization of Alkaloids in the Bulbs of In Vitro-Regenerated Snake's Head Fritillary (*Fritillaria meleagris* L.): The Effect of a Temperature Regime

Marija Marković, Milana Trifunović-Momčilov, Olga Radulović, Danijela M. Paunović, Dragana D. Antonić Reljin, Slađana Jevremović and Branka Uzelac *

Department of Plant Physiology, Institute for Biological Research "Siniša Stanković"—National Institute of the Republic of Serbia, University of Belgrade, Bulevar despota Stefana 142, 11108 Belgrade, Serbia; marija.nikolic@ibiss.bg.ac.rs (M.M.); milanag@ibiss.bg.ac.rs (M.T.-M.); olga.radulovic@ibiss.bg.ac.rs (O.R.); danijela.paunovic@ibiss.bg.ac.rs (D.M.P.); dragana.antonic@ibiss.bg.ac.rs (D.D.A.R.); sladja@ibiss.bg.ac.rs (S.J.)

* Correspondence: branka@ibiss.bg.ac.rs

Abstract: The genus *Fritillaria* is characterized by the production of alkaloids, primarily of the isosteroidal type. The aim of this preliminary study was to perform a histolocalization screening aimed at revealing the presence of alkaloids in the bulbs of *F. meleagris*, cultured in vitro under different growth regimes and at different developmental stages, as a powerful tool to identify preferable in vitro plant material for alkaloid isolation. Histochemical localization of alkaloids was performed using Wagner's and Dragendorff's reagents in fresh sections of bulbs cultured at 24 °C or 7 °C for 4 weeks, as well as those cultured at 24 °C following the 4-week chilling treatment, which were sampled at the beginning of sprouting. A positive reaction was observed with both reagents and was particularly intense in cold-treated bulbs. Alkaloids were mainly distributed in the bulbs and partly in the sprouts of the in vitro grown *F. meleagris*. The most intense staining, indicative of high alkaloid content, was observed in the bulb scales of pre-chilled bulbs that sprouted at 24 °C, rendering them preferable as in vitro plant material for alkaloid isolation. The results suggest that alkaloid production in the bulbs of *F. meleagris* can be improved by manipulating growth in the microenvironment of in vitro cultures, in order to meet the increasing industrial demand for medicinally or commercially important metabolites used as traditional medicines and herbal remedies.

Keywords: snake's head fritillary; bulbs; alkaloids; histolocalization

1. Introduction

Fritillaria meleagris L. (Liliaceae) is a valuable horticultural species that is mainly distributed in the temperate zones of the northern hemisphere. It is often propagated vegetatively using its underground storage organs (bulbs). Bulbous plants, such as fritillary, evade unfavorable conditions by going dormant as bulbs underground during the winter months and germinating the following year, often requiring a period of low temperatures to overcome dormancy. The absence of low-temperature exposure leads to very slow plant growth and causes various anomalies in flower formation and/or development [1]. The very low natural propagation rate of *Fritillaria* species renders their mass production and cultivation generally problematic [2]. *F. meleagris* is considered a rare and threatened species throughout its geographic range [3]. A comparison of the current status of *F. meleagris* populations in Serbia with the status 50 years ago shows that the populations have declined to such an extent that they could be considered endangered [4].

In addition to their great horticultural importance, fritillaries also have potential for medicinal use due to their phytochemical properties, with a rich pharmacological history [5]. *Fritillaria* species, which are most commonly consumed as dried bulbs [Bei Mu in Chinese], exhibit pharmacological effects due to the presence of alkaloids, terpenoids, and many other metabolites, including antitussive, expectorant, antiasthmatic, and antitumor activity [6]. Fritillaries have been used to treat hypotension, sensory disorders, various lung infections including tuberculosis, respiratory disorders, myocardial irritation, scrofulous swellings, and chest nodules [6,7]. Alkaloids, the major compounds isolated from the crude extracts of the bulbs and identified as the active ingredients in *Fritillaria*, are considered to be useful agents that reduce coronavirus-associated lung injury [8]. To date, 120 alkaloids have been isolated from plants of the genus *Fritillaria* [9]. The alkaloids extracted from *Fritillaria* species belong to the isosteroidal and steroidal types [10]. Most alkaloids belong to the isosteroidal type (72.7%), and their contents and structure types differ among species [11, 12].

Natural products from *Fritillaria* species have many advantages over conventional drugs [13]. Recently, a very thorough investigation was conducted as a strategy to produce new drugs from *Fritillaria* [14]. In view of the traditional use of the taxa belonging to *Fritillaria*, several review articles addressed the phytochemistry, chemotaxonomy, molecular phylogeny, and pharmacology of *Fritillaria* species [6,15]. Many Chinese *Fritillaria* sp. have been extensively studied, especially *Fritillariae thunbergii* bulbus [16], but the species distributed outside China have been greatly neglected in scientific research. Beyond China, the bulbs of *Fritillaria* are also approved as active ingredients in medicinal preparations in Oceania, North America, and Asia [17]. However, the disappearance of most *Fritillaria* species, including *F. meleagris*, from their natural habitats, and the decline in the number of wild-growing plants, have resulted in diminishing sources of potentially valuable medicinal compounds.

Reproduction of fritillaries under natural conditions is very slow [18]. The entire growth period from seed to commercial form takes at least three years, so in vitro propagation can speed up the process of obtaining valuable plant material [19]. Propagation of *F. meleagris* bulbs in vitro can be enhanced by various tissue culture protocols. Dormant bulbs and those that begin to sprout after the cold period exhibit various physiological and morphological alterations [20]. In vitro-regenerated bulbs of *F. meleagris* also require chilling in order to overcome dormancy and continue their growth and development [21]. The physiology and biochemistry of dormancy, a developmental stage without visible morphogenesis, is not yet fully elucidated.

In recent years, great efforts have been made to investigate the in vitro culture of medicinal plants [22]. Concerning snake's head fritillary, research has focused on optimizing protocols for dormancy breaking, sprouting, and overall improvement of plant growth. A number of our previous studies [21,23–25] reported different techniques for the propagation and regeneration of *F. meleagris*. Additional physiological and biochemical studies need to be carried out to increase the production of metabolites of therapeutic interest in bulbs of in vitro-cultured *F. meleagris*. Therefore, we considered necessary a histolocalization screening aimed at revealing bioactive metabolites in the bulbs of *F. meleagris* under different growth regimes and at different developmental stages before performing in-depth phytochemical studies.

The aim of this preliminary study was to investigate the presence of alkaloids in the bulbs of *F. meleagris*, to localize the alkaloids using specific histochemical tests, and to investigate possible differences in their distribution under different growth regimes and at different developmental stages. The results may provide a new perspective for improving alkaloid production and increasing alkaloid accumulation in fritillary bulbs. Further research on this species, focusing on analytical techniques for the chemical evaluation of alkaloids, could be important from an economic point of view. This work is a preliminary study for a larger comprehensive investigation of the effects of different physical, nutritional, and hormonal factors, plant age, and bulb size, on alkaloid production in *F. meleagris*.

2. Materials and Methods

2.1. Plant Material and Culture Conditions

Bulb cultures of *Fritillaria meleagris* L. were established according to the procedure modified after Petrić et al. [25]. Briefly, seeds of *F. meleagris* were washed with running water for 1 h, surface sterilized with 0.8% (w/v) sodium hypochlorite in sterile deionized water for 20 min, and then rinsed three times with sterile deionized water. Mature zygotic embryos were isolated aseptically and used as explants. Five explants were placed in each Petri dish filled with 15 mL Murashige and Skoog (MS) medium [26] supplemented with 30 g L^{-1} sucrose, 7 g L^{-1} agar, 250 mg L^{-1} casein hydrolysate, 250 mg L^{-1} L-proline, 80 mg L^{-1} adenine sulfate, and 1 mg L^{-1} thidiazuron (TDZ). After 4 weeks of cultivation, bulblet formation was observed [25]. Obtained bulblets were cut longitudinally into four scale sections and cultured on MS medium as previously described, with the exception of 0.1 mg L^{-1} TDZ used instead of 1 mg L^{-1}, for an additional 4 weeks, where new bulblets were formed [25] and used to establish stock cultures.

Bulblet scale explants were used as primary explants for in vitro propagation of *F. meleagris*. The stock cultures were maintained on MS medium supplemented with 30 g L^{-1} sucrose, 7 g L^{-1} agar, and 0.1 mg L^{-1} TDZ for shoot and bulb multiplication. Four bulb scale explants were cultured per Petri dish. After four weeks, regenerated bulbs of about 80–100 mg were used for further experiments. The pH of all culture media was adjusted to 5.8 with 1 N NaOH before autoclaving at 121 °C for 25 min. All cultures were maintained in a growth chamber under a 16 h photoperiod, with a photon flux density of 40 µmol m^{-2} s^{-1} provided by cool white fluorescent lamps, at 24 ± 2 °C.

Bulbs regenerated in vitro were transferred to fresh MS medium without plant growth regulators in 100 mL Erlenmayer flasks containing 50 mL of medium. Five explants were placed in each Erlenmayer flask and cultured at different temperatures. Four replicates, each consisting of five explants, were used per treatment (n = 20). Bulbs were subjected to one of two chilling treatments (at 7 °C), performed in a cold chamber (Snaigė): (1) a chilling treatment for 4 weeks, after which the bulbs were sampled immediately; sprouting rarely occurred at this temperature, but if it did, sprouted bulbs were sampled at that time point, and (2) a chilling treatment for 4 weeks, followed by cultivation at 24 °C; after transfer to 24 °C, bulbs started to sprout and were sampled as soon as the sprouting began. Bulbs cultured at 24 °C for 4 weeks were used as a control (Figure 1).

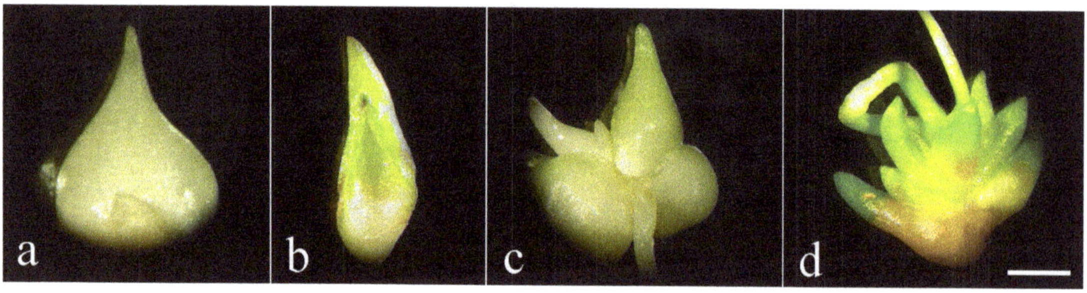

Figure 1. In vitro-cultured bulbs of *F. meleagris* collected at the beginning of sprouting, at different time points of different treatments, which were used for localization of alkaloids. (**a**) Sprouted bulb after 7 days of chilling; (**b**) bulb sprouted after 4 weeks of chilling; (**c**) control bulb sprouted after 7 days at 24 °C; (**d**) bulb sprouted after 4 weeks at 24 °C. Not all bulbs or all treatments are shown. Scale bar = 5 mm.

2.2. Determination of Alkaloids

The presence of alkaloids was detected by Wagner's or Dragendorff's reagents [27], which produce a golden-brown- or reddish-brown-colored reaction in response to alkaloids, respectively (the intensity of staining depends on the quantity of alkaloids). Wagner's reagent (iodine potassium iodide solution) was prepared by dissolving 2 g I_2 and 6 g KI in

100 mL of distilled water. The solution was left at room temperature for 24 h. Dragendorff's reagent was prepared by dissolving 0.85 g bismuth subnitrate in 10 mL glacial acetic acid and 40 mL distilled water, followed by thorough mixing with 20 mL of 40% KI aqueous solution.

Bulbs as well-as sprouted bulbs were cut with a scalpel and dipped into one of the solutions for 20 min. After staining, the longitudinal- and cross sections of the bulbs were photographed using a photomicroscope (Leica wild M3Z). The reaction using Wagner's reagent produced a golden-brown to brownish color in response to alkaloids and a blue/purple color in response to starch. Dragendorff's reagent produced an orange-brown-colored reaction with alkaloids.

2.3. Morpho-Anatomical Study

Bulbs cultured for 4 weeks at 7 °C and 24 °C, as well as sprouted bulbs, were fixed in FAA fixative (5 mL of 40% formaldehyde, 5 mL of glacial acetic acid, and 90 mL of 70% ethanol) at 4 °C for 24 h, according to the procedure of Jensen [28]. The fixed material was washed, dehydrated in a graded ethanol series, and embedded in paraffin. Cross sections (5–7 μm thick) were stained with Alcian blue and photographed using a Leitz DMRB photomicroscope (Leica, Wetzlar, Germany).

3. Results

F. meleagris bulbs grown for 4 weeks in a cold chamber (which did not sprout) showed alkaloids scarcely distributed in the whole bulb (Figure 2a). When the bulbs started to sprout, alkaloids were mostly observed in the bulb scales, as revealed by the more intense golden-brown staining, and also occasionally displaced between the reserve parenchyma cells, which contained an abundance of amyloplasts (Figure 2b). These amyloplasts could conceal a positive reaction for alkaloids, because they stained dark blue in response to the Wagner's reagent and dark brown in response to the Dragendorff's reagent. Alkaloids were also detected in the newly formed sprouts. The strongest positive reaction for alkaloids was observed in the bulb scales of the sprouted bulb at the end of cold treatment (after 4 weeks), indicative of a high alkaloid content (Figure 2c).

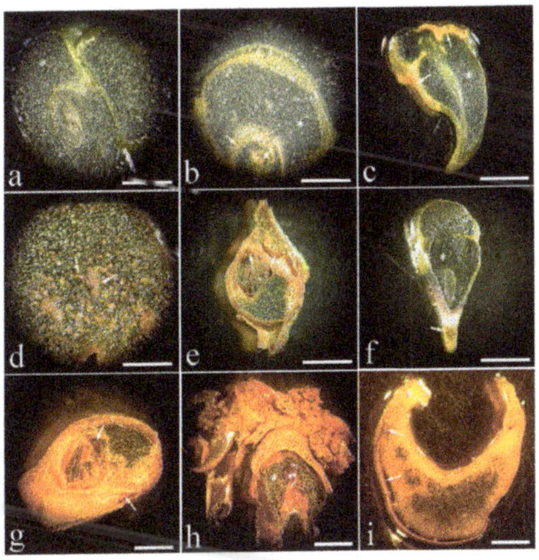

Figure 2. In situ histochemical localization of alkaloids in the bulbs of in vitro-regenerated *Fritillaria meleagris* cultured at different temperature regimes. (**a–c**) Bulbs cultured at 7 °C after (**a**) 4 weeks; bulb did not sprout (scale bar = 1 mm); (**b**) 7 days; beginning of sprouting (scale bar = 1.7 mm); (**c**) 4 weeks;

bulb sprouted (scale bar = 5 mm); (**d–f**) Control bulbs, cultured at 24 °C after: (**d**) 4 weeks; bulb did not sprout (scale bar = 3.5 mm); golden-brown colored drops (*arrows*) indicate alkaloids; (**e**) 10 days; beginning of sprouting (scale bar = 5 mm); (**f**) 14 days; bulb sprouted (scale bar = 5 mm); (**g–i**) *F. meleagris* bulbs pre-chilled for 4 weeks in cold chamber then cultured at 24 °C where they sprouted after (**g**) 7 days (scale bar = 5 mm); (**h**) 3 weeks (scale bar = 1.7 mm); (**i**) bulb scale of the bulb sprouted after 2 weeks at 24 °C (scale bar = 1 mm). Alkaloids are mostly detected as orange- or golden-brown staining (*arrows*) in bulb scales, whereas the dark-blue staining indicates starch granules (*asterisks*).

Control bulbs (non-chilled) that did not sprout showed a more intense golden-orange staining in cross sections (Figure 2d) compared to the bulbs cultured in the cold chamber. At the beginning of sprouting (after 10 days of sprouting), alkaloids were mainly detected in the bulb scales and the newly formed sprout, while the parenchyma cells were greenish blue in color, indicating the absence of alkaloids (Figure 2e). Non-chilled, sprouted bulbs (after 14 days of sprouting) showed almost no alkaloids in the center of the bulb, while the thickened, outer parts of the bulb and sprout remained faintly colored, suggesting lower levels of alkaloids (Figure 2f).

Bulbs cultivated in the cold chamber for 4 weeks and then transferred to 24 °C, where they sprouted, showed the highest alkaloid content (Figure 2g–i), already at the beginning of sprouting after 7 days (Figure 2g). After 3 weeks of sprouting, most of the starch was found in the sprout, while the bulb scales mostly showed no staining for starch but a positive reaction for alkaloids (Figure 2h). Compared to the non-sprouted bulbs, the scales of the sprouted pre-chilled bulbs showed a very intense brown staining after 2 weeks of sprouting at 24 °C (Figure 2i).

Histoanatomical analysis did not reveal a distinctive difference between chilled and non-chilled bulbs (Figure 3). Microscopic observations showed that the scales of both non-chilled (Figure 3a) and chilled bulbs (Figure 3b) contained densely packed parenchyma cells. Amyloplasts with numerous starch granules were clearly visible in the scales, but they were more abundant in the chilled bulbs (Figure 3b). The scales of bulbs sprouted after cold treatment showed ruptured cells with a lower number of amyloplasts (Figure 3c), as revealed by the absence of blue-colored starch in the cross sections (Figure 2i).

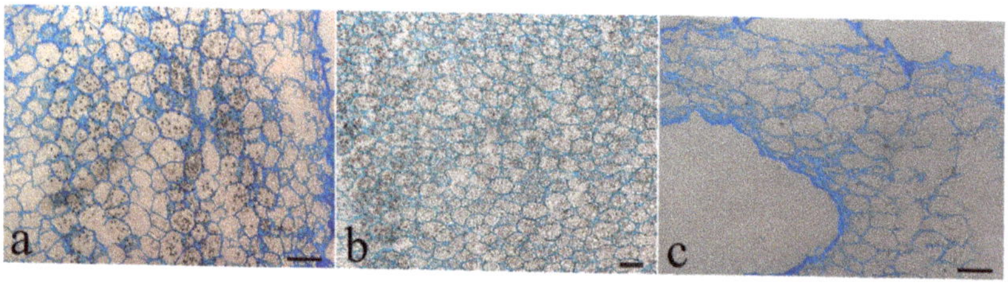

Figure 3. Histoanatomical analysis of chilled and non-chilled bulbs of *F. meleagris* cultured on medium without plant growth regulators. (**a**) Non-chilled bulb (control); (**b**) bulb cultured for 4 weeks at 7 °C; (**c**) bulb scale of the bulb sprouted after 4 weeks at 7 °C. Scale bars = 100 μm.

4. Discussion

All studied *Fritillaria* species contain alkaloids, which are essential constituents of the bulbs [29] and are responsible for the alleviation and treatment of pulmonary diseases. Nevertheless, *Fritillaria* species differ in their phytochemical composition and thus in their pharmacological activity [6]. We therefore hypothesized that *F. meleagris* might contain some specific valuable alkaloids that remain to be determined. Since *F. meleagris* is very rare and endangered and is already extinct in some places in Serbia [4], in vitro propagation techniques provide a very effective method for its large-scale propagule generation, thus expanding the source of the sought alkaloids. In the present study, we relied on our previous studies on in vitro propagation of *F. meleagris* and the possibilities of overcoming dormancy by cold treatment [23,25]. Our results showed that the highest alkaloid accumulation was observed in the bulbs after cold treatment, particularly in those that sprouted. The positive reactions to Dragendorff's and Wagner's reagents were localized in the bulb scales as well as in the newly formed sprouts. The strongest positive reaction for alkaloids, indicative of high alkaloid content, was observed in the bulb scales of pre-chilled bulbs that sprouted at 24 °C. Histological analysis showed that chilled and non-chilled bulbs differed only in the number of amyloplasts and that the number of amyloplasts with starch granules was greatly reduced after the onset of sprouting. Li et al. [30] also showed that alkaloid accumulation in the bulbs of *F. thunbergii* depends on the season. Therefore, for potential alkaloid production, it is preferable to cultivate the bulbs at low temperatures and use them for alkaloid isolation.

In vitro-derived bulblets of *F. cirrhosa* showed the presence of peimissine, the main isosteroidal alkaloid detected in commercial crude drug samples [31]. Lower but significant amounts of this alkaloid were also found in the callus, the concentration depending on light conditions during cultivation [31]. Considering that bulbs are the main repository for the alkaloids, the higher number of bulbs obtained in vitro is a significant result, since under natural conditions about one bulb per plant develops in about one year [32]. Moreover, the bulbs take several years to reach the desired size as they are geophytes and have a long life cycle of 270–280 days below ground and 80–90 days above ground [33]. The levels of isosteroidal alkaloids in the bulbs of *F. cirrhosa* are strongly influenced by environmental conditions and fluctuate with the age and reproductive stage of the plant [34]. The above results are very similar to our observations for the bulbs of *F. meleagris*. The environmental conditions led to a strong accumulation of alkaloids in the bulbs, which is in agreement with the findings that the occurrence of alkaloid-containing plants and the amount of alkaloids contained in their vegetative tissues depend on their growth habit and ecogeographical distribution [35]. Genetic studies also showed that in vitro-regenerated bulbs of *F. cirrhosa* had a higher alkaloid content than bulbs under natural conditions [36]. The differences in alkaloid content could be the result of evolutionary progress and the defensive role of alkaloids as barriers against fungal, microbial, and insect attack. The need to allocate defensive compounds to a particular plant part primarily depends on its susceptibility to predators but also varies with the developmental stage of the plant and environmental conditions [37]. The biosynthesis of alkaloids is often affected by temperature, which also influences respiration and photosynthesis [38]. Enhanced respiration and photosynthesis may lead to elevated protein synthesis and alkaloid production [38]. Despite the seasonal variation, Konchar et al. [34] pointed out that the reproductive stage of the plant is more consistent with the observed differences in alkaloid accumulation than any variation in the environment. However, in vitro-cultured plants are by definition considered as juvenile-stage plants [39]. Therefore, it is also possible that alkaloid production in *F. meleagris* bulbs grown at low temperatures is less influenced by environmental conditions than by genetic factors, as the developmental stage is triggered by temperature conditions.

Further improvement of in vitro techniques for bulblet propagation, such as liquid medium in bioreactors, could help to increase the number of new bulblets for the commercial production of alkaloids. The logical next step for further experimentation is to investigate the effects of additional factors that could enhance alkaloid production in the bulbs of *F. meleagris*, such as plant developmental stage, bulb size, in vitro culture conditions, plant growth regulators, and combinations thereof. The next stage of research must be the determination of alkaloids and their pharmaceutical activity. An interesting follow-up study would be to investigate the alkaloid shift in *F. meleagris* at different developmental stages and to sample more frequently at different time points. Histolocalization remains a powerful tool for solving the problem of cost-effective isolation of natural products, whose content varies according to the growth phase, and for optimizing research in the field of phytochemistry and phytotherapy.

5. Conclusions

This preliminary investigation revealed that alkaloids were mainly distributed in the bulbs and partly in the sprouts of in vitro grown *F. meleagris*. The most intense staining, indicative of high alkaloid content, was observed in the bulb scales of pre-chilled sprouted bulbs, rendering them preferable as in vitro plant material for alkaloid isolation.

Author Contributions: Conceptualization, M.M. and S.J.; Methodology, M.M. and S.J.; Investigation, M.M., M.T.-M., O.R., D.M.P., D.D.A.R. and B.U.; Writing—Original Draft Preparation, M.M. and B.U.; Writing—Review and Editing, M.M. and B.U.; Visualization—M.M., M.T.-M., O.R., D.M.P., D.D.A.R. and B.U.; Supervision, M.M.; Project Administration, M.M. All authors have read and agreed to the published version of the manuscript.

Funding: This work was supported by the Ministry of Science, Technological Development and Innovation of the Republic of Serbia, contract number 451-03-47/2023-01/200007. All experiments were conducted at the Department of Plant Physiology, Institute for Biological Research "Siniša Stanković"—National Institute of the Republic of Serbia, Belgrade, Serbia.

Data Availability Statement: Data are contained within the article

Conflicts of Interest: The authors declare no conflict of interest.

References

1. De Hertogh, A.A.; Le Nard, M. Physiological and biochemical aspects of flower bulbs. In *The Physiology of Flower Bulbs*; De Hertogh, A.A., Le Nard, M., Eds.; Elsevier: Amsterdam, The Netherlands, 1993; pp. 53–69.
2. Carasso, V.; Hay, F.R.; Probert, R.J.; Mucciarelli, M. Temperature control of seed germination in *Fritillaria tubiformis* subsp. *moggridgei* (Liliaceae) a rare endemic of the South-west Alps. *Seed Sci. Res.* **2011**, *21*, 33–38.
3. Tatarenko, I.; Walker, K.; Dyson, M. Biological Flora of Britain and Ireland: *Fritillaria meleagris*. *J. Ecol.* **2022**, *110*, 1704–1726. [CrossRef]
4. Tomović, G.; Vukojičić, S.; Niketić, M.; Zlatković, B.; Stevanović, V. *Fritillaria* (Liliaceae) in Serbia: Distribution, habitats and some taxonomic notes. *Phytol. Balc.* **2007**, *13*, 359–370.
5. Kukulczanka, K.; Kromer, K.; Czastka, B. Propagation of *Fritillaria meleagris* L. through tissue culture. *Acta Hortic.* **1989**, *251*, 147–153. [CrossRef]
6. Hao, D.-C.; Gu, X.-J.; Xiao, P.-G.; Peng, Y. Phytochemical and biological research of *Fritillaria* medicine resources. *Chin. J. Natl. Med.* **2013**, *11*, 330–344. [CrossRef]
7. Rashid, I.; Yaqoob, U. Traditional uses, phytochemistry and pharmacology of genus *Fritillaria*—A review. *Bull. Natl. Res. Cent.* **2021**, *45*, 124–161. [CrossRef]
8. Majnooni, M.B.; Fakhri, S.; Shokoohinia, Y.; Kiyani, N.; Stage, K.; Mohammadi, P.; Gravandi, M.M.; Farzaei, M.H.; Echeverria, J. Phytochemicals: Potential therapeutic interventions against coronavirus-associated lung injury. *Front. Pharmacol.* **2020**, *11*, 588467. [CrossRef]
9. Xiao, P.G.; Jiang, H.R.; Li, P.; Luo, Y.B.; Liu, Y. The botanical origin and pharmacophylogenetic treatment of Chinese material medica Beimu. *Acta Phytotaxon. Sin.* **2007**, *45*, 473–487. [CrossRef]
10. Li, H.; Jiang, Y.; Li, P. Chemistry, bioactivity and geographical diversity of steroidal alkaloids from the Liliaceae family. *Nat. Prod. Rep.* **2006**, *23*, 735–752. [CrossRef]
11. Li, S.L.; Lin, G.; Chan, S.W.; Ho, Y.P. Simultaneous determination of seven major isosteroidal alkaloids in bulbs of *Fritillaria* by gas chromatography. *J. Chromatogr. A* **2000**, *873*, 221–228. [CrossRef]
12. Lin, G.; Li, P.; Li, S.L.; Chan, S.W. Chromatographic analysis of *Fritillaria* isosteroidal alkaloids, the active ingredients of Beimu, the antitussive traditional Chinese medicinal herb. *J. Chromatogr. A* **2001**, *935*, 321–338. [CrossRef] [PubMed]
13. De Marco, M.B.; Accordini, S.; Cerveri, I.; Corsico, A.; Anto, J.M.; Künzli, N.; Janson, C.; Sunyer, J.; Jarvis, D.; Chinn, S.; et al. Incidence of chronic obstructive pulmonary disease in a cohort of young adults according to the presence of chronic cough and phlegm. *Am. J. Respir. Crit. Care Med.* **2007**, *175*, 32–39. [CrossRef] [PubMed]
14. Wang, C.Y.; Bai, X.Y.; Wang, C.H. Traditional Chinese medicine: A treasured natural resource of anticancer drug research and development. *Am. J. Chin. Med.* **2014**, *42*, 543–559. [CrossRef] [PubMed]
15. Wang, Y.; Hou, H.; Ren, Q.; Hu, H.; Yang, T.; Li, X. Natural drug sources for respiratory diseases from *Fritillaria*: Chemical and biological analyses. *Chin. Med.* **2021**, *16*, 40. [CrossRef] [PubMed]
16. Li, H.; Hung, A.; Li, M.; Yang, A.W.H. Fritillariae thunbergii bulbus: Traditional uses, phytochemistry, pharmacodynamics, pharmacokinetics and toxicity. *Int. J. Mol. Sci.* **2019**, *20*, 1667. [CrossRef]
17. Cunningam, A.B.; Brinckmann, J.A.; Pei, S.J.; Luo, P.; Schippmann, U.; Long, X.; Bi, J.F. High altitude species, high profits: Can the trade in wild harvested *Fritillaria cirrhosa* (Liliacea) be sustained? *J. Ethnopharmacol.* **2018**, *223*, 142–151. [CrossRef]
18. Paek, K.Y.; Murthy, H.N. High frequency of bulblet regeneration from bulb scale section of *Fritillaria thunbergii*. *Plant Cell Tissue Organ Cult.* **2002**, *68*, 247–252. [CrossRef]

19. Subotić, A.; Trifunović, M.; Jevremović, S.; Petrić, M. Morpho-histological study of direct somatic embryogenesis in endangered species *Fritillaria meleagris*. *Biol. Plant.* **2010**, *54*, 592–596. [CrossRef]
20. Atif, J.M.; Ahanger, M.A.; Amin, B.; Ghani, M.I.; Ali, M.; Cheng, Z. Mechanism of Allium crops bulb enlargement in response to photoperiod: A review. *Int. J. Mol. Sci.* **2020**, *21*, 1325. [CrossRef]
21. Petrić, M.; Jevremović, S.; Trifunović, M.; Tadić, V.; Milošević, S.; Dragićević, M.; Subotić, A. The effect of low temperature and GA_3 treatments on dormancy breaking and activity of antioxidant enzymes in *Fritillaria meleagris* bulblets cultured in vitro. *Acta Physiol. Plant.* **2013**, *35*, 3223–3236. [CrossRef]
22. Park, S.Y.; Paek, K.Y. *Bioreactor Culture of Shoots and Somatic Embryos of Medicinal Plants for Production of Bioactive Compounds*; Springer: Berlin/Heidelberg, Germany, 2014; pp. 337–368.
23. Marković, M.; Trifunović-Momčilov, M.M.; Uzelac, B.; Jevremović, S.; Subotić, A. Bulb dormancy in vitro—*Fritillaria meleagris*: Initiation, release and physiological parameters. *Plants* **2021**, *10*, 902. [CrossRef] [PubMed]
24. Petrić, M.; Subotić, A.; Trifunović, M.; Jevremović, S. Morphogenesis in vitro of *Fritillaria* spp. Invited review. In *Floriculture and Ornamental Biotechnology 6, Special Issue 1 Bulbous Ornamentals*; Tuyl, J.V., Arens, P., Eds.; Global Science Books: London, UK, 2012; pp. 78–89.
25. Petrić, M.; Subotić, A.; Jevremović, S.; Trifunović, M. Somatic embryogenesis and bulblet regeneration in snakehead fritillary (*Fritillaria meleagris* L.). *Afr. J. Biotechnol.* **2011**, *10*, 16181–16188.
26. Murashige, T.; Skoog, F. A revised medium for rapid growth and bioassays with tobacco culture. *Physiol. Plant.* **1962**, *15*, 473–497. [CrossRef]
27. Zhang, N.; Wang, M.; Li, Y.; Zhou, M.; Wu, T.; Cheng, Z. TLC–MS identification of alkaloids in Leonuri Herba and Leonuri Fructus aided by a newly developed universal derivatisation reagent optimised by the response surface method. *Phytochem. Anal.* **2021**, *32*, 242–251. [CrossRef] [PubMed]
28. Jensen, W.A. *Botanical Histochemistry: Principles and Practices*; WH Freeman and Company: San Francisco, CA, USA, 1962.
29. Kitajima, J.; Komori, T.; Kawasaki, T. Studies on the constituents of the crude drug "Fritillariae bulbus". III. On the diterpenoid constituents of fresh bulbs of *Fritillaria thunbergii* Miq. *Chem. Pharm. Bull.* **1982**, *30*, 3922–3931. [CrossRef]
30. Li, S.; Li, P.; Bian, Y.; Zhu, J.; Wang, M.; Lin, G. Studies on variations of alkaloids in zhebeimu (*Fritillaria thunbergii* Miq.) by improved colorimetric determination. *China J. Chin. Mater. Med.* **1999**, *24*, 592–594.
31. Chang, H.C.; Xie, H.M.; Lee, M.R.; Lin, C.Y.; Yip, M.K.; Agrawal, D.C.; Tsay, H.S. In vitro propagation of bulblets and LC-MS/MS analysis of isosteroidal alkaloids in tissue culture derived materials of Chinese medicinal herb *Fritillaria cirrhosa* D. Don. *Bot. Stud.* **2020**, *61*, 9. [CrossRef]
32. Paek, K.Y.; Sung, N.S.; Park, C.H. Several factors affecting bulblet regeneration from the culture of scale segment and node-bud in fritillary as medicinal bulbous plant. *Acta Hortic.* **1996**, *440*, 499–503. [CrossRef]
33. Kumar, P.; Partap, M.; Ashrita; Rana, D.; Kumar, P.; Warghat, A.R. Metabolite and expression profiling of steroidal alkaloids in wild tissues compared to bulb derived in vitro cultures of *Fritillaria roylei*—High value critically endangered Himalayan medicinal herb. *Ind. Crops Prod.* **2020**, *145*, 111945. [CrossRef]
34. Konchar, K.; Li, X.L.; Yang, Y.P.; Emshwillerr, E. Phytochemical variation in *Fritillaria cirrhosa* D. Don (*Chuan Bei Mu*) in relation to plant reproductive stage and timing of harvest. *Econ. Bot.* **2011**, *65*, 283–294. [CrossRef]
35. Levin, D.A. Alkaloid-bearing plants: An ecogeographic perspective. *Am. Nat.* **1976**, *110*, 261–284. [CrossRef]
36. Zhao, Q.; Li, R.; Zhang, Y.; Huang, K.; Wang, W.; Li, J. Transcriptome analysis reveals in vitro-cultured regeneration bulbs as a promising source for targeted *Fritillaria cirrhosa* steroidal alkaloid biosynthesis. *3 Biotech* **2018**, *8*, 191. [CrossRef] [PubMed]
37. McKey, D. Adaptive patterns in alkaloid physiology. *Am. Nat.* **1974**, *108*, 305–320. [CrossRef]
38. da Silva, G.C.; de-la-Cruz-Chacón, I.; Honório, A.B.M.; Martin, B.C.; Sousa, M.C.; Campos, F.G.; Boaro, C.S.F.; Ferreira, G. Temperature and GA3 as Modulating Factors in the Biosynthesis of Alkaloids during Imbibition and Early Development of Annona x atemoya Mabb. cv. 'Gefner' Seedlings. *Horticulturae* **2022**, *8*, 766. [CrossRef]
39. Croteau, R.; Felton, M.; Karp, F.; Kjonaas, R. Relationship of camphor biosynthesis to leaf development in sage (*Salvia officinalis*). *Plant Physiol.* **1981**, *67*, 820–824. [CrossRef]

horticulturae

MDPI

Communication

Optimization of In Vitro Regeneration of *Pinus peuce* (Gris.)

Dragana Stojičić [1], Snežana Budimir [2], Vlado Čokeša [3] and Branka Uzelac [2,*]

1 Department of Biology with Ecology, Faculty of Sciences and Mathematics, University of Niš, Višegradska 33, 18000 Niš, Serbia; dragana.stojicic@pmf.edu.rs
2 Department of Plant Physiology, Institute for Biological Research "Siniša Stanković"-National Institute of the Republic of Serbia, University of Belgrade, Bulevar despota Stefana 142, 11108 Belgrade, Serbia; budimir@ibiss.bg.ac.rs
3 Institute of Forestry, Kneza Višeslava 3, 11030 Belgrade, Serbia; vlado.cokesa@gmail.com
* Correspondence: branka@ibiss.bg.ac.rs

Abstract: *Pinus peuce* (Macedonian pine) is considered a valuable ornamental tree that is frequently planted in parks and gardens, especially in Western Europe. This endemic pine is one of the most valuable conifer species in its native range, which currently consists of only two disjunct populations restricted to small mountainous areas of the Balkans and is listed as a near-threatened species. The reproduction of Macedonian pine by seed is limited, so in vitro propagation methods have emerged as a promising tool for large-scale propagation. The objective of this study was to develop an improved system for the micropropagation of *P. peuce* from juvenile plant material using a short-term liquid cytokinin pulse. For that, explants derived from 4-week-old seedlings were pulse-treated with different concentrations of N^6-benzyladenine (BA) for 1 or 2 h to stimulate the induction of axillary buds. The highest axillary shoot formation was achieved with 222 μM BA pulse treatment, with an average number of ~six shoots per explant. Elongated shoots (≥10 mm) were detached from the explants and pulse-treated with 0.27 or 1.08 mM α-naphthaleneacetic acid (NAA) or 0.25 or 0.98 mM indole-3-butyric acid (IBA) for 1 or 2 h. IBA was more effective than NAA and led to a maximum rooting percentage (up to 40%) and the highest number of acclimatized plants (15–20%). Rooted plants were successfully transferred to ex vitro conditions.

Keywords: cytokinin pulse treatment; axillary bud induction; Macedonian pine; micropropagation

1. Introduction

Pinus peuce (Gris.), or Macedonian pine, was first discovered in 1939 by the German botanist August Grisebach on Mount Baba in Macedonia as a new species of five-needled pine that had not previously been recorded on the Balkan Peninsula [1]. *P. peuce* is phylogenetically isolated among the Balkan species of the genus *Pinus* and belongs to the subgenus *Strobus* according to the classification proposed by Gernandt et al. [2].

The Macedonian pine is a Tertiary relict and an endemic species of the Balkan Peninsula. Its current range is restricted to small mountainous areas of the Balkans between the northern latitudes of 41° and 43° and consists of two disjunct populations separated by the valley of the Vardar River [3,4]. The species is classified by Farjon [5] and IUCN [6] as "Near Threatened" under the criteria B2a.

P. peuce occurs on some of the highest Mediterranean and sub-Mediterranean mountainous regions of North Macedonia, Serbia, Montenegro, Bulgaria, Albania and Greece at altitudes ranging from 800–900 m to 2300–2400 m, with the optimum usually being 1600–1900 m [3]. This pine species favors cold mountain climates and high air humidity, occurring mainly on silicate soils and less frequently on carbonate ones. In its native range, it grows on soils usually poor in nutrients and derived from acid materials, but it can also be found on serpentine [7]. However, it is highly adaptable to different ecological sites and can therefore be found in a wide altitudinal range [8].

The Macedonian pine is one of the most valuable conifer species in its native range, where it mostly forms pure stands, but it can also form mixed stands with other conifers (Bosnian pine, black pine, fir) and deciduous species (beech). In the past, Macedonian pine forests used to form a spacious and powerful forest belt in the Balkans; however, this belt has been reduced to a limited, natural present-day range due to the negative anthropogenic impact [9]. The most extensive stands in its native range are found on north- and northwest-facing slopes [10]. The tree grows from the lower border of the submountain belt to the upper border of the subalpine forest belt [3,11]. This exceptional adaptability to harsh mountain climate conditions makes the species highly useful for afforestation at high altitudes and for protection against erosion [3,11].

P. peuce is considered a valuable, 35–40 m tall ornamental tree that is frequently planted in parks and gardens, particularly in Europe [4,10]. The durable wood of the tree is highly valued, and its resin is used in the chemical, optical and pharmaceutical industries, while the local populations use it to cure wounds, stomach diseases and other ailments [3]. Despite its slow juvenile growth, its basal area increment may later exceed that of other pines. Macedonian pine could be an important species for afforestation, as it appears to be less susceptible to biotic and abiotic hazards and is highly stable due to its deeply penetrating lateral root system, which renders it valuable for watershed and avalanche protection [3,4].

The natural reproduction of *P. peuce* by seeds starts in 5–6-year-old trees and is limited, as cones are formed every 2–4 years, while seeds exhibit low germination rates (~0.1%) unless stratified. The seeds of the Macedonian pine are deeply dormant and require chilling treatment at 3–5 °C for an average of 30 weeks to overcome dormancy [12,13]. The hard seed coat covered with resin appears to be an efficient barrier to water entry into the seed, which is necessary for the onset of germination, and a mechanical barrier to root emergence. The seeds of *P. peuce* are also characterized by undifferentiated or underdeveloped embryos [14].

In vitro propagation methods can help to overcome the problems encountered with classical methods of reproduction. Since the first report on the plantlet regeneration by organogenesis from mature zygotic embryos of *Pinus palustris* [15], considerable progress has been made in exploring opportunities for the use of in vitro vegetative propagation of conifers in forestry. Clonal propagation by somatic embryogenesis or organogenesis is still difficult in many conifer species and is often limited to the use of juvenile explants [16]. Another possibility is regeneration by propagating or promoting the axillary bud development of existing meristems. In *Pinus* species, axillary shoots arise from pre-existing quiescent meristems located adaxial to juvenile needles, in the axils of cotyledons, in the normally dormant short shoots surrounded by needle primordia, and adaxial to scale leaves (cataphylls) in older plant material [17–19].

Shoot proliferation via axillary meristems has several advantages over the de novo organogenic route. Besides rapidity and its time- and space-saving nature, the method of clonal propagation through axillary shoot formation is less prone to the risk of genetic instability [18]. Axillary shoot induction is a promising method for *P. peuce* reproduction in order to restore its former range and increase forest resilience through species diversification. The aim of the present study was to develop an improved system for the shoot propagation of this horticulturally valuable and potentially profitable pine species via axillary meristems from mature zygotic embryos using a short-term liquid cytokinin pulse.

2. Materials and Methods

2.1. Explant Source, Culture Medium and Culture Conditions

Cones of *Pinus peuce* (Gris.) were collected from open-pollinated trees in a seed orchard located on Mučanj Mountain (Serbia) (Figure 1). Prior to the experiments, the seeds were removed from the cones, washed under running tap water for 24 h, surface-disinfected with 25% (*v*/*v*) commercial bleach, "Snežnik" (Panonija AD, Pančevo, Serbia), in sterile deionized water (1% (*w*/*v*) sodium hypochlorite) for 25 min and rinsed three times with sterile distilled water. Seed coats were then removed, and the mature embryos were excised from the surrounding gametophytic tissue and placed vertically in test tubes on 1/4-strength Murashige and Skoog (MS) culture medium [20] supplemented with 2% (*w*/*v*) sucrose and solidified with 0.7% (*w*/*v*) agar (Torlak, Belgrade, Serbia). The pH of the medium was adjusted to 5.7 with 0.1 N NaOH before autoclaving at 114 °C for 25 min. The cultures were maintained under cool white fluorescent tubes with a photon flux density of 45 μmol m^{-2} s^{-1} and a 16 h photoperiod at 25 ± 2 °C.

Germinated embryos produced seedlings with upright hypocotyls, elongated cotyledons and brown radicles (Figure 2A). Four weeks after embryo germination, the radicle and part of the hypocotyl were removed to obtain 1.5–2.0 cm long explants (consisting of the stem apex, the cotyledons and the apical part of the hypocotyl) for further experiments.

Figure 1. *Pinus peuce*. (**A**) Adult trees of the Macedonian pine in a seed orchard on the Mučanj Mountain (Serbia). (**B**) Uppermost branches of cone-bearing Macedonian pine tree. (**C**) Immature (*white arrow*) and mature (*black arrow*) cones collected from open-pollinated *P. peuce* trees. (**D**) Seeds isolated from mature cones containing mature zygotic embryos that were used as initial explants.

2.2. Axillary Shoot Induction

Explants were transferred to 100 mL Erlenmeyer flasks containing 6-benzyladenine (BA) purchased from Sigma-Aldrich (Saint Louis, MO, USA) and dissolved in 0.1% (*v*/*v*) dimethyl sulfoxide (DMSO) (Sigma-Aldrich, Saint Louis, MO, USA) at the following concentrations: 4.4, 11.5, 22.5, 44.0, 220.0 and 440.0 μM. Control explants were treated with 0.1% (*v*/*v*) DMSO aqueous solution. Control and BA pulse-treated explants were agitated on a rotary shaker at 100 rpm for 1 h or 2 h at 25 °C. After cytokinin pulse treatment, the explants were deposited on a sheet of sterile filter paper placed inside an empty sterile Petri dish and left to dry in a laminar flow hood. After drying, the basal ends of the explants were immersed in Gresshoff and Doy culture medium [21], as modified by Sommer et al. [15] (GDmS medium), supplemented with 3% (*w*/*v*) sucrose, 0.5% (*w*/*v*) activated charcoal (Duchefa Biochemie BV, Haarlem, The Netherlands) and solidified with 0.7% (*w*/*v*) agar in 100-mL Erlenmeyer flasks. Explants were subcultured at 4-week intervals on a fresh medium of the same composition and grown for a total of 12 weeks, under the same light and temperature regime as used for culture initiation.

After 12 weeks of cultivation on a medium without plant growth regulators (PGRs), the number of explants developing shoots, the number of axillary shoots per explant, the shoot length and the number of shoots longer than 10 mm were recorded.

2.3. In Vitro Rooting

In order to induce adventitious rooting, a total of 207 elongated axillary shoots (≥10 mm) were isolated and transferred to 100 mL Erlenmeyer flasks containing auxins: α-naphthaleneacetic acid (NAA) at the concentrations of 0.27 or 1.08 mM or indole-3-butyric acid (IBA) at the concentrations of 0.25 or 0.98 mM, both purchased from Sigma-Aldrich (Saint Louis, MO, USA) and dissolved in 0.1% (*v*/*v*) DMSO. The control shoots were treated with 0.1% (*v*/*v*) DMSO aqueous solution. The control shoots and the auxin-pulse-treated shoots were agitated on a rotary shaker at 100 rpm for 1 h or 2 h at 25 °C. After auxin pulse treatment, the shoots were deposited on a sheet of sterile filter paper placed inside an empty sterile Petri dish and left to dry in a laminar flow hood. After drying, the shoots were transferred to test tubes containing 10 mL of half-strength GDmS medium, supplemented with 2% (*w*/*v*) sucrose and solidified with 0.7% (*w*/*v*) agar. Auxin-pulse-treated shoot cultures were maintained under the same environmental conditions as used in previous experiments for at least eight weeks.

2.4. Acclimatization

The rooted shoots were removed from the test tubes, washed with water to remove adhering culture medium and transferred to plastic pots with the substrate Substral® (Evergreen Garden Care Deutschland GmbH, Meinz, Germany), and then enclosed in a thin, transparent plastic cover to reduce water loss and maintained in a greenhouse with high humidity (70–80%). The plantlets were gradually acclimatized over a 4-week period by removing the thin cover daily at increasing intervals until final acclimatization. Plant survival was assessed after 16 weeks.

2.5. Experimental Design and Data Analysis

The axillary shoot induction experiment was conducted in a completely randomized block design. Five replicates, each consisting of six explants, were used per treatment ($n = 30$). The frequency of axillary shoot formation and the number of axillary shoots per explant were recorded with the aid of a Leica Wild M3Z stereomicroscope (Leica, Wetzlar, Germany) after 12 weeks of culture. To take into account both the mean number and frequency of axillary shoot formation, the bud-forming capacity (BFC) index was calculated as follows:

$$\text{BFC} = (\text{mean number of buds per explant}) \times (\%\text{ of explants forming buds})/100$$

For the root induction experiment, ten or eleven replicates were used per treatment, each consisting of a single axillary shoot immersed vertically into the medium, and the experiment was repeated (n = 20 or 21).

Percentage data were subjected to angular transformation prior to analysis. Statistical analyses were performed using Statistica 10 software (StatSoft, Hamburg, Germany). Data were subjected to standard analysis of variance (ANOVA), and means were separated using Duncan's test at a confidence level of $p < 0.05$. All results are presented as means ± standard errors.

3. Results and Discussion

BA-pulse-treated explants of *P. peuce* started to elongate after being transferred to a PGR-free medium, which was due to elongation of the hypocotyl rather than the shoot axis. Hypocotyl elongation was inversely proportional to the BA concentration used. In explants treated with higher BA concentrations, the shoot axes either failed to elongate or only elongated to a very small extent.

Two weeks after BA pulse treatment of the explants and their subsequent transfer to PGR-free medium, the first axillary buds were discerned as swollen protrusions beneath the stem tip of the explants. These axillary buds eventually elongated into axillary shoots. Both the 1 and 2 h control treatments exhibited a high frequency of responsive explants but a relatively low number of shoots per explant, which is why their BFC indices were the lowest (Table 1). The percentage of explants forming shoots in any of the BA pulse treatments did not differ significantly from that of the respective controls. However, for both BA pulse treatment durations, a significant increase in the number of shoots per explant was recorded for BA concentrations higher than 11.5 μM compared to the controls. The highest number of axillary shoots and the maximum BFC index were obtained with 222 μM BA pulse treatment, regardless of the treatment duration (Table 1).

Table 1. Effects of BA concentrations and exposure time on axillary shoot formation.

Treatment Duration	BA (μM)	Explants Forming Shoots (%)	Number of Shoots per Explant	Shoot Length (mm)	BFC Index	Number of Shoots ≥ 10 mm
1 h	-	90.00 ± 5.77 ᵃ	0.22 ± 0.10 ᵃ	5.17 ± 0.40 ᵃ	0.19	0
	4.4	83.33 ± 3.33 ᵃ	0.40 ± 0.14 ᵃ	8.00 ± 0.58 ᵇᶜ	0.33	2
	11.5	80.00 ± 5.77 ᵃ	0.88 ± 0.22 ᵃ	8.29 ± 0.39 ᵇᶜ	0.70	5
	22.5	86.67 ± 3.33 ᵃ	1.65 ± 0.25 ᵇ	**9.12 ± 0.32 ᶜ**	1.43	21
	44.0	83.33 ± 8.82 ᵃ	4.52 ± 0.34 ᵈ	**9.57 ± 0.30 ᶜ**	3.77	53
	222.0	80.00 ± 5.77 ᵃ	**5.50 ± 0.40 ᵉ**	6.69 ± 0.24 ᵃᵇ	**4.40**	19
	444.0	76.67 ± 6.67 ᵃ	3.17 ± 0.31 ᶜ	5.34 ± 0.50 ᵃ	2.43	6
2 h	-	83.33 ± 3.33 ᵃ	0.28 ± 0.11 ᵃ	4.43 ± 0.62 ᵃ	0.23	0
	4.4	86.67 ± 3.33 ᵃ	0.65 ± 0.18 ᵃ	7.12 ± 0.41 ᶜ	0.63	0
	11.5	83.33 ± 8.82 ᵃ	0.76 ± 0.19 ᵃ	7.84 ± 0.62 ᶜ	0.63	2
	22.5	80.00 ± 5.77 ᵃ	2.13 ± 0.33 ᵇ	**9.78 ± 0.36 ᵈ**	1.70	24
	44.0	83.33 ± 3.33 ᵃ	4.92 ± 0.36 ᶜ	**10.11 ± 0.28 ᵈ**	4.10	**63**
	222.0	83.33 ± 6.67 ᵃ	**5.96 ± 0.37 ᵈ**	6.85 ± 0.21 ᵇᶜ	**4.97**	17
	444.0	80.00 ± 5.77 ᵃ	2.63 ± 0.26 ᵇ	5.41 ± 0.31 ᵃᵇ	2.10	0

Values represent means ± SE, n = 30. Different letters within columns indicate significant differences between treatments according to Duncan's Multiple Range Test ($p \leq 0.05$). Maximum values are in bold.

For all treatments, the number of axillary shoots developed per explant was lower than the number of adventitious shoots previously reported to develop from cytokinin-induced adventitious buds of the Macedonian pine [22]. However, the axillary shoots were more vigorous and developed more rapidly, so the axillary buds elongated to shoots with average lengths of 9.57 mm (1 h pulse treatment) and 10.11 mm (2 h pulse treatment) within 12 weeks. Interestingly, BA stimulated shoot elongation at all but the highest concentrations tested, but even then, it had no inhibitory effect on shoot elongation (Table 1). However, explants treated with the highest BA concentrations (222 and 444 μM) produced a bundle of needles that were occasionally fasciated. In vitro culture conditions are often reported to lead to the formation of plantlets with abnormal morphology and physiology, but these abnormalities can be recovered after transfer to ex vitro conditions [23].

Axillary shoot multiplication is generally considered more difficult in conifers than in angiosperm trees [24]. Although conifers are particularly recalcitrant to clonal propagation, successful adventitious regeneration and axillary shoot multiplication have been described for a number of conifer species [25–32]. The main mechanism determining axillary meristem growth is the balance between cytokinin and auxin. One way to influence this balance is the addition of cytokinins or cytokinin-like PGRs to the tissue culture medium [32,33]. An alternative method of cytokinin application in bud induction and shoot formation is cytokinin administration via liquid pulse treatment [34]. A number of studies have reported improved organogenesis using the liquid pulse method [34–36]. The effect of liquid BA pulse treatment on the development of axillary shoots has also been described in several *Pinus* species [37,38], in which formed axillary shoots developed more rapidly and were more vigorous than their adventitious counterparts [27,37].

Shoots longer than 10 mm were used for the rooting experiment (Figure 2B). A small proportion (5%) of the non-pulsed shoots formed roots when maintained on a PGR-free medium for 8 weeks. IBA was more effective than NAA and resulted in a maximum rooting percentage and the highest number of acclimatized plants (Table 2). The IBA pulse significantly improved rooting compared to the control, regardless of the duration of the pulse treatment. NAA at either concentration did not significantly affect rooting percentage for any treatment duration.

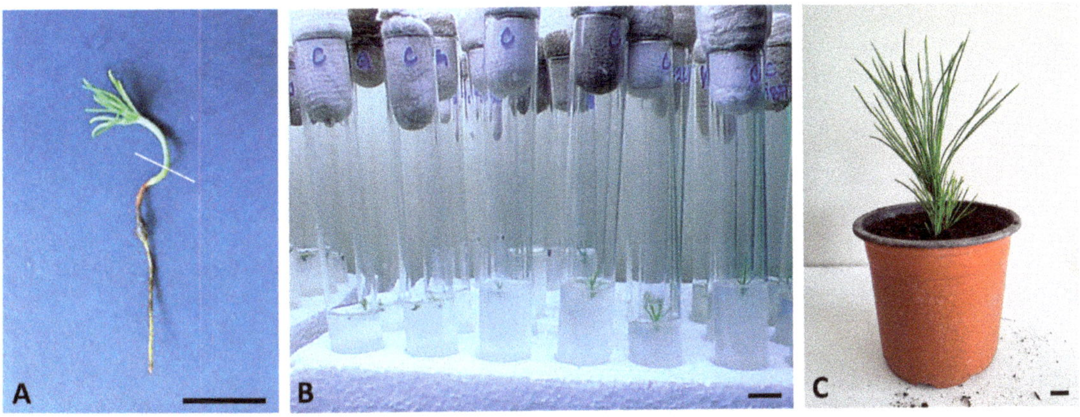

Figure 2. Overview of micropropagation of *Pinus peuce* by axillary shoots. (**A**) Four-week-old seedling, from which the radicle and part of the hypocotyl were removed (*white incision line*) to obtain explants for cytokinin pulse treatment. (**B**) Elongated axillary shoots detached from explants immediately after auxin pulse treatment and transfer to PGR-free medium. (**C**) *P. peuce* plant produced with the established micropropagation protocol after 21 months of acclimatization. Scale bar = 1 cm.

Table 2. Effects of auxin concentrations and exposure time on rooting and acclimatization effectiveness.

Treatment Duration	Auxins (mM)		Number of Shoots per Treatment	Rooting Percentage (%)	Number of Acclimatized Plants
	NAA	IBA			
1 h	-	-	20	5.00 ± 2.07 [a]	1
	0.27	-	21	19.05 ± 2.93 [abc]	2
	1.08	-	21	19.05 ± 2.93 [abc]	1
	-	0.25	21	33.33 ± 3.16 [bcd]	**4**
	-	0.98	20	**40.00 ± 2.15 [d]**	3
2 h	-	-	20	5.00 ± 2.07 [a]	1
	0.27	-	21	14.29 ± 2.93 [ab]	1
	1.08	-	21	23.81 ± 2.67 [abcd]	2
	-	0.25	21	**38.10 ± 2.67 [cd]**	**4**
	-	0.98	21	28.57 ± 2.07 [bcd]	**4**

Values represent means ± SE, $n = 20$. Different letters within columns indicate significant differences between treatments according to Duncan's Multiple Range Test ($p \leq 0.05$). Maximum values are in bold.

The formation of adventitious roots is a critical step in vegetative propagation that is essential for achieving an economically efficient production system as required in clonal forestry [39]. This complex developmental process, which is controlled by many environmental and endogenous factors, is considered a difficult step in conifers. The use of exogenous auxin is usually required for the induction of adventitious roots in difficult-to-root species. In most conifers, including *Pinus* spp., IBA and NAA are the most commonly used auxins [40]. However, the effectiveness of auxin pulses varies among species. A similar approach in the seedling explant culture of *Pinus elliottii* resulted in a very low rooting frequency (less than 1%) [37]. Humánez et al. [30] obtained similar rooting percentages in maritime pine *Pinus pinaster* shoots pulse-treated with significantly higher auxin concentrations (NAA 25 μM + IBA 4.4 μM) and a longer exposure time (48 h) at 4 °C before their transfer to potting mix in the greenhouse, whereas the application of auxin pulses before the transfer to a PGR-free medium or the presence of NAA (1 μM) in the rooting medium induced much lower rooting percentages (less than 15%).

Rooted plantlets of *P. peuce* were transferred to pots in the greenhouse, where they continued to grow as phenotypically normal plants (Figure 2C). Plants derived from IBA-pulse treated explants had higher acclimatization rates (15–20%) than those treated with NAA pulse (Table 2). Similar results were obtained in *P. elliottii* plantlets, where plants obtained on media containing IBA or IBA + NAA had higher acclimatization percentages than those obtained on NAA-containing media [41].

4. Conclusions

The results of this research demonstrate that short-term liquid cytokinin pulse treatment and a PGR-free medium provide a starting point for developing an effective method for axillary shoot proliferation in *P. peuce* that is suitable for true-to-type plant regeneration and large-scale micropropagation for the sustainable management of its unique ecosystem. The effectiveness of auxin pulses in adventitious root formation requires further optimization, possibly by including a cold pretreatment followed by ex vitro rooting, which could lead to a higher survival rate of acclimatized plants.

Author Contributions: Conceptualization, D.S. and S.B.; methodology, D.S. and V.Č.; validation, D.S., S.B. and B.U.; formal analysis, D.S. and V.Č.; investigation, D.S. and S.B.; resources, D.S. and V.Č.; writing—original draft preparation, D.S. and B.U.; writing—review and editing, B.U., D.S. and S.B.; visualization, B.U.; supervision, D.S.; project administration, D.S. All authors have read and agreed to the published version of the manuscript.

Funding: This research was funded by the Ministry of Science, Technological Development and Innovation of the Republic of Serbia, contract numbers 451-03-47/2023-01/200124, 451-03-47/2023-01/200007 and 451-03-47/2023-01/200027. The APC was funded by the Institute for Biological Research "Siniša Stanković"—the National Institute of the Republic of Serbia, University of Belgrade.

Data Availability Statement: Data are contained within the article.

Conflicts of Interest: The authors declare no conflicts of interest. The funders had no role in the design of the study; in the collection, analyses, or interpretation of data; in the writing of the manuscript; or in the decision to publish the results.

References

1. Fukarek, P. Otkriće i današnja rasprostranetost molike (*Pinus peuce* Gris.) (in Macedonian) [Discovery and present-day distribution of Macedonian pine (*Pinus peuce* Gris.)]. In Proceedings of the Symposium on Pinus peuce, Skopje, North Macedonia, 2–6 September 1969; pp. 17–25.
2. Gernandt, D.S.; López, G.G.; García, S.O.; Liston, A. Phylogeny and classification of *Pinus*. *Taxon* **2005**, *54*, 29–42. [CrossRef]
3. Alexandrov, A.H.; Andonovski, V. *Pinus peuce*—EUFORGEN Technical guidelines for genetic conservation and use for Macedonian pine (*Pinus peuce*). *Pinus peuce*—EUFORGEN European forest genetic resources programme. In Proceedings of the Bioversity International, Rome, Italy, 17 October 2011.
4. Savill, P.; Mason, B. *Pinus peuce* Griseb., Macedonian or Balkan pine. *Q. J. For.* **2015**, *109*, 245–252. Available online: https://rfs.org.uk/wp-content/uploads/2021/06/97-savill-mason-pinus-peuce-oct-2015.pdf (accessed on 20 November 2023).
5. Farjon, A. *Pinus peuce*. The IUCN Red List of Threatened Species 2017, e.T34193A95751594. Available online: https://www.iucnredlist.org/species/34193/95751594 (accessed on 20 November 2023).
6. IUCN (2022-2). The IUCN Red List of Threatened Species. Version 2022-2. Available online: https://www.iucnredlist.org (accessed on 20 November 2023).
7. Vidaković, M. *Četinjače. Morfologija i Varijabilnost [Conifers. Morphology and Variability, in Serbocroat]*; Yugoslav Academy of Science and Arts: Zagreb, Croatia, 1982.
8. Janković, M.; Bogojević, R. Neke karakteristike mikroklime munikovih šuma (*Pinetum heldreichii—Seslerietum autumnalis* M. Jank. Et R. Bog.) na Ošljaku, Šarplanina (in Serbian) [Some characteristics of microclimate in *Pinus heldreichii* forests—*Pinetum heldreichii-Seslerietum autumnalis* M. Jank. Et R. Bog.—on Ošljak, Šarplanina]. In Proceedings of the Symposium on Pinus heldreichii Christ., Dečani, Serbia,, 4–7 September 1972; pp. 134–145.
9. Janković, M. Neki problemi ekologije, cenologije i rasprostranjenja endemoreliktne balkanske vrste *Pinus peuce* [Some problems of ecology, cenology and distribution of the endemorelic Balkan pine species *Pinus peuce*]. In Proceedings of the Symposium on Pinus peuce, Skopje, North Macedonia, 2–6 September 1969; pp. 173–178.
10. Holzer, K. Intrinsic qualities and growth potential of *Pinus cembra* and *Pinus peuce* in Europe. In *Biology of Rust Resistance in Forest Trees: Proceedings of a NATO-IUFRO Advanced Study Institute, 17–24 August 1969*; U.S. Department of Agriculture, National Agricultural Library: Washington, DC, USA, 1972; US Department of Agriculture Miscellaneous Publication 1221. Available online: https://www.biodiversitylibrary.org/ (accessed on 1 December 2023).

11. Jovanović, B. *Dendrologija sa Osnovama Fitocenologije [Dendrology with Basics Phytocenology, in Serbian]*; Naučna knjiga: Belgrade, Serbia, 1971; pp. 123–127.
12. Stilinović, S. *Semenarstvo šumskog i Ukrasnog Drveća i žbunja [Seed Production of Forest and Ornamental Trees and Shrubs, in Serbocroat]*; Intitute of Forestry, Faculty of Forestry: Belgrade, Serbia, 1985.
13. Gosling, P. *Raising Trees and Shrubs from Seed*; Forestry Commission: Edinburgh, UK, 2007.
14. Đorđeva, M.; Ničota, B.; Stamenkov, M. Rtlivosta na semeto odmolika (*Pinus peuce* Gris.) i pojava na nerazvien inedorazvien embrion kajnego (in Macedonian) [Fertility of the seedof Macedonian pine (*Pinus peuce* Gris.) and the appearance of undeveloped and underdeveloped embryos]. In Proceedings of the Symposium on *Pinus peuce*, Skopje, North Macedonia, 2–6 September 1969; pp. 119–126.
15. Sommer, H.E.; Brown, C.L.; Kormanik, P.P. Differentiation of plantlets in longleaf pine (*Pinus palustris* Mill) tissue cultured in vitro. *Bot. Gaz.* **1975**, *136*, 196–200. [CrossRef]
16. Bonga, J.M.; Klimaszewska, K.K.; von Aderkas, P. Recalcitrance in clonal propagation, in particular of conifers. *Plant Cell Tissue Organ Cult.* **2010**, *100*, 241–254. [CrossRef]
17. Toribio, M.; Pardos, J.A. Scots pine (*Pinus sylvestris* L.). In *Trees II. Biotechnology in Agriculture and Forestry*; Bajaj, Y.P.S., Ed.; Springer: Berlin/Heidelberg, Germany, 1989; Volume 5, pp. 479–506. [CrossRef]
18. Schwarz, O.J.; Schlarbaum, S.E.; Burns, J.A. Tissue culture micropropagation of conifers. In Proceedings of the Southern Regional Information Exchange Group Biennial Symposium on Forest Genetics "Applications of Vegetative Propagation in Forestry", Huntsville, AL, USA, 8–10 July 1992; pp. 1–22. Available online: https://www.srs.fs.usda.gov/pubs/gtr/gtr_so108.pdf (accessed on 29 November 2023).
19. de Oliveira, L.F.; Quoirin, M.; Koehler, H.S.; Amano, E.; Higa, A.R.; Ribas, L.L.F. Propagation from axillary buds and anatomical study of adventitious roots of *Pinus taeda* L. *Afr. J. Biotechnol.* **2013**, *12*, 5413–5422. [CrossRef]
20. Murashige, T.; Skoog, F. A revised medium for rapid growth and bioassays with tobacco tissue cultures. *Physiol. Plant.* **1962**, *15*, 473–497.
21. Gresshoff, P.M.; Doy, C.H. Development and differentiation of haploid *Lycopersicon esculentum* (tomato). *Planta* **1972**, *107*, 161–170.
22. Stojičić, D.; Janošević, D.; Uzelac, B.; Čokeša, V.; Budimir, S. Micropropagation of *Pinus peuce*. *Biol. Plant.* **2012**, *56*, 362–364.
23. Pospíšilová, J.; Tichá, L.; Kadleček, P.; Haisel, D.; Plzáková, Š. Acclimatization of micropropagated plants to ex vitro conditions. *Biol. Plant.* **1999**, *42*, 481–497. [CrossRef]
24. Burrows, G.E. Gymnosperm Resprouting—A Review. *Plants* **2021**, *10*, 2551. [CrossRef]
25. Nour, K.A.; Yeung, E.C.; Thorpe, T.A. Shoot Bud Histogenesis from Mature Embryos and Shoots of Eastern White Cedar (*Thuja occidentalis* L.) Cultured In vitro. *Int. J. Plant Sci.* **1993**, *154*, 378–385. [CrossRef]
26. Capuana, M.; Giannini, R. In Vitro Plantlet Regeneration from Embryonic Explants of *Pinus pinea* L. *Vitro Cell. Dev. Biol.-Plant* **1995**, *31*, 202–206.
27. Stojičić, D.; Budimir, S.; Ćulafić, L. Micropropagation of *Pinus heldreichii*. *Plant Cell Tissue Organ Cult.* **1999**, *59*, 147–150. [CrossRef]
28. Kowalski, B.; van Staden, J. Micropropagation of *Podocarpus henkelii* and *P. elongatus*. *S. Afr. J. Bot.* **2001**, *67*, 362–366. [CrossRef]
29. Renau-Morata, B.; Ollero, J.; Arrillaga, I.; Segura, J. Factors influencing axillary shoot proliferation and adventitious budding in cedar. *Tree Physiol.* **2005**, *25*, 477–486. [CrossRef]
30. Humánez, A.; Blasco, M.; Brisa, C.; Segura, J.; Arrillaga, I. Thidiazuron enhances axillary and adventitious shoot proliferation in juvenile explants of Mediterranean provenances of maritime pine *Pinus pinaster*. *Cell. Dev. Biol.-Plant* **2011**, *47*, 569–577. [CrossRef]
31. Xiong, Y.P.; Chen, S.Y.; Guo, B.Y.; Niu, M.Y.; Zhang, X.H.; Li, Y.; Wu, K.L.; Zheng, F.; da Silva, J.A.T.; Zeng, S.J.; et al. An efficient micropropagation protocol for *Metasequoia glyptostroboides* Hu et Cheng from shoot segments of 2-year-old trees. *Trees* **2020**, *34*, 307–313. [CrossRef]
32. Pereira, C.; Montalbán, I.A.; Pedrosa, A.; Tavares, J.; Pestryakov, A.; Bogdanchikova, N.; Canhoto, J.; Moncaleán, P. Regeneration of *Pinus halepensis* (Mill.) through Organogenesis from Apical Shoot Buds. *Forests* **2021**, *12*, 363. [CrossRef]
33. Wilms, H.; De Bièvre, D.; Longin, K.; Swennen, R.; Rhee, J.; Panis, B. Development of the first axillary in vitro shoot multiplication protocol for coconut palms. *Sci. Rep.* **2021**, *11*, 18367. [CrossRef]
34. Bornman, C.H. Possibilities and constraints in the regeneration of trees from cotyledonary needles of *Picea abies* in vitro. *Physiol. Plant.* **1983**, *57*, 5–16.
35. Goldfarb, B.; Howe, G.T.; Baily, L.M.; Strauss, S.H.; Zaerr, J.B. A liquid cytokinin pulse induces adventitious shoot formation from Douglas-fir cotyledons. *Plant Cell Rep.* **1991**, *10*, 156–160. [CrossRef] [PubMed]
36. Drake, P.M.W.; John, A.; Power, J.B.; Davey, M.R. Cytokinin pulse-mediated shoot organogenesis from cotyledons of Sitka spruce (*Picea sitchensis* (Bong.) Carr.) and high frequency in vitro rooting of shoots. *Plant Cell Tissue Organ Cult.* **1997**, *50*, 147–151. [CrossRef]
37. Burns, J.A.; Schwarz, O.J.; Schlarbaum, S.E. Multiple shoot production from seedling explants of slash pine (*Pinus elliottii* Engelm.). *Plant Cell Rep.* **1991**, *10*, 439–443. [CrossRef] [PubMed]
38. Stojičić, D.; Budimir, S. Cytokinin-mediated axillary shoot formation in *Pinus heldreichii*. *Biol. Plant.* **2004**, *48*, 477–479. [CrossRef]
39. George, E.F.; Debergh, P.C. Micropropagation: Uses and Methods. In *Plant Propagation by Tissue Culture*, 3rd ed.; George, E.F., Hall, M.H., De Klerk, G.J., Eds.; Springer Press: Dordrecht, The Netherlands, 2008; Volume 1, pp. 29–64.

40. Ragonezi, C.; Klimaszewska, K.; Castro, M.R.; Lima, M.; de Oliveira, P.; Zavattieri, M.A. Adventitious rooting of conifers: Influence of physical and chemical factors. *Trees* **2010**, *24*, 975–992. [CrossRef]
41. Nunes, S.; Sousa, D.; Pereira, V.T.; Correia, S.; Marum, L.; Santos, C.; Dias, M.C. Efficient protocol for in vitro mass propagation of slash pine. *Cell. Dev. Biol.-Plant* **2018**, *54*, 175–183. [CrossRef]

horticulturae

MDPI

Article

Metabolism of Fluorinated Topolin Cytokinins in Micropropagated *Phalaenopsis amabilis*

Nino Murvanidze [1,*], Karel Doležal [2,3], Lenka Plačková [3] and Stefaan P. O. Werbrouck [1,*]

1 Laboratory for Applied In Vitro Plant Biotechnology, Ghent University, Valentin Vaerwyckweg 1, 9000 Ghent, Belgium
2 Department of Chemical Biology, Faculty of Science, Palacký University, Šlechtitelů 27, 78371 Olomouc, Czech Republic; karel.dolezal@upol.cz
3 Laboratory of Growth Regulators, Faculty of Science, Palacký University & Institute of Experimental Botany AS CR, Šlechtitelů 27, 78371 Olomouc, Czech Republic; lenka.plackova@upol.cz
* Correspondence: nina.murvanidze@gmail.com (N.M.); stefaan.werbrouck@ugent.be (S.P.O.W.)

Abstract: Fluorinated cytokinins have emerged as promising alternatives to traditional cytokinins in Phalaenopsis plant tissue culture, offering enhanced stability and bioactivity. However, their metabolic fate and impact on endogenous cytokinin profiles remain largely unexplored. This study builds upon previous research to investigate the comparative metabolism of the traditional cytokinin 6-Benzylaminopurine (BA) with the successful alternatives 6-(3-fluorobenzylamino)purine (FmT) and 6-(3-fluorobenzylamino)purine 9-riboside (FmTR). Additionally, this study examines the impact of another crucial factor, the use of ventilated versus closed containers, on metabolic processes. The results revealed the distinct metabolic profiles associated with each treatment, highlighting the complex interplay between exogenous and endogenous cytokinin levels. This study is the first to investigate the effects of these stable, synthetic, and exogenous cytokinins on the naturally occurring cytokinin levels and their metabolites in micropropagated Phalaenopsis. Additionally, we proposed an alternative inactivation pathway involving the conversion of FmTR and BA to pT and pTR. These findings provide valuable insights into the intricate relationship between cytokinin metabolism and plant growth under in vitro conditions.

Keywords: *Phalaenopsis*; orchid; cytokinin metabolism; in vitro tissue culture; micropropagation

1. Introduction

Cytokinins (CKs), a class of essential plant hormones, play a pivotal role during plant growth in vitro. These phytohormones also regulate numerous developmental processes, including meristem maintenance, cell division, shoot formation, and leaf senescence [1]. During in vitro micropropagation, the exogenous application of CKs is crucial for achieving optimal plant multiplication and quality [2,3]. Plants have evolved complex mechanisms to maintain homeostasis between the exogenous supply and endogenous levels of CKs through intricate metabolic pathways, essential for overall plant growth and development [4]. Despite their widespread use in micropropagation, the comprehensive metabolic profiles of CKs have only been elucidated for a limited number of plant species. Moreover, the relative abundance of various cytokinin metabolites is influenced by the specific type of applied CKs and exhibits interspecies variability [5,6].

Orchid micropropagation through plant tissue culture techniques has been extensively studied and applied over the years [7]. Phalaenopsis micropropagation is crucial for commercial production since traditional propagation methods, such as cutting off-shoots, are characterized by a low multiplication rate [8]. However, the in vitro clonal propagation of *Phalaenopsis* still faces challenges for large-scale production. The choice of CKs significantly influences shoot proliferation, in addition to the culture system and media components. Common CKs used in orchid tissue culture include BA (N6-benzylaminopurine), kinetin,

2iP (2-isopentyladenine), and TDZ (thidiazuron), often combined with auxins like 2,4-D (2,4-dichlorophenoxyacetic acid), NAA (α-naphthaleneacetic acid), and IAA [7,9,10]. In our study, we showed that supplementing in vitro media with fluorinated CKs improved the micropropagation rate compared to traditionally applied BA [11]. Our recent research has demonstrated that fluorinated CKs significantly enhance shoot proliferation in orchids grown in vitro, particularly in hermetically sealed containers [11]. However, the mechanisms behind this phenomenon remain elusive. To explore this subject, we performed a study focusing on the metabolites of the traditionally used 6-benzylaminopurine (BA)-type cytokinin and "second"-generation topolins 6-(3-fluorobenzylamino)purine (FmT) and 6-(3-fluorobenzylamino)purine 9-riboside (FmTR) with supposed improved stability and bioactivity [12]. Understanding the metabolic response of orchids to fluorinated CKs is crucial for developing optimized micropropagation protocols, not only for orchids but potentially for other recalcitrant plant species as well.

Upon absorption from the culture medium, CKs are subjected to extensive metabolic transformations within plant tissues, resulting in a dynamic array of metabolites with varying biological activities [13]. They encompass active forms, predominantly free bases like BA, meta-topolin (mT), 2iP, and zeatin (Z), which directly influence plant growth and development. These active forms are often converted into less active transport forms, including ribosides and ribotides, facilitating their movement within the plant [14]. Additionally, CKs can be reversibly stored as O-glucosides, which are inactive conjugates that can be readily hydrolyzed back to their active free base forms when required. Conversely, N-glucosides (such as N7 glucosides and N9-glucosides) are generally considered to be irreversible and completely inactive storage forms [14,15]. This intricate metabolic network allows plants to tightly regulate cytokinin levels in response to developmental and environmental cues, a critical factor in maintaining plant homeostasis. In the context of in vitro culture, where exogenous cytokinin application and the absence of natural environmental cues can disrupt this delicate balance, understanding and manipulating cytokinin metabolism is important for achieving the optimal conditions for de novo organogenesis and successful micropropagation [15–18].

As fluorinated CKs have shown enhanced shoot proliferation in phalaenopsis, in this study, we aim to understand the response of in vitro plants to these fluorinated CKs. We therefore investigated the metabolic profile of applied CKs. These insights offer a deeper understanding of the intricate relationship between cytokinin metabolism and plant growth in a controlled environment, which could lead to optimized micropropagation protocols (not only for this valuable ornamental plant but also for other challenging crops). We demonstrate the balance between endogenous and exogenous CKs and interaction with the culture conditions.

2. Materials and Methods

2.1. Plant Material and Growth Conditions

An unnamed cultivar of *Phalaenopsis amabilis* served as the model plant for this study. The initial plant material was propagated on a 2.14 g/L MS (Murashige and Skoog, 1962) basal medium supplemented with 10 μM FmTR. Following a two-month growth period, these shoots were transferred to the fresh medium of the same composition, but supplemented with one of three different CKs: FmT, FmTR (Olchemim, Olomouc, Czech Republic), or BA, each at a fixed concentration of 10 μM. Axillary shoots (1 cm) with two young leaves were separated from in vitro grown plants. Five explants were transferred to each 565 mL polypropylene container containing 100 mL of medium. The lids of these recipients were tightly closed or contained a 'Large' filter ('L' = #10 filter) (Microbox™, Eco2box, Geraardsbergen, Belgium). The cultures were maintained for 10 weeks at 22 ± 2 °C in a 16/8 light/dark photoperiod (40 μmol m^{-2} S^{-1} PAR) provided by white fluorescence tubes.

2.2. Cytokinin Analyses

Five plants were randomly selected from each treatment group and weighed before lyophilization. Additionally, leaves from a potted Phalaenopsis plant were collected to serve as a control/reference sample for comparison. All samples were stored at −80 °C before CK analysis. Cytokinin metabolites were quantified using a modified protocol by Novák et al. [19]. In brief, lyophilized samples were finely ground in liquid nitrogen and 5 mg of dry weight was aliquoted for analysis. Subsequently, 1 µL of cold extraction solution (Modified Bieleski buffer MeOH/H_2O/HCOOH (15/4/1), Bieleski, 1964) and 23 deuterium-labelled cytokinin (CK) internal standards (Olchemim, Olomouc, Czech Republic) were added to each sample. The internal standards, applied at concentrations of 1 pmol (free bases) and 2 pmol (O-glucoside and nucleotides) per sample, served to assess the recovery of target CKs during purification and validate their quantification. Following homogenization in a vibration mill (27 Hz, 3 min), sonication (4 °C, 5 min), and agitation (4 °C, 30 min), samples were centrifuged (20,000 rpm, 10 min, 4 °C) and the resulting pellets were re-extracted using the same procedure. The collected supernatants were then subjected to a multi-step purification process involving octadecyl C18 (100 mg/1 mL) speed SPE cartridges, Oasis MCX (30 mg/1 mL) columns, and immunoaffinity chromatography (IAC). Blank samples were prepared in parallel using an identical procedure.

Detailed purification and quantification procedures of CKs and its metabolites were previously described by Plačková [20]. Briefly, samples were analyzed using an Acquity UPLC™ system (Waters, Milford, CT, USA) coupled with a Xevo™ TQ MS™ (Waters, Milford, CT, USA) triple quadrupole mass spectrometer equipped with an electrospray ionization (ESI) interface. CK quantification was achieved via the multiple reaction monitoring (MRM) of [M + H]+ and the appropriate product ions. The optimal MRM conditions (dwell time, cone voltage, and collision energy) were determined for each CK. The ratio of CKs to their corresponding labelled standards was used to quantify CK levels based on the known internal standard concentrations [16]. All CK quantifications were performed in triplicate.

2.3. Data Analysis

The experiment followed a completely randomized design with three repetitions for each sample to assess CKs. Each treatment included three replicates, and each replicate consisted of 10 explants per culture vessel. CKs were measured in the leaves of plants grown in vitro for 10 weeks. An independent two-sample *t*-test was conducted across all means using Python 3.10 to compare the means of different sample groups. Significant difference was set at $p < 0.05$. This statistical approach was chosen to determine if there were significant differences between the independent sample groups under various conditions. All tests were performed assuming unequal variances.

3. Results

Effect of Exogenously Administered CKs on Total Quantified Cytokinin Levels and Their Metabolic Profile

The total content of CKs (expressed as pmol/1 g FW) was determined in eight-week-old in vitro shoots of *Phalaenopsis* grown in the presence of 10 µM BA, 10 µM FmT, and 10 µM FmTR, and was compared between plants grown in ventilated or hermetically sealed containers. A total of 48 different CKs and their metabolites were detected in varying concentrations and are listed in Table 1. The concentrations of different cytokinin types vary depending on the treatment and container type; both aromatic CKs (such as BA, ortho-topolin (oT), meta-topolin (mT), para-topolin (pT), and their derivatives) and isoprenoid CKs (such as trans-zeatin (tZ), cis-zeatin (cZ), dihydrozeatin (DHZ), isopentenyladenine (iP), and its derivatives) were detected, with diverse abundance across treatments and container types.

The presence and abundance of specific cytokinin metabolites varied significantly under different treatment and culture conditions. For instance, tZ7G, present in trace amounts under BA treatment, was completely absent in plants grown in closed containers and supplemented with FmT or FmTR. Additionally, the abundance of cZ and its riboside form, cZR, exhibited distinct patterns depending on the treatment and container type. While cZ remained detectable at minimal concentrations in BA-treated plants, it fell below the limit of detection (LOD) in FmT- or FmTR-treated plants. Intriguingly, cZR was consistently abundant across all treated plants.

Table 1. Concentrations of CKs (pmol/g FW) and its metabolites in *Phalaenopsis* shoots grown in ventilated and closed containers in vitro.

	Cytokinin Treatment					
	BA		FmT		FmTR	
Cytokinin	Container Type					
Metabolites	Ventilated	Closed	Ventilated	Closed	Ventilated	Closed
tZ	<LOD	<LOD	<LOD	<LOD	<LOD	<LOD
tZOG	<LOD	<LOD	<LOD	<LOD	<LOD	<LOD
tZR	<LOD	<LOD	<LOD	<LOD	<LOD	<LOD
tZROG	<LOD	<LOD	<LOD	<LOD	<LOD	<LOD
tZ7G	20.2 ± 2.07 a	4.9 ± 0.2 b	1.6 ± 0.05 c	<LOD	<LOD	<LOD
tZ9G	<LOD	<LOD	<LOD	<LOD	<LOD	<LOD
tZR5′MP	<LOD	<LOD	<LOD	<LOD	<LOD	<LOD
cZ	20.8 ± 2.5 a	11.09 ± 1.4 b	<LOD	<LOD	<LOD	<LOD
cZOG	<LOD	<LOD	<LOD	<LOD	<LOD	<LOD
cZR	2059.5 ± 96.3 a	1296.1 ± 9.8 b	743.5 ± 56.5 c	893.7 ± 77.0 d	1362.8 ±129.3 b	270.5 ± 21.8 e
cZROG	26.1 ± 1.7 a	11.3 ± 0.5 b	<LOD	3.85 ± 0.2 d	<LOD	4.1 ± 0.1 c
cZ7G	<LOD	<LOD	<LOD	<LOD	<LOD	<LOD
cZ9G	<LOD	<LOD	<LOD	8.6 ± 0.2 a	<LOD	6.1 ± 0.4 b
cZR5′MP	<LOD	74.6 ± 5.3 a	<LOD	40.7 ± 3.6 c	<LOD	66.1 ± 3.9 b
DHZ	<LOD	<LOD	<LOD	<LOD	<LOD	<LOD
DHZOG	<LOD	<LOD	<LOD	<LOD	<LOD	<LOD
DHZR	5.7 ± 0.8 a	3.8 ± 0.3 b	3.4 ± 0.06 b	1.5 ± 0.2 c	<LOD	<LOD
DHZROG	164.1 ± 4.8 b	<LOD	296.6 ± 20.1 a	<LOD	34.9 ± 1.1 c	<LOD
DHZ7G	<LOD	3.3 ± 0.07 b	<LOD	2.4 ± 0.2 c	<LOD	3.8 ± 0.1 a
DHZ9G	<LOD	<LOD	<LOD	<LOD	<LOD	<LOD
DHZR5′MP	<LOD	<LOD	<LOD	<LOD	<LOD	<LOD
iP	8.6 ± 0.1 a	5.2 ± 0.1 b	1.8 ± 0.1 d	3.1 ± 0.2 c	3.1 ± 0.3 c	2.5 ± 0.2 c
iPR	272.9 ± 4.7 a	125.8 ± 5.3 b	53.2 ± 1.1 d	22.1 ± 0.4 f	87.07 ± 3.8 c	23.8 ±1.2 e
iP7G	<LOD	<LOD	<LOD	<LOD	<LOD	<LOD
iP9G	<LOD	<LOD	<LOD	<LOD	<LOD	<LOD
iPR5′MP	<LOD	21.5 ± 2.03 a	<LOD	14.4 ± 1.3 b	<LOD	20.01 ± 0.8 a
BA	1398.3 ± 151.1 b	4159.68 ± 497.2 a	9.03 ± 0.6 d	15.4 ± 3.1 c	<LOD	<LOD
BAR	84.04 ± 2.9 a	61.4 ± 1.3 b	0.5 ± 0.09 e	<LOD	1.1 ± 0.08 d	3.1 ± 0.3 c
BA7G	<LOD	<LOD	<LOD	<LOD	<LOD	<LOD
BA9G	106.5 ± 3.4 a	37.3 ± 1.3 b	<LOD	<LOD	<LOD	<LOD
BAR5′MP	<LOD	19.1 ± 0.4	<LOD	<LOD	<LOD	<LOD
mT	30.8 ± 1.9 a	16.1 ± 0.3 b	<LOD	9.1 ± 1.0 d	<LOD	19.4 ± 1.3 c
mTR	67.2 ± 6.3 a	17.5 ± 0.9 e	27.1 ± 1.3 c	7.8 ± 0. 2 f	32.2 ± 2.0 b	26.6 ± 1.8 d
mT7G	<LOD	<LOD	<LOD	<LOD	<LOD	<LOD
mT9G	<LOD	<LOD	<LOD	16.8 ± 0,1 a	<LOD	15.1 ± 1.3 b
oT	<LOD	<LOD	<LOD	<LOD	<LOD	<LOD
oTR	17.1 ± 1.5 a	4.7 ± 0.4 b	<LOD	<LOD	<LOD	<LOD
oT7G	<LOD	<LOD	<LOD	<LOD	<LOD	<LOD
oT9G	<LOD	<LOD	<LOD	<LOD	<LOD	<LOD
pT	147.6 ± 9.4 b	27.8 ± 2.3 c	140.7 ± 10.6 b	<LOD	196.5 ± 6.1 a	<LOD
pTR	1242.2 ± 27.8 b	251.9 ± 17.8 d	1142.1 ± 24.9 c	21.6 ± 0.8 f	1346.1 ± 74.4 a	25.8 ±1.3 e
pT7G	<LOD	<LOD	<LOD	<LOD	<LOD	<LOD
pT9G	<LOD	<LOD	<LOD	<LOD	<LOD	<LOD
K	<LOD	<LOD	<LOD	<LOD	<LOD	<LOD
KR	<LOD	<LOD	<LOD	<LOD	<LOD	<LOD
K9G	<LOD	<LOD	<LOD	<LOD	<LOD	<LOD
FmT	99.2 ± 11.2 f	140.8 ± 11.4 e	1952.3 ± 220.9 c	7836.4 ± 713.3 a	326.1 ± 25.4 d	2350.7 ± 360.5 b
FmTR	844.4 ± 107.2 c	804.2 ± 88.6 d	627.4 ± 44.5 e	490.7 ± 41.03 f	4478.6 ± 150.7 b	12,156.5 ± 1712.8 a

Values are presented as mean ± SD; (n = 3) LOD = below detection limit. Means in the same row followed by different letters are significantly different ($p \leq 0.05$) according to the t-test.

Endogenous isoprenoid metabolites, specifically the reversible storage form dihydrozeatin-O-glucoside riboside (DHZROG) and the transport form isopentenyladenine riboside (iPR), exhibited higher concentrations in ventilated containers. BA treatment generally led to elevated concentrations of BA, BAR, and their derivatives, while FmT and FmTR treatments led to the accumulation of fluorinated CKs. Notably, less glucosylation of BA occurred in closed containers, with higher levels of free BA observed. Further, small amounts of FmT were transformed into BA in both ventilated and closed containers, unlike FmTR. pT and pTR were observed as metabolites of mF and mFR, implying that mFR might degrade into BA and BAR, and then further convert into pT and pTR.

To better understand how treatment and culture conditions influence Phalaenopsis development in vitro, we analyzed the absolute cytokinin concentrations across different treatments. The highest total cytokinin level was found in FmTR-treated plants grown in closed containers. This suggests that the added FmTR remained stable and constituted the largest portion of the total cytokinin pool across all treatments (Figure 1a). Interestingly, culture conditions not only affected the total cytokinin levels but also the ratio between isoprenoid CKs (iP and zeatins) and aromatic CKs (BA and topolins) [21] (Figure 1b). The concentrations of isoprenoid and aromatic CKs varied depending on the treatment. For example, the highest isoprenoid cytokinin concentration, making up 38.97% of the total cytokinin content, was found in BA-treated shoots grown in ventilated containers, while the lowest concentration, about 2.65% of total CKs, was observed in the FmTR-treated plants grown in closed containers.

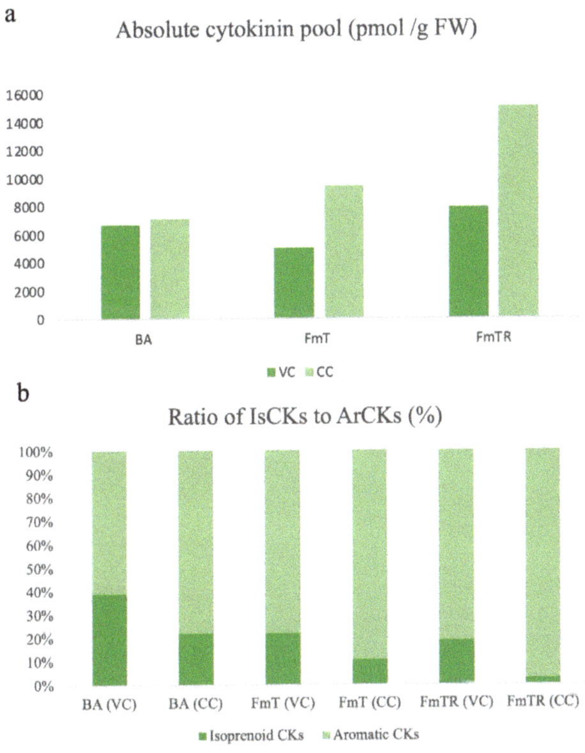

Figure 1. Cytokinin pool in *Phalaenopsis* axillary shoots under different treatments. Shoot grown on media supplemented with BA, FmT, and FmTR, with two types of culture vessels: closed container (CC) and ventilated container (VC). (**a**) Absolute cytokinin pool and (**b**) the ratio (%) of isoprenoid CKs (IsCKs) to aromatic CKs (ArCKs).

Figure 2 illustrates the distribution of CK metabolites in *Phalaenopsis* under different treatments and container types. In filter containers, ribosides constituted the majority of the total CK content across all treatments, consistent with their role as long-distance transport forms capable of interconversion into active forms [1,14]. Free bases represented the second largest proportion, with varying percentages depending on the applied cytokinin (BA: 26%; FmT: 42%; and FmTR: 7%). In closed containers, free bases predominated, except in the case of FmTR treatment.

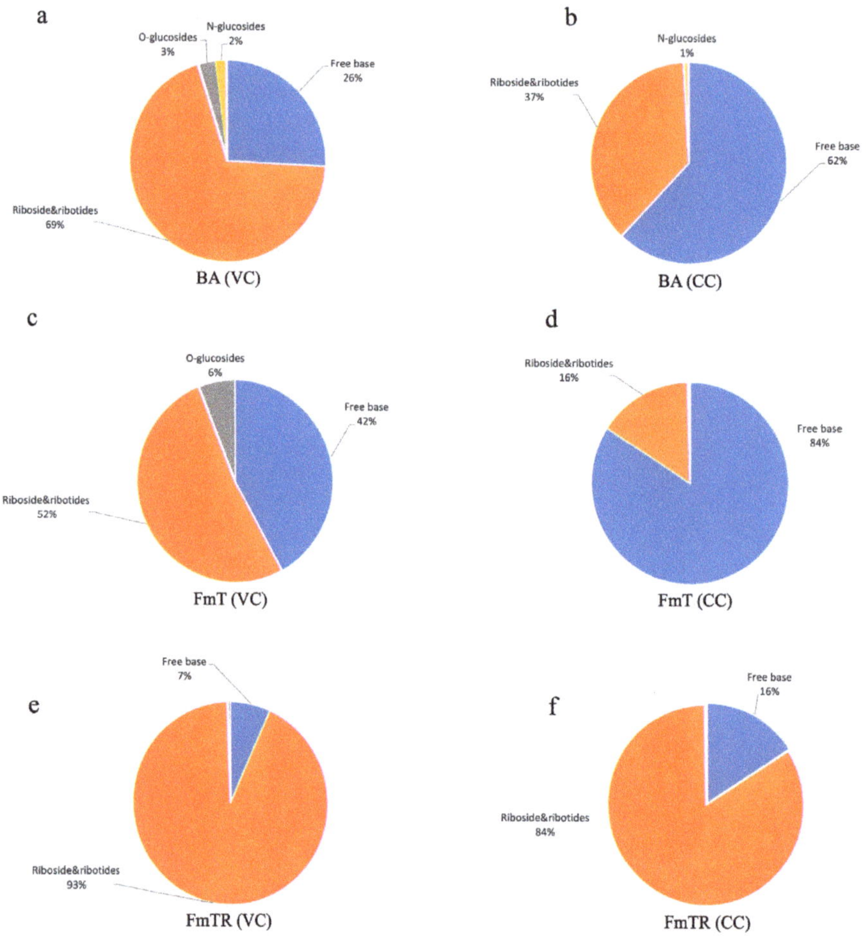

Figure 2. Distribution of cytokinin metabolites in Phalaenopsis shoots treated with BA, FmT, or FmTR and grown in either closed containers (CCs) or ventilated containers (VCs). The metabolites are grouped into free bases (cZ, IP, BAP, mT, oT, pT, and FmT), ribosides (FmTR, pTR, oTR, mTR, BAR, iPR, cZR, and DHZR), and O-glucoside ribosides and N-glucosides (DHZROG, tZ7G, DHZ7G, cZ9G BAP9G, and mT9G). Panels (**a**), (**c**), and (**e**) show the distribution in ventilated containers for BA, FmT, and FmTR treatments, respectively. Panels (**b**), (**d**), and (**f**) illustrate the distribution in closed containers for BA, FmT, and FmTR treatments, respectively.

O-glucosides, detectable only in FmT- and BA-treated plants, accounted for a small proportion of the total CK pool (3% with BA and 6% with FmT). These reversible storage forms can contribute to the active CK pool upon hydrolysis [22–24]. In closed containers, only negligible amounts of cZROG were found. N-glucosides were largely below the detection limit, except in BA-treated plants where tZ7G and BAP9G were present. The predominance of BAP9G suggests that 9-glycosylation is a primary BA deactivation pathway in *Phalaenopsis*. Notably, FmT-treated plants exhibited extremely low levels of N-glucosides (Table 1).

4. Discussion

Our study reveals a complex interplay between exogenous cytokinin application, headspace aeration, and the resulting endogenous cytokinin profiles in *Phalaenopsis*. The differential accumulation of specific metabolites under varying treatments and container types underscores the dynamic nature of cytokinin metabolism in this orchid species. Notably, the absence of tZ7G in plants treated with FmT or FmTR, particularly in closed containers, suggests the potential inhibitory effect of these fluorinated CKs on the N7-glucosylation pathway. Furthermore, the unexpected abundance of cZR across all treatments, despite the low or undetectable levels of its precursor cZ, raises questions surrounding the biological significance of these compounds in *Phalaenopsis*. cis-Zeatin was previously considered biologically inactive, potentially as a byproduct of tRNA degradation; however, recent studies challenge this notion and suggest its involvement in various stages of plant development [25]. Notably, neither cZ nor its metabolites were detected in control leaves from a potted plant. Li et al. [26] described the dynamics of endogenous CKs in TDZ-treated *Phalaenopsis* plants, demonstrating that TDZ promotes the biosynthesis of CK-ribosides, particularly trans-zeatin riboside (tZR) and isopentenyladenosine riboside (iPR), with a small increase in the level of cis-zeatin riboside (cZR). Contrary to their findings, our study observed an increase in cZR levels under fluorinated cytokinin treatment. These findings indicate that not only can the metabolism of endogenous CKs be strongly influenced by exogenous CKs, but also that endogenous CK biosynthesis can vary between species of the same genus. While in Phalaenopsis aphrodite, TDZ promoted shoot proliferation via CK-ribosides, in *P. amabilis*, we did not observe an increase in tZR or iPR, and cZR was not detected in control *P. amabilis*.

Interestingly, isoprenoid metabolites DHZROG and iPR were found at higher concentrations in filter pots compared to closed containers, contrary to the observation that closed containers typically yield more shoots [11]. This finding suggests that the fluorinated CKs FmT and FmTR, as well as headspace aeration, may modulate isoprenoid cytokinin biosynthesis or metabolism. Additionally, the de-fluorination of FmT to BA was proposed, putatively observed in both ventilated and closed containers but not measured with FmTR, highlights the potential for metabolic interconversions between different cytokinin types and underscores the importance of considering the stability and metabolic fate of synthetic CKs in plant tissue culture.

The abundance of total cytokinin content in closed containers suggests that a hermetically sealed environment might prevent the degradation of certain CKs, potentially explaining the earlier reported increased shoot proliferation under closed containers [11]. However, the exact mechanism behind this action remains unclear. Interestingly, we also observed that within the same treatment group, a higher concentration of aromatic CKs was linked to an increased number of shoots, particularly in plants grown in closed containers. This finding suggests that aromatic CKs may have a positive impact on shoot development in *Phalaenopsis*. Our previous work [11] showed that in vitro *Phalaenopsis* treated with FmTR produced more shoots (average shoot per explant = 25.3) compared to FmT (average shoots per explant = 14.6) and BA (average shoots per explant = 7), with smaller leaves when treated with FmTR or FmT, compared to BA treatment. In addition, while BA and FmT treatment did not affect rooting, FmTR completely prevented root formation during micropropagation.

Notably, the hydroxylated metabolites of BA, such as mT, oT, and pT [27–29], were observed in our study, despite not being detected in greenhouse-grown *Phalaenopsis*. Since para-topolin and its derivatives are not very biologically active [29], this pathway could be an alternative inactivation method, as aromatic CKs are known to be less easily broken down by CKX enzymes [30]. pT and pTR have also been found in *Handroanthus guayacan*, *Tabebuia rosea*, and *Pistacia vera* L. [17,31], though as metabolites of mTR. However, high-resolution mass spectrometry is needed to confirm the identity of these products. The decline of pT and pTR levels in closed containers may be linked to altered metabolic flux under these conditions. The presence of FmT and FmTR in BA-treated plants is not surprising, as their previous subculture involved FmTR, and its stability allows for carryover. These observations underscore the importance of understanding the dynamics of CK metabolism and the effects of exogenous CKs (such as TDZ, BA, FmT, or FmTR), which can help optimize micropropagation protocols. By manipulating CK levels through genetic or chemical means, it is possible to enhance shoot proliferation, thereby improving the efficiency of orchid propagation.

5. Conclusions

In conclusion, our investigation into the CK metabolism of *Phalaenopsis* under varying exogenous CK treatments and culture conditions has revealed a complex and dynamic system. The differential accumulation of specific metabolites highlights the intricate interplay between exogenous and endogenous CK levels, influenced by both the chemical structure of the applied CK and the culture environment.

The observed metabolic alterations, such as the unexpected abundance of cZR and the proposed de-fluorination of FmT, underscore the need for a deeper understanding of cytokinin metabolism in orchids. The identification of a potential novel metabolic pathway involving pT and pTR further emphasizes the complexity of this system.

These findings have significant implications for optimizing *Phalaenopsis* micropropagation protocols. By tailoring cytokinin treatments and culture conditions to specific metabolic profiles, we can potentially enhance shoot proliferation, improve plant quality, and overcome recalcitrance in orchid tissue culture. Moreover, our results contribute to a broader understanding of CK metabolism in plants and its role in regulating growth and development. Further research is warranted to elucidate the precise mechanisms underlying these metabolic processes and to explore their potential applications in plant biotechnology.

Author Contributions: Conceptualization, N.M. and S.P.O.W.; methodology, N.M. validation, S.P.O.W. and K.D.; endogenous cytokinin analysis: L.P.; formal analysis, N.M. and K.D.; investigation, N.M. and L.P.; resources, S.P.O.W.; writing—original draft preparation, N.M.; writing—review and editing, N.M. and S.P.O.W.; visualization, resources, and supervision, S.P.O.W.; All authors have read and agreed to the published version of the manuscript.

Funding: This research received no external funding.

Data Availability Statement: Data are contained within the article.

Conflicts of Interest: The authors declare no conflicts of interest.

References

1. Kieber, J.J.; Schaller, G.E. Cytokinin signaling in plant development. *Development* **2018**, *145*, dev149344. [CrossRef] [PubMed]
2. Singh, J.; Sengar, R.S.; Prasad, M.; Kumar, A. Effect of Different Plant Growth Regulators on In-vitro Micropropagation of Banana Cultivar Grand Naine (*Musa* spp.). *J. Adv. Biol. Biotechnol.* **2024**, *27*, 90–98. [CrossRef]
3. Mazri, M.A. Role of CKs and physical state of the culture medium to improve in vitro shoot multiplication, rooting and acclimatization of date palm (*Phoenix dactylifera* L.) cv. *Boufeggous*. *J. Plant Biochem. Biotechnol.* **2015**, *24*, 268–275. [CrossRef]
4. Zürcher, E.; Müller, B. Cytokinin synthesis, signaling, and function—Advances and new insights. *Int. Rev. Cell Mol. Biol.* **2016**, *324*, 1–38. [PubMed]
5. Aremu, A.O.; Bairu, M.W.; Szüčová, L.; Doležal, K.; Finnie, J.F.; Van Staden, J. Genetic fidelity in tissue-cultured 'Williams' bananas–The effect of high concentration of topolins and benzyladenine. *Sci. Hortic.* **2013**, *161*, 324–327. [CrossRef]
6. Ahmad, M.Z.; Hussain, I.; Roomi, S.; Zia, M.A.; Zaman, M.S.; Abbas, Z.; Shah, S.H. In vitro response of cytokinin and auxin to multiple shoot regeneration in *Solanum tuberosum* L. *Am. Eurasian J. Agric. Environ. Sci.* **2012**, *12*, 1522–1526.
7. Lal, N.; Madhulika, S. Prospects of plant tissue culture in orchid propagation: A review. *Indian J. Biol.* **2020**, *7*, 103–110.
8. Khatun, K.; Nath, U.K.; Rahman, M.S. Tissue culture of Phalaenopsis: Present status and future prospects. *J. Adv. Biotechnol. Exp. Ther.* **2020**, *3*, 273–285. [CrossRef]
9. Parthibhan, S.; Rao, M.V.; Kumar, T.S. In vitro regeneration from protocorms in Dendrobium aqueum Lindley–An imperiled orchid. *J. Genet. Eng. Biotechnol.* **2015**, *13*, 227–233. [CrossRef]

10. Zanello, C.A.; Duarte, W.N.; Gomes, D.M.; Cardoso, J.C. Micropropagation from inflorescence nodal segments of Phalaenopsis and acclimatization of plantlets using different substrates. *Horticulturae* **2022**, *8*, 340. [CrossRef]
11. Murvanidze, N.; Doležal, K.; Werbrouck, S.P. Fluorine containing topolin CKs for *Phalaenopsis Amabilis* (L.) blume micropropagation. *Propag. Ornam. Plants* **2019**, *19*, 48–51.
12. Plíhalová, L. Synthesis and chemistry of meta-Topolin and related compounds. In *Meta-topolin: A Growth Regulator for Plant Biotechnology and Agriculture*; Springer: Singapore, 2021; pp. 11–22.
13. Mok, M.C. CKs and plant development—An overview. In *Cytokinins—Chemistry, Activity, and Function*; Mok, D.W.S., Mok, M.C., Eds.; CRC Press: Boca Raton, FL, USA, 1994; pp. 155–166.
14. Zhao, J.; Wang, J.; Liu, J.; Zhang, P.; Kudoyarova, G.; Liu, C.-J.; Zhang, K. Spatially Distributed CKs: Metabolism, Signaling, and Transport. *Plant Commun.* **2024**.
15. Mok, D.W.S.; Mok, M.C. Cytokinin metabolism and action. *Annu. Rev. Plant Biol.* **2001**, *52*, 89–118. [CrossRef] [PubMed]
16. Klemš, M.; Plačková, L.; Doležal, K.; Bettaieb, T.; Werbrouck, S.P. Changes in cytokinin levels and metabolism in tobacco (*Nicotiana tabacum* L.) explants during in vitro shoot organogenesis induced by trans-zeatin and dihydrozeatin. *Plant Growth Regul.* **2011**, *65*, 427–437. [CrossRef]
17. Abdouli, D.; Plačková, L.; Doležal, K.; Bettaieb, T.; Werbrouck, S.P. Topolin CKs enhanced shoot proliferation, reduced hyperhydricity and altered cytokinin metabolism in Pistacia vera L. seedling explants. *Plant Sci.* **2022**, *322*, 111360. [CrossRef] [PubMed]
18. Aremu, A.O.; Bairu, M.W.; Szüčová, L.; Doležal, K.; Finnie, J.F.; Van Staden, J. Assessment of the role of meta-topolins on in vitro produced phenolics and acclimatization competence of micropropagated 'Williams' banana. *Acta Physiol. Plant.* **2012**, *34*, 2265–2273. [CrossRef]
19. Smykalova, I.; Plačková, L.; Doležal, K.; Bettaieb, T.; Werbrouck, S.P. The role of cytokinin regulation in micropropagation. *Plant Cell Rep.* **2019**, *38*, 123–132.
20. Novák, O.; Hauserová, E.; Amakorová, P.; Doležal, K.; Strnad, M. Cytokinin profiling in plant tissues using ultra-performance liquid chromatography–electrospray tandem mass spectrometry. *Phytochemistry* **2008**, *69*, 2214–2224. [CrossRef]
21. Plačková, L.; Hrdlička, J.; Smýkalová, I.; Cvečková, M.; Novák, O.; Griga, M.; Doležal, K. Cytokinin profiling of long-term in vitro pea (*Pisum sativum* L.) shoot cultures. *Plant Growth Regul.* **2015**, *77*, 125–132. [CrossRef]
22. Kamada-Nobusada, T.; Sakakibara, H. Molecular basis for cytokinin biosynthesis. *Phytochemistry* **2009**, *70*, 444–449. [CrossRef]
23. Werbrouck, S.P.O.; van der Jeugt, B.; Dewitte, W.; Prinsen, E.; Van Onckelen, H.A.; Debergh, P.C. The metabolism of benzyladenine in *Spathiphyllum floribundum* Schott 'Petite' in relation to acclimatisation problems. *Plant Cell Rep.* **1995**, *14*, 662–665. [CrossRef] [PubMed]
24. Jameson, P.E. Cytokinin Translocation to, and Biosynthesis and Metabolism within, Cereal and Legume Seeds: Looking Back to Inform the Future. *Metabolites* **2023**, *13*, 1076. [CrossRef] [PubMed]
25. Schäfer, M.; Plačková, L.; Doležal, K.; Bettaieb, T.; Werbrouck, S.P. The role of cis-zeatin-type CKs in plant growth regulation and mediating responses to environmental interactions. *J. Exp. Bot.* **2015**, *66*, 4873–4884. [CrossRef] [PubMed]
26. Li, Y.-Y.; Hao, Z.-G.; Miao, S.; Zhang, X.; Li, J.-Q.; Guo, S.-X.; Lee, Y.-I. Profiles of CKs Metabolic Genes and Endogenous CKs Dynamics during Shoot Multiplication In Vitro of Phalaenopsis. *Int. J. Mol. Sci.* **2022**, *23*, 3755. [CrossRef] [PubMed]
27. Werbrouck, S.P.; Doležal, K.; Strnad, M.; Van Staden, J. Meta-topolin, an alternative to benzyladenine in tissue culture? *Physiol. Plant.* **1996**, *98*, 291–297. [CrossRef]
28. Van Staden, J.; Drewes, F.E. The stability and metabolism of benzyladenine glucosides in soybean callus. *J. Plant Physiol.* **1992**, *140*, 92–95. [CrossRef]
29. Holub, J.; Plačková, L.; Doležal, K.; Strnad, M. Biological activity of CKs derived from ortho-and meta-hydroxybenzyladenine. *Plant Growth Regul.* **1998**, *26*, 109–115. [CrossRef]
30. Galuszka, P.; Popelková, H.; Werner, T.; Frébortová, J.; Pospíšilová, H.; Mik, V.; Köllmer, I.; Schmülling, T.; Frébort, I. Biochemical characterization of cytokinin oxidases/dehydrogenases from Arabidopsis thaliana expressed in *Nicotiana tabacum* L. *J. Plant Growth Regul.* **2007**, *26*, 255–267. [CrossRef]
31. Grira, M.; Prinsen, E.; Werbrouck, S. New Understanding of Meta-Topolin Riboside Metabolism in Micropropagated Woody Plants. *Plants* **2024**, *13*, 1281. [CrossRef]

horticulturae

MDPI

Article

Mycorrhization of Black Crowberry (*Empetrum nigrum* L.) After In Vitro Propagation with Mineral Fertilizers

Ivan Nechiporenko [1,2,*], Svetlana Akimova [2,*] and Natalia Semenova [3]

1 State Scientific Institution "All-Russian Research Institute of Phytopathology", St. Institute, Own 5, Bolshie Vyazemy, Odintsovo District, Moscow Region, 143050 Big Vyazyomy, Russia
2 Institute of Horticulture and Landscape Architecture, Russian State Agrarian University—Moscow Timiryazev Agricultural Academy, Timiryazevskaya St., 49, 127434 Moscow, Russia
3 Prokhorov General Physics Institute of the Russian Academy of Sciences, Vavilov Str. 38, 119991 Moscow, Russia; natalia.86@inbox.ru
* Correspondence: vannechiporenko@gmail.com (I.N.); akimova@rgau-msha.ru (S.A.)

Abstract: *Empetrun nigrum* L. is a rare berry crop with a high content of biologically active substances, which are of great interest in medicine. In order to obtain sufficient transplants of this crop, in vitro propagation is preferred over other methods. It is known that a large percentage of propagated plants can be lost or damaged, not only at the stage of adaptation to non-sterile conditions, but also in the post-adaptation period. Plants may have weak growth due to poor root development and a lack of nutrients. Therefore, to improve the understanding of the plant requirements in the post-adaptation period of *E. nigrum*, clonal micropropagation was performed by combining mycorrhizal preparations containing (1) mycelium of *Glomus* sp., and humic substances, and (2) mycelium of *Glomus* sp., *Trichoderma harzianum*, and other mycorrhiza formation microorganisms with different types of mineral fertilizer ($N_{20}P_{20}K_{20}$, $N_{20}P_{16}K_{10}(S_5)$, and $N_{19}P_9K_{10}$ + 2MgO). Analysis of the growth dynamics of ex vitro plants over 98 days of cultivation in containers under greenhouse conditions showed the effectiveness of mineral fertilizer and mycorrhizal preparations treatment. The total root length increased by 30–50% and the total shoot length by 40–80%. The survival was 95.8–100%.

Keywords: *Empetrum nigrum*; micropropagation; growing ex vitro plants; *Glomus*; microorganisms; root system; *Trichoderma*

1. Introduction

Crowberry (*Empetrum* sp.) is a rare berry crop which belongs to the *Empetraceae* family [1–3]. Due to the similarity of morphological characteristics and common habitats with plants such as *Erica carnea* L. and *Phyllodoce empetriformis* (Sm.) D. Don., it is sometimes classified as belonging to the *Ericaceae* family [4,5].

This botanical genus is divided into two large groups of species with different habitats. Plants in the northern hemisphere of Eurasia mainly have black fruits and are grouped under the main species *E. nigrum* L. Meanwhile, American species have fruits of various colors: pink (*E. eamesii* Fern. & Wieg.), red (*E. rubrum* Vahl ex Willd), purple, or violet (*E. atropurpureum* Fern. & Wieg.) [4].

Black crowberry (*E. nigrum* L.) is a groundling, creeping, evergreen shrub belonging to a widespread species whose habitat is associated with both forested and treeless mountain and arctic tundra zones [6]. It is highly tolerant of its habitat; however, it reacts extremely negatively to environmental pollution [7]. It is known that in the struggle for competition and resources, crowberry is capable of allelopathic action, suppressing various flora in its habitat [8,9]. It has been found that species of the genus *Empetrum* release phenolic compounds from their leaves into the soil, causing chemical effects and suppressing other plants. The main phenolic compound responsible for this effect is the allelopathic compound batatasin-III [10]. It is known that crowberry can grow on soils of a wide range of acidity (pH 2.5–7.5), but the crop cannot survive in areas with alkaline soils with a pH of 8.0 and more [11] because of the presence of various mycorrhizal symbiotic fungi on the roots, mainly ericoid mycorrhiza, which die in such conditions [9,12].

Crowberry belongs to a group of wild berry plants classified as "superfoods," which also includes such plants as bilberry (*Vaccinium myrtillus* L.), lingonberry (*Vaccinium vitis-idaea* L.), and cloudberry (*Rubus chamaemorus* L.), all of which have a high content of biologically active substances. The indigenous peoples of Scandinavia, Russia, and Canada use berries of *E. nigrum* L. as raw materials for the production of various preparations (jams, pastilles, beverages), but only in the northern regions of natural growth are the fruits a commercial crop [13–18]. Crowberries are also used as a natural dye [19]. Since this berry plant is rare in horticulture, there is no information about the existence of its industrial plantations.

In urban areas, crowberry has broad prospects as a medicinal plant raw material containing a large amount of flavonoids, widely used in the pharmaceutical industry [20,21]. Therefore, interest in this berry and medicinal plant is currently growing.

E. nigrum is usually propagated vegetatively by softwood and hardwood cuttings [22]. However, the rate of propagation using these methods is ineffective for reproducing a sufficient amount of plant material. Seed propagation is used in forest nurseries to restore forest ecosystems. However, this method does not preserve the characteristics of the original plant [2], and it is slower than in morphologically similar plants of the *Ericaceae* family, such as the *Calluna* and *Erica* genera [9]. The problem of prolonged seed germination is associated with the hardening of the endocarp, while only a small number of seeds germinate the following spring after stratification under natural conditions [4,23,24].

Thus, there is currently a lack of planting material or black crowberry, and demand for developing accelerated methods of vegetative propagation is increasing. For plants that are difficult to propagate, the most effective technology for multiplication is clonal micropropagation (in vitro), which successfully develops both in industrial propagation and scientific research [25–27]. When using this in vitro technology, it is possible to propagate and obtain genetically uniform high-quality planting material in the required quantity [28–30].

The key stage in obtaining high-quality planting material after in vitro technology and their further sale in a nursery is the of growing, completion, or post-adaptation of ex vitro plants [31]. There is a lack of information on how plants develop after clonal micropropagation in vitro during ex vitro adaptation and when grown in containers in greenhouse conditions before planting ex vitro plants in the field.

At the same time, there is evidence showing that ex vitro microplants during adaptation in non-sterile conditions may experience increased stress. The latter can cause slowed growth, suppression, and even death of ex vitro plants, ultimately affecting further development during container cultivation. Survival rates after adaptation can vary from 35 to 90%, depending on the species [32–34].

One of the reasons for the slow growth of black crowberry when grown in containers may be a lack of nutrition and poor nutrient uptake due to an underdeveloped root system [35]. To eliminate unfavorable factors for better survival and root system stimulation, it is advisable to add special preparations to the substrate which contain various additives capable of stimulating vegetative growth, increasing immunity, for example, agrobacteria, bacterial derivatives (protein preparations), and spores of mycorrhizal fungal [32,36–41].

The natural habitat of ericaceous plants is associated with acidic soils and symbiotic relationships with mycorrhizae; thus, preparations containing mycorrhizal fungi may be the most preferable plant cultivation after in vitro conditions [42,43].

These fungi are able to stimulate the formation of new roots and improve the absorption of applied mineral fertilizers due to the fact that the produce dissolves enzymes that convert insoluble minerals into soluble bioavailable forms that are easily used by the plant [44–47]. They produce extracellular phosphatases and proteolytic enzymes that provide access to difficult-to-decompose organic compounds by decomposing and absorbing the released products. These can then be transferred to the host plant, thereby increasing the supply of nitrogen and phosphorus. This contributes to the potential competitive advantage of ericaceous plants, as they have access to soil organic matter [48].

It is noted that ambient temperature can also affect the absorption of nutrients such as nitrogen [49]. In particular, the carbon/nitrogen ratio shifts due to the warming of carbon being supplied to mycorrhizal fungi. Later, the ratio changes to the opposite. Ultimately, a lower carbon/nitrogen exchange ratio may allow *Empetrum* sp. plants to redirect carbon to other functions—growth, respiration, etc. [50]. The plants are also known to accumulate trace elements in various parts of plants that are normally inaccessible [51].

Mycorrhizal fungi are divided into various groups. There are ericoid mycorrhizal fungi (ErM), which are found only in acidophilic plants of the *Ericaceae* family that grow in the wild [52,53].

Ericoid mycorrhizal fungi are associated with the root surface of the host plant, forming a loose network of hyphae. These hyphae colonize the cortical cells of the host plant roots and form characteristic intracellular spirals. Only a few taxonomically diverse species of ericoid mycorrhizal fungi have been identified as ericoid mycorrhizae. Most fungi that form ericoid mycorrhizae remain unidentified because they are sterile. The best-known species is *Hymenoscyphus ericae*, which is believed to colonize both the *Ericaceae* and *Cupressaceae* families [9,54–56].

Arbuscular mycorrhizal fungi (AMF) belong to the *Glomeromycota* family, where the species diversity is limited to 350 species described to date. The most common AM fungi belong to the *Glomus* genus and form a symbiotic relationship with 80% of land plant species, the growth of which depends on them [57,58]. *Rhizophagus irregularis* is the main species used to study AM and is a key component of various mycorrhizal plant biostimulants, being widely used in horticulture and plant cultivation [59]. Symbiosis with AM is characterized by their penetration of fungi into root the cells of the root cortex with the formation of microscopic, branched structures called arbuscular structures. These structures increase the efficiency of metabolite exchange between the plant and the fungus. The extraradical hyphae of AM create a surface for colonization by functional populations of bacteria [60].

Several studies have reported on the interactions between nitrogen-fixing fungi and bacteria that mobilize phosphorus and nitrogen, as well as the impact of nitrogen fixation on the structure of bacterial community [61,62]. It has been noted that the artificial infection of young, actively growing seedlings with mycorrhizal fungi improves rooting and early plant growth, as there is a need to increase the above-ground system during this growth period [63–65]. However, this cannot be achieved without increasing the absorption capacity of the roots and mineral nutrition improvement [66,67]. This is due to the fact that ericaceous plants do not have root hairs at their root tips and are able to absorb nutrients only through a symbiotic relationship with mycorrhizal fungi [68,69]. Therefore, for better plant development, it is necessary to direct all agricultural techniques to create conditions that stimulate the development of mycorrhizal fungi and optimize the conditions for the development of perennial plantings [70].

In view of the rise in prices of mineral fertilizers, the combined use of mycorrhizal fungi can help to reduce fertilizer consumption and improve its absorption [71,72].

Therefore, the aim of our research was to investigate the effectiveness of adding mycorrhizal preparations and various types of fertilizers (water-soluble and slow-released) using different application methods when growing black crowberry plants in greenhouses after in vitro propagation.

2. Materials and Methods

2.1. In Vitro Culture Conditions

The experiments were carried out threefold in 2024–2025 at the Federal State Budgetary Scientific Establishment the All-Russian Scientific Research Institute of Phytopathology (VNIIF), in the Department of the Laboratory for Recovery and Research on the Adaptive Potential of Cultures and Plants, and at the Russian State Agrarian University, Moscow Timiryazev Agricultural Academy, in the department of biotechnology and berry crops of the Edelstein Educational Scientific and Production Center for Horticulture and Vegetable Growing.

Crowberry microplants were obtained using in vitro propagation. At the stage of initiation of in vitro culture, cut young shoots of *E. nigrum* L. cv. 'Irland' (European variety) were used (Figure 1). They were washed under running water together with a soap solution. Shoots were sterilized in a laminar box using 3% sodium hypochlorite (NaClO) with an addition of 0.01% polysorbate (Tween 80) for 15 min and 70% ethanol for 1 min. Then the shoots were then washed with bidistilled water threefold, cut into 1.0 cm pieces, and planted on Woody Plant Medium (Lloyd & McCown, 1980) [73] with an addition of 0.2 μM 6-BA (6-benzylaminopurine) (Duchefa Biochemie, Haarlem, The Netherlands).

Figure 1. The process of obtaining adapted ex vitro plants of *E. nigrum* cv. 'Irland': (**A**) explant introduced into in vitro culture; (**B**) in vitro microcuttings at the multiplication stage; (**C**) microcuttings at the in vitro rooting stage; (**D**) ex vitro microplants in the adaptation process; (**E**) adapted ex vitro microplant with roots; (**F**) ex vitro plants in the process of growing in a greenhouse.

At the stage of multiplication, three sequential passages were produced on the nutritive growth medium based on WPM enriched with the following substances (mg/L): thiamine hydrochloride (B1); pyridoxine hydrochloride (B6); nicotinamide (PP), 0.5; meso-inositol, 100; and sucrose, 30.000, with added cytokinin of Zeatin (2-Methyl-4-(9H-purin-6-ylamino)-2-buten-1-ol) (HiMedia Laboratories LLC, Maharashtra, India) at a concentration of 2.2 μM and 0.8% *w*/*v* agar-agar (American-type bacteriological) (Dia-m, Moscow, Russia), with the pH adjusted to 4.5 with 1M NaOH or HCl. The plants in vitro were cultivated in a controlled environment room with a photosynthetic photon flux density (PPFD) of 26.62 $\mu mol \cdot s^{-1} \cdot m^{-2}$, a 16/8 h (light/dark) photoperiod, and temperature of 20 ± 2 °C. The subcultivation period was 60 days.

At the rooting stage, in vitro microcuttings were planted on the nutritive growth medium based on ½WPM (reduction of macronutrients and sucrose) with added IAA (indolyl-3-acetic acid) (Duchefa Biochemie, Haarlem, The Netherlands) at a concentration of 12.3 μM. Rooting in vitro plants was subcultivated with a PPFD of 26.62 $\mu mol \cdot s^{-1} \cdot m^{-2}$, with a 16/8 h (light/dark) photoperiod and 20 ± 2 °C temperature. The subcultivation period was 60 days.

2.2. Adaptation Conditions

At the adaptation stage, microplants were replaced in industrial chambers—dome-shaped mini-greenhouses with holes (for ventilation) (with temperature conditions of 22 s 2 °C, humidity 90%)—and located under Zěma ZML-0160 LED phytolamps (Ecolight, Moscow, Russia) with a level PPFD of 120 $\mu mol \cdot s^{-1} \cdot m^{-2}$ and a 16/8 h (light/dark) photoperiod for 45 days. After adaptation to non-sterile conditions, the plants were kept in a greenhouse box (temperature 22–30 °C, humidity 70–75%).

2.3. Experimental Treatments

In the first ten days of March, the experimental adapted crowberry plants were replaced in 0.5 L containers (square pot 0.5 L 9 × 9 × 10 cm) in high-moor peat substrate (pH ≤ 4.0), to which mycorrhiza-forming preparations containing arbuscular mycorrhizal fungi of the genus *Glomus* were added according to the experimental design. The mycorrhizal preparations 'Kormilitsa' (BashInkom, Ufa, Russia) and 'Mycofriend' (Organic Line, Lytkarino, Russia) were used.

In addition, to ensure better plant survival, growth and development mineral fertilizers were applied with mycorrhiza combination and without it (Table 1).

'Plantafol' (Valagro®, Chieti, Italy)—chelated $N_{20}P_{20}K_{20}$ fertilizer with microelements, which has increased solubility and complete absorption—was applied for foliar treatment seven times with 14-day intervals (pH of 1% solution: 4.5).

'Rastvorin' (Buyskiy Himicheskiy Zavod, Buy, Russia)—water-soluble fertilizer for heather was used for root nutrition $N_{20}P_{16}K_{10}(S_5)$ seven times, once every 14 days (pH of 1% solution—3.0).

'Osmocote PRO' 3–4M (ICL Group, Glebe, The Netherlands)—a slow-release fertilizer $N_{19}P_9K_{10}$ + 2MgO with trace elements—was used to enrich the substrate.

Thus, by combining mineral fertilizers and mycorrhizal additives and using them separately, we obtained 12 experimental variants (Table 2).

pH measurements were performed using an Expert-001 pH meter-ionomer (Ekoniks-Expert, Moscow, Russia).

Morphometric indicators of ex vitro plant development were recorded seven times, every two weeks, on the 14th, 28th, 42nd, 56th, 70th, 84th, and 98th days of cultivation. The following parameters were recorded: number (pcs.) and length of shoots (cm), total shoot length (cm), and, at the end of the last data recording, number (pcs.) and length of roots (cm), as well as total root length (cm).

Table 1. Composition, method of application, and concentration of mycorrhiza preparations and mineral fertilizers.

Mycorrhiza and Fertilizers	Composition	Application Method	Concentration (g/L)
	Mycorrhiza forming additives		
Kormilitsa	Granular mycorrhizae containing mycelium and spores of fungi of the genus *Glomus*, colonized root fragments, natural organic fillers, humic substances	Substrate application	6
Mycofriend	Powdered preparation containing fungi (*Glomus* sp., *Trichoderma harzianum*); microorganisms that support the formation of mycorrhizae and plant rhizosphere (*Streptomyces* sp., *Pseudomonas fluorescens*); phosphate-solubilizing bacteria with fungicidal and bactericidal properties (*Bacillius megaterium* var. *phosphaticum*, *Bacillus subtilis*, *Bacillus muciloginosus*, *Enterobacter* sp.); total number of viable cells (1.0–1.5) $\times$ 10^8 CFU/g; biologically active substances (phytohormones, vitamins, amino acids)	Substrate application	0.4
	Mineral fertilizers		
Plantafol	N—20% (N-NO_3—4%, N-NH_4—2%, N-NH_2—14%), P_2O_5—20%, K_2O—20%, microelement chelates with EDTA (B-0.02%; Fe-0.01%; Mn-0.05%; Cu-0.005)	Foliar application	2
Rastvorin	N—20%, P_2O_5—16%, K_2O—10%, S-5.5%, Microelements: B-0.01%; Mn-0.1%; Cu-0.01%; Zn-0.01%; Mo-0.001%	Direct application to substrate	2
Osmocote PRO	N—19% (N-NO_3—6.6%, N-NH_4—9%, N-NH_2—1.4%), P_2O_5—9%, K_2O—10%, MgO-2%, Microelements: B-0.01%; Fe-0.007%; Mn-0.04%; Cu-0.023%; Zn-0.011%; Mo-0.01%,	Substrate application	2

Table 2. Experimental treatment variants.

Treatment Variant	Mycorrhiza		Mineral Fertilizers		
	Kormilitsa	Mycofriend	Plantafol	Rastvorin	Osmocote PRO
T1					
T2		+			
T3	+				
T4			+		
T5					+
T6				+	
T7		+	+		
T8		+			+
T9		+		+	
T10	+		+		
T11	+				+
T12	+			+	

'+' means that it is used in this treatment variant.

2.4. Data Analysis

The experiments were repeated three times, with eight plants used for each repetition. The statistical data processing was first performed using multi-way analysis of variance (ANOVA), after which significant differences between the mean values for variant were assessed using Duncan's multiple range test ($p \leq 0.05$). This confirmed the reliability of the research results. The data are presented as mean values and the standard error of the mean (M ± SEM).

3. Results

During the ex vitro cultivation of *E. nigrum* 'Irland' plants in a greenhouse, the survival and morphometric indicators of plant development were recorded every 14 days. It was found that the survival of plants during 98 days of cultivation was 100% in the control variant (T1) without fertilizers or mycorrhiza, in variants with the mycorrhiza-forming preparation separately T2 or with $N_{20}P_{16}K_{10}(S_5)$ (T9) fertilizer, in variants with the $N_{20}P_{20}K_{20}$ fertilizer (T4) or with mycorrhiza containing fertilizer (T10), and in variants with the mycorrhiza-forming preparation with the $N_{20}P_{16}K_{10}(S_5)$ fertilizer (T12) application (Table 3).

Table 3. Survival of *E. nigrum* 'Irland' plants ex vitro during greenhouse cultivation in containers. Values represent mean SEM (n = 8). Letters indicate significant differences among treatments, ns—indicates non-significant differences according to Duncan's test ($p \leq 0.05$).

Variant	Plant Survival, %						
	14th Day	28th Day	42nd Day	56th Day	70th Day	84th Day	98th Day
T1	100.0 ± 0^{ns}	100.0 ± 0^{c}	100.0 ± 0^{d}	100.0 ± 0^{e}	100.0 ± 0^{e}	100.0 ± 0^{e}	100.0 ± 0^{e}
T2	100.0 ± 0^{ns}	100.0 ± 0^{c}	100.0 ± 0^{d}	100.0 ± 0^{e}	100.0 ± 0^{e}	100.0 ± 0^{e}	100.0 ± 0^{e}
T3	100.0 ± 0^{ns}	95.8 ± 4.2^{b}	95.8 ± 4.2^{cd}	95.8 ± 4.2^{de}	95.8 ± 4.2^{de}	95.8 ± 4.2^{de}	95.8 ± 4.2^{de}
T4	100.0 ± 0^{ns}	100.0 ± 0^{c}	100.0 ± 0^{d}	100.0 ± 0^{e}	100.0 ± 0^{e}	100.0 ± 0^{e}	100.0 ± 0^{e}
T5	100.0 ± 0^{ns}	79.2 ± 8.5^{a}	58.3 ± 10.3^{b}	45.8 ± 10.4^{b}	41.6 ± 10.1^{b}	37.5 ± 10.1^{b}	37.5 ± 10.1^{b}
T6	100.0 ± 0^{ns}	95.8 ± 4.2^{b}	95.8 ± 4.2^{cd}	95.8 ± 4.2^{de}	95.8 ± 4.2^{de}	95.8 ± 4.2^{de}	95.8 ± 4.2^{de}
T7	100.0 ± 0^{ns}	95.8 ± 4.2^{b}	91.7 ± 5.8^{c}	91.7 ± 5.8^{d}	91.7 ± 5.8^{d}	91.7 ± 5.8^{d}	91.7 ± 5.8^{d}
T8	100.0 ± 0^{ns}	79.2 ± 8.5^{a}	37.5 ± 10.1^{a}	37.5 ± 10.1^{a}	33.3 ± 9.8^{a}	29.2 ± 9.5^{a}	29.2 ± 9.5^{a}
T9	100.0 ± 0^{ns}	100.0 ± 0^{c}	100.0 ± 0^{d}	100.0 ± 0^{e}	100.0 ± 0^{e}	100.0 ± 0^{e}	100.0 ± 0^{e}
T10	100.0 ± 0^{ns}	100.0 ± 0^{c}	100.0 ± 0^{d}	100.0 ± 0^{e}	100.0 ± 0^{e}	100.0 ± 0^{e}	100.0 ± 0^{e}
T11	100.0 ± 0^{ns}	95.8 ± 4.2^{b}	95.8 ± 4.2^{cd}	83.3 ± 7.8^{c}	83.3 ± 7.8^{c}	75.0 ± 9.0^{c}	75.0 ± 9.0^{c}
T12	100.0 ± 0^{ns}	100.0 ± 0^{c}	100.0 ± 0^{d}	100.0 ± 0^{e}	100.0 ± 0^{e}	100.0 ± 0^{e}	100.0 ± 0^{e}

The lowest survival (plants on the 98th day of growth: 29.2–75.0%) were observed in variants with the introduction of the slow-release fertilizer $N_{19}P_9K_{10}$ + 2MgO into the substrate (T5), both separately and in combination with mycorrhiza-forming preparations (T8, T11). The best survival of plants was observed in the variant with the combination of the mycorrhiza-forming preparation T3, T10, T12.

The survival of crowberry plants in the other variants of the experiment on the 98th day was 91.7–95.8% (Table 3).

An analysis of seven counts and observations showed that mycorrhizal preparations and mineral fertilizers have a significant impact on the morphometric parameters of adapted plants.

Significant differences were found when measuring the height of *E. nigrum* experimental plant shoots on the 14th day of cultivation compared to the control in all variants without mycorrhizae (2.8–3.3 cm compared to 1.6 cm in the control T1). Significant differences were also found in the T11 variant where $N_{19}P_9K_{10}$ + 2MgO fertilizers and mycorrhiza-forming preparation were added to the substrate (3.5 cm compared to 2.5 cm in T1) (Figure 2B).

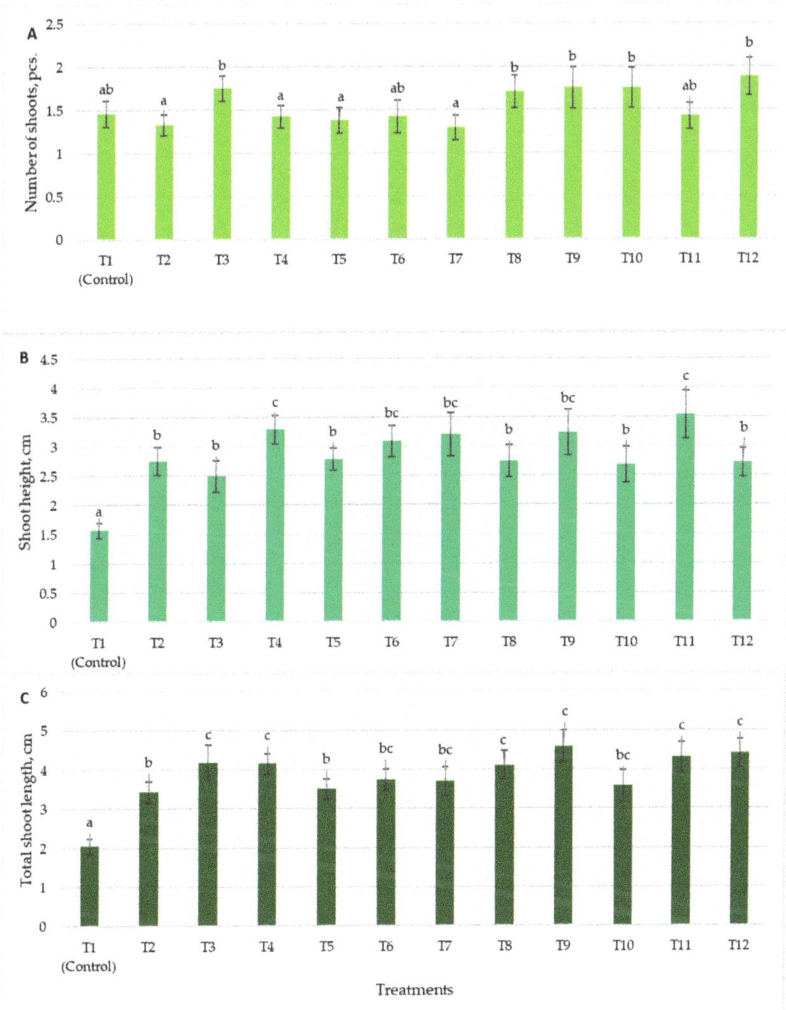

Figure 2. Morphometric indicators ((**A**) number of shoots, (**B**) shoot height, (**C**) total shoot length) of ex vitro development of *E. nigrum* 'Irland' plants on the 14th day of cultivation. Values represent mean SEM (n = 8). Letters indicate significant differences among treatments according to Duncan's test ($p \leq 0.05$).

Taking the total shoot length into account, significant differences were also found with the control in all variants without mycorrhizal preparations (3.5–4.2 cm). Additionally, significant differences were observed in all variants of the experiment with the control without mycorrhizae, except for those with foliar applications using $N_{20}P_{20}K_{20}$ fertilizer (Figure 2C).

On the 28th day of cultivation, the obtained results did not demonstrate reliable differences with the control of all the studied the experimental variants. This is probably due to the biological characteristics of crowberry plants, which in natural habitat conditions grow mainly on soils with a poor composition of nutrients. Due to the fact that our plants were multiplied using the technology of clonal micropropagation and were planted in substrates enriched with mycorrhizae and nutrients, it can be assumed that on the 28th of cultivation, according to development indicators, they were approximately at the same level. The data are presented in Supplementary Figure S1.

However, it should be noted that on the 28th day of cultivation, the $N_{19}P_9K_{10}$ + 2MgO fertilizer was found to have a detrimental effect on the plants, both when used separately (T5) and when used in combination with the mycorrhiza-forming preparation (T8). The plants' survival was 79.2% at this point and subsequently declined steadily to 29.2–37.5% by the end of the observation period. The data are presented in Supplementary Figure S1.

Records of the number of shoots taken on the 42nd day of cultivation showed significant differences between variants T8 and T9 (1.8–2.1 pcs). Additionally, in variants with the addition of mycorrhizae to the substrate, there were significant differences observed in both the variants without fertilizers (T3) (2.2 pcs) and in combination with the $N_{20}P_{16}K_{10}(S_5)$ fertilizer (T12) (2.0 pcs).

Average shoot height observations showed significant differences from the control for all fertilizer application options in experimental variants without mycorrhizae—T4, T5, and T6 (4.5–5.6 cm). The same was observed in variants T10 and T11 (4.8–5.9 cm compared to 3.6 cm in the control, T1).

Data analysis of the total shoot height (Figure S2) showed significant differences from T1 in all variants with the mycorrhizae application to the substrates T4, T10, T11, and T12 (6.8–8.0 cm). The same was observed in the T6, T9, and T12 variants with the addition of the $N_{20}P_{16}K_{10}(S_5)$ fertilizer (6.3–8.0 cm).

On the 56th day of cultivation, it was observed that foliar fertilizer $N_{20}P_{20}K_{20}$ significantly affected shoot height (6.5 cm compared to 4.2 cm in T1) and total shoot length (10.5 cm compared to 5.9 cm in T1) in variants without mycorrhizae (T4). The $N_{20}P_{16}K_{10}(S_5)$ fertilizer had the best effect on total shoot length in T9 and T12 variants with mycorrhiza-forming preparations (10.6–10.6 cm). A significant increase in the parameter was also observed in the T3, T11, and T12 variants. The total shoot length was 10.5–10.6 cm in these variants (Figure S3).

On the 70th day of cultivation, the advantage of the previously identified effective treatment options was preserved (Figure S4). The $N_{20}P_{20}K_{20}$ fertilizer with preparation mycorrhizae had a significant effect on shoot length in the T4 variant (9.3 cm compared to 6.3 cm in T1) and total shoot length without mycorrhizae in the T4 variant (14.5 cm compared to 9.8 cm in T1). The $N_{20}P_{16}K_{10}(S_5)$ fertilizer had the best effect on total shoot length in variants with in the T9 and T12 variants (15.4–15.6 cm).

On the 84th day of cultivation, the addition of $N_{20}P_{20}K_{20}$ fertilizer to variants without mycorrhizae (T4) had a significant effect on both the number of shoots (2.8 pcs. compared to 1.8 pcs. in T1) and the total shoot length (20.3 cm compared to 11.8 cm in T1). The $N_{20}P_{16}K_{10}(S_5)$ fertilizer produced the best results in terms of total shoot length in the T9 and T12 variants treated with mycorrhizae (20.5–20.8 cm compared to 14.8 cm in the control without mycorrhizae).

Significant advantages were found in variants with the addition of the mycorrhiza-forming preparation KM to the substrate, both with T10 and T12 and without T3 fertilizer (19.9–20.9 cm compared to 11.8 cm in T1) (Figure S5).

Morphometric indicators of ex vitro *E. nigrum* plants on the 98th day of cultivation showed that the advantages of the previously selected variants were preserved. The effectiveness of adding $N_{20}P_{20}K_{20}$ fertilizer without mycorrhiza-forming preparations (T4) was demonstrated, with 5.1 pcs. shoots compared to 1.8 pcs. in T1 and a total shoot length of 25.8 cm compared to 14.9 cm. At the same time, the highest values for shoot length were in the T7 variant (10.1 cm).

In direct application to substrate, the $N_{20}P_{16}K_{10}(S_5)$ fertilizer shows significant advantages in total shoot length. This is evident in both variants, with and without mycorrhizae.

In the variant without mycorrhizae (T6), the total shoot length was 20.3 cm, compared to 14.9 cm in T1. In the variant with preparations mycorrhizae (T9 and T12), the total shoot length was 25.4–26.1 cm, compared to 14.9 cm in the control variant.

Significant advantages were also confirmed for the T3 and T12 variants with the addition of the mycorrhiza-forming preparation to the substrate, in terms of total shoot length (25.2 cm compared to 14.9 cm in T1) (Figures 3 and 4).

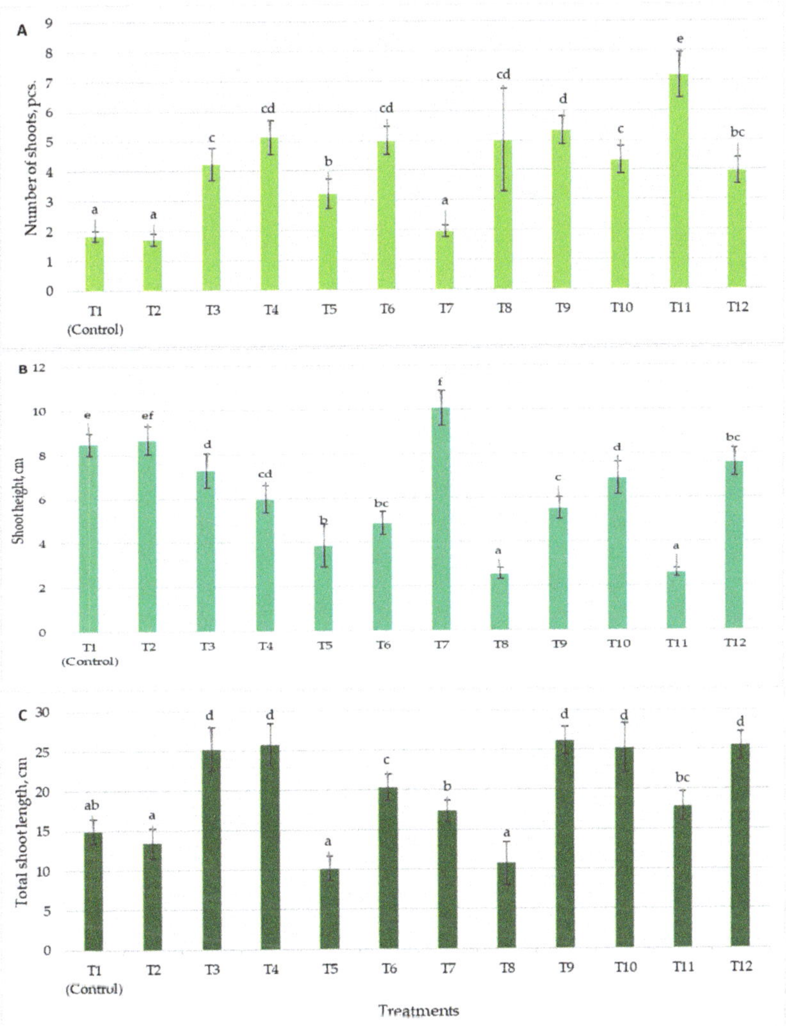

Figure 3. Morphometric indicators ((**A**) number of shoots, (**B**) shoot height, (**C**) total shoot length) of ex vitro development of *E. nigrum* 'Irland' plants on the 98th day of cultivation. Values represent mean SEM (n = 8). Letters indicate significant differences among treatments according to Duncan's test ($p \leq 0.05$).

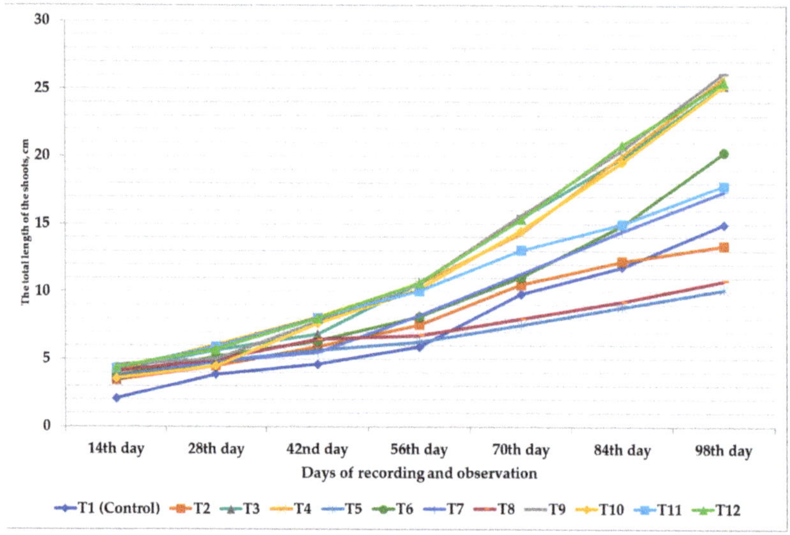

Figure 4. The dynamics of changes in the total shoot length of experimental *E. nigrum* 'Irland' plants, depending on the addition of mycorrhizal preparations and mineral fertilizers application.

On the 98th day of cultivation of *E. nigrum* 'Irland' plants, the morphometric indicators of the root system development of the experimental plants were recorded after completing the accounting and observations of the above-ground system.

The results confirmed the effectiveness of the previously selected variants and demonstrated the positive impact of foliar application with $N_{20}P_{20}K_{20}$ fertilizer in the absence of mycorrhizae (T4). This treatment increased root length to 6.9 cm and total root length to 33.3 cm. $N_{20}P_{16}K_{10}(S_5)$ fertilizer positively affected total root length in the T6 variant without mycorrhizae (32.7 cm compared to 24.8 cm in T1). It also had a positive effect on total root length in the T9 and T12 variants with mycorrhizae (37.1–37.2 cm compared to 24.8 cm in the control group T1).

The positive effect of the mycorrhiza-forming agent KM on root length (6.5 cm compared to 5.4 cm in T1) and total root length in the no-fertilizer and mineral fertilizer $N_{20}P_{16}K_{10}(S_5)$ conditions in the T3 and T12 variants was also revealed to be 33.8–37.6 cm compared to 24.8 cm in the control (Figures 5 and 6).

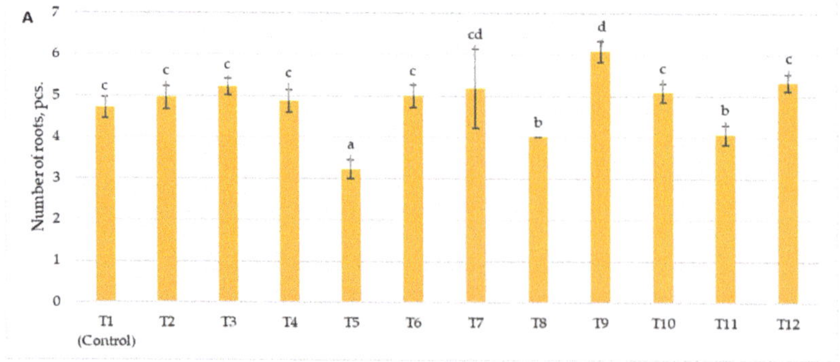

Figure 5. *Cont.*

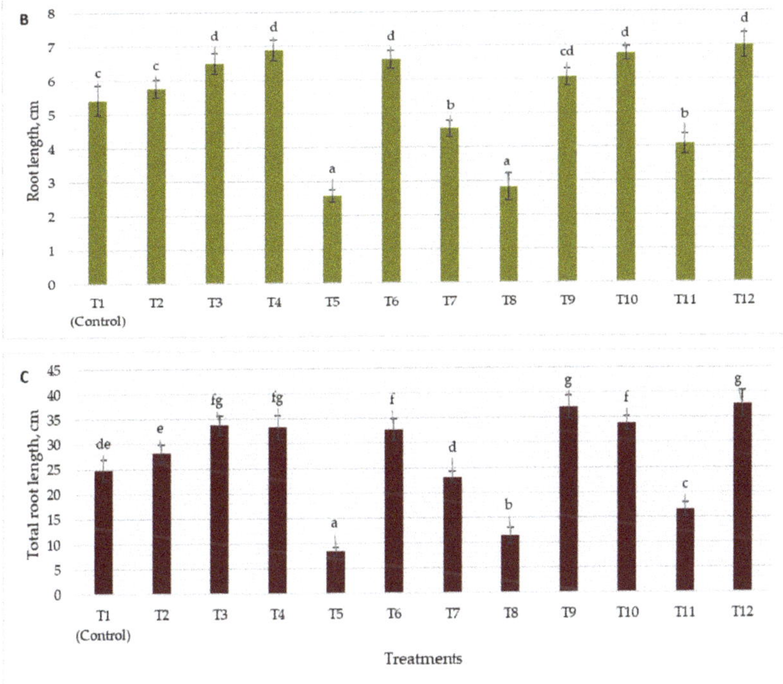

Figure 5. Morphometric indicators ((**A**) number of roots, (**B**) root height, (**C**) total root length) of *E. nigrum* 'Irland' plants on the 98th day of cultivation. Values represent mean SEM (n = 8). Letters indicate significant differences among treatments according to Duncan's test ($p \leq 0.05$).

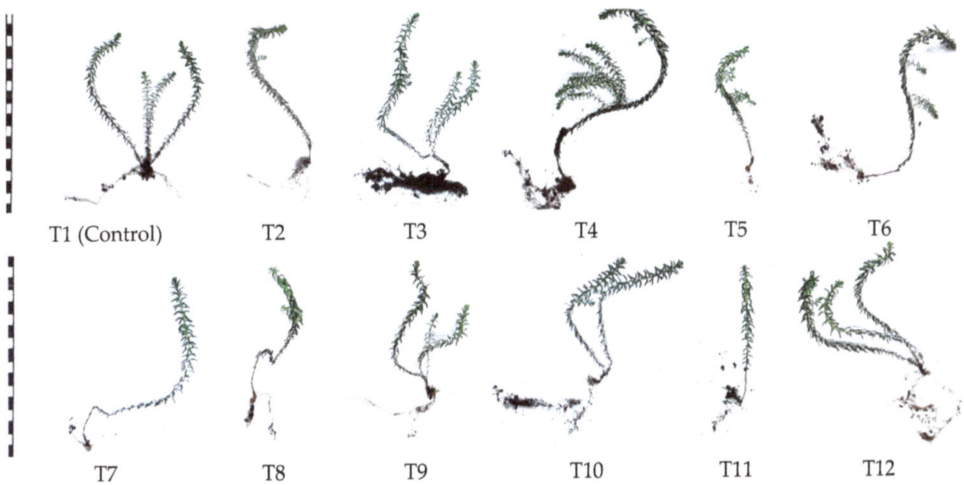

Figure 6. The effect of adding mycorrhiza-forming preparations and mineral fertilizers to the substrate on the morphometric indicators of the development of experimental *E. nigrum* 'Irland' plants on the 98th day of cultivation.

4. Discussion

Due to the increase in the price of mineral fertilizer, some plant nurseries are already using the practice of combining mycorrhizal preparations and mineral fertilizers, which allow for more rational and effective use of fertilizers [74–77]. Therefore, it was important to study the peculiarities of the influence of mycorrhizal preparations on the ex vitro cultivation of *E. nigrum* plants in containers under greenhouse conditions with optimally selected mineral nutrition.

It is known that *E. nigrum* has a predominantly taproot-type root system that is later replaced by creeping lateral roots without root hairs, located in the upper 5–10 cm soil layer [5,23]. For the epidermal cells of *E. nigrum*, for the absorption of water and nutrients, symbiosis with ericoid or arbuscular mycorrhizal fungi is necessary [78–81].

In habitat, crowberry grows in nutrient-poor areas, which determines the level of ash elements. Additionally, nitrogen, phosphorus, and potassium are present in peat bog organic matter in a form that is difficult to assimilate [82,83]. At the same time, phosphorus is unavailable in acidic soils with a large amount of undecomposed and semi-decomposed plant residues (high-moor peat) as it precipitates together with iron and aluminum [84–86].

Symbiosis with mycorrhizal fungi develops better when the soils contain low amounts of soluble phosphorus and nitrogen. Therefore, symbiosis between mycorrhizae and plants is of great importance for the adaptation of *E. nigrum* plants to conditions of low levels of available phosphorus, sulfur, nitrogen, and micronutrients in the soil. In this case, ammonium nitrogen is considered the most suitable source, since it is more available in acidic soils than the nitrate form. Furthermore, it was found that ammonium nitrogen allows for better vegetative growth and root system development in *E. nigrum* [50,87].

AM is widely used for introduction into the substrate when growing plants of the genus *Vaccinium* L., since it is capable of increasing adaptive capacity and may have powerful stress-protective properties capable of stimulating development on the root and, subsequently, above-ground parts of experimental plants [88]. This is probably associated with the effectiveness of the experimental treatment options with mycorrhiza-forming preparations of MF and KM, which contain fungi of the genus *Glomus*.

The high stress-protective and stimulating effect of the mycorrhiza-forming preparation (T3), containing mycelium and spores of fungi of the genus *Glomus*, on the survival, safety, and development of experimental ex vitro *E. nigrum* should also be noted. Therefore, during all the recorded observations of the variants, with the inclusion of the mycorrhizal preparations in the substrate composition, both independently and in combination with the fertilizers $N_{19}P_9K_{10}$ + 2MgO and $N_{20}P_{16}K_{10}(S_5)$, reliable differences were revealed in all the considered indicators.

The humic substances included in the mycorrhizal-forming preparation (T3) are characterized by a high capacity of cation and anion exchange, the ability to interact with soil enzymes, and vitamins. These activate plant immunity and development [89].

New research shows that not only mycorrhizal fungi, but also some root-symbiotic bacteria, are capable of stimulating ericaceous plants to become resistant to unfavorable environmental conditions [90]. *Trichoderma harzianum*, which is part of the 'Mycofriend' preparation, is a well-known biological agent that occurs in nature and is an endophytic symbiont. Similarly, the genus *Bacillus* is considered useful in agriculture due to its high level of antagonism towards various phytopathogenic microorganisms [91–95]. *Pseudomonas fluorescens* is a Gram-negative bacterium with antibacterial properties that has attracted the attention of researchers as an alternative to chemical bactericide agents [37,96,97]. This is probably related to the high survival of crowberry plants in variants using the mycorrhiza-forming preparation (T2, T7, T9).

In order for a plant to grow and develop well, it is necessary to create conditions that stimulate the development of mycorrhizal fungi and optimize the absorption of nutrients by plants. Therefore, in our research, we studied combinations of mycorrhizal preparations with different types of fertilizer (water-soluble and slow-release) with various methods of application (substrate application, direct application to the substrate, and foliar application).

It is known that the slow-release fertilizers $N_{19}P_9K_{10}$ + 2MgO are included in the group of fertilizers with a coating based on organic polymers—they have a multi-layer, porous coating-based alkyd resins. When entering the soil, moisture, under the action of osmotic pressure, penetrates through the pores into the granule, as a result of which it stretches, and through the expanded pores, nutrients enter in their dissolved form [98,99].

It is known that the nutrients of slow-release fertilizers are unaffected by soil salt content, pH level, microbial activity, or water quality. The only important factor is a temperature above +21 °C; above this, the release of nutrients is faster, and below it, it is slower.

Many studies conducted in recent years have shown that acidic or alkaline environments have virtually no effect on the release of nutrients from the polymer coatings of slow-release fertilizers [100,101]. However, our studies revealed no advantages of the slow-release fertilizer $N_{19}P_9K_{10}$ + 2MgO for the ex vitro survival of *E. nigrum* plants, which had a survival of between 29.2% and 75.0% on the 98th day of cultivation.

We assume that due to the release of phenolic substances by crowberry roots, in particular Batatasin-III, there could be an accelerated release of nutrients contained in the fertilizer granules. For this reason, young roots could get burned, and the plants could die. It is known that phenolic compounds can significantly slow down the rate of nutrient decomposition, impact the ability to accumulate organic matter, and also change the circulation of nitrogen and suppress the saprophytic activity of the soil [102,103].

Nevertheless, other researchers have demonstrated that the pH of the environment in which the release occurs significantly affects the release of nutrients from slow-release fertilizers, particularly when pH-sensitive polymers are used as the coating material [104]. Furthermore, some polymers may undergo structural changes or changes in properties in response to pH changes [105,106].

Depending on the nature of the functional group (e.g., a carboxyl or amino group) and its state (protonation or deprotonation in response to pH changes), electrostatic repulsion between chains may occur. This can lead to expansion or contraction of the chains, which affects polymer swelling. This can potentially either promote or inhibit moisture penetration and nutrient dissolution [102,107].

For example, studies on the release of nitrogen from urea coated with a bio-nanocomposite hydrogel in solutions with different pH levels have shown that nitrogen release increases as the pH rises. Within 24 h, 20.0%, 25.6%, and 34.9% of nitrogen was released in solutions with pH levels of 4.0, 7.0, and 10.0, respectively [108].

Foliar application with $N_{20}P_{20}K_{20}$ leaf fertilizer containing microelements in a chelated form, which increases solubility and absorption, proved effective without the need for mycorrhiza-forming preparations. When used on the 98th day of cultivation, plant survival was 100%, total root length increased by 1.3 times compared to the control, and the total shoot length increased by 1.7 times. This is probably because foliar application results in the rapid absorption of nutrients due to the fertilizer interacting most effectively with the physiological processes of the foliage [109,110].

The best results were obtained by direct application to substrate-experimental *E. nigrum* plants with a water-soluble fertilizer for heathers containing sulfur $N_{20}P_{16}K_{10}(S_5)$, both separately and in combination with mycorrhiza-forming preparations. The survival on the 98th day of cultivation was 95.8–100%, and the total root and shoot length increased by 1.4–1.5 times and 1.4–1.8 times, respectively, compared to the control.

In forest and berry plant nurseries, sulfur fertilizers are used to acidify the soil [111, 112]. The importance of sulfur (S) is due to the fact that it is a component of protein and non-protein amino acids, it plays a crucial structural role in cells in the form of disulfide bonds in proteins, and it participates in enzyme regulation and provides protection against oxidative stress with the help of glutathione. However, plant species vary greatly in their sulfur requirements. The assimilation of sulfur and nitrogen is closely interrelated [113]. Sulfur also plays an important role in protecting plants against disease [111,114]. Various plant species prevent fungal infections by depositing elemental sulfur in the xylem parenchyma [115].

However, sulfur deficiency can lead to reduced root development and a decrease or change in root exudates, affecting the community of microorganisms that use them as a carbon source [116–118]. Plant root activity affects the physical and chemical properties of soil by releasing organic compounds (rhizodeposition), accounting for 15–30% of photosynthetically produced carbon. This process provides soil organisms with energy to perform their functions. Thus, there is a synergistic effect between mycorrhizal preparations and sulfur-containing fertilizers.

5. Conclusions

The results obtained have improved our understanding of the conditions for ex vitro cultivation black crowberry of the 'Irland' variety using various mycorrhiza-forming preparations and different types of fertilizers.

On the 98th day of cultivation, the advantage of adding the mycorrhizal preparation granular mycorrhizae containing only mycelium of the genus *Glomus* to the substrate, both separately and with the mineral fertilizer $N_{20}P_{16}K_{10}(S_5)$, was revealed (the total root length of the plant increased by 43–50%, and the total shoot length increased by 70%). Also, when using the mycorrhizal preparation KM, the survival on the 98th day of cultivation was 95.8%, and when combined with the water-soluble fertilizer $N_{20}P_{16}K_{10}(S_5)$, it was 100%. In addition, it was found that it is effective to apply $N_{20}P_{16}K_{10}(S_5)$ fertilizer every 14 days, both independently and in combination with mycorrhiza-forming preparations (increase in total root length by 31–51% and total shoot length by 43–88%).

Without mycorrhizal preparations, it is effective to use foliar applications of the fertilizer $N_{20}P_{20}K_{20}$ every 14 days, which increases the total root length by 34% and the total shoot length by 72%, at the same time, and the survival on the 98th day of cultivation is 100%.

At the same time, the worst variants were found where the slow-release fertilizer $N_{19}P_9K_{10}$ + 2MgO was applied in the substrate. In T5 and T8 variants, the survival on the 98th day of cultivation of experimental plants was 29.2–37.5%.

This study highlights the importance of using mycorrhizal preparations in combination with different types of fertilizers during the post-adaptation period of *E. nigrum* plants to obtain more vigorous and stress-resistant plants for cultivation in plant nurseries, and then in industrial plantations for fruit.

Supplementary Materials: The following supporting information can be downloaded at https://www.mdpi.com/article/10.3390/horticulturae11091063/s1, Figure S1. Morphometric indicators (A—number of shoots, B—shoot height, C—total shoot length) of ex vitro development of *E. nigrum* 'Irland' plants on the 28th day of cultivation. Figure S2. Morphometric indicators (A—number of shoots, B—shoot height, C—total shoot length) of ex vitro development of *E. nigrum* 'Irland' plants on the 42nd day of cultivation. Figure S3. Morphometric indicators (A—number of shoots, B—shoot height, C—total shoot length) of ex vitro development of *E. nigrum* 'Irland' plants on the 56th day of cultivation. Figure S4. Morphometric indicators (A—number of shoots, B—shoot height, C—total shoot length) of ex vitro development of *E. nigrum* 'Irland' plants on the 70th day of cultivation. Figure S5. Morphometric indicators (A—number of shoots, B—shoot height, C—total shoot length) of ex vitro development of *E. nigrum* 'Irland' plants on the 84th day of cultivation.

Author Contributions: Conceptualization, I.N.; methodology, I.N. and S.A.; formal analysis, I.N. and N.S.; investigation, I.N. and S.A.; resources, I.N.; data curation, I.N., S.A. and N.S.; writing—original draft preparation, I.N.; writing—review and editing, I.N. and S.A.; visualization, I.N.; supervision, S.A.; project administration, S.A. and N.S. All authors have read and agreed to the published version of the manuscript.

Funding: This research received no external funding.

Data Availability Statement: Data is contained within the article and Supplementary Materials.

Conflicts of Interest: The authors declare no conflicts of interest.

References

1. Muravnik, L.E.; Shavarda, A.L. Leaf glandular trichomes in *Empetrum nigrum*: Morphology, histochemistry, ultrastructure and secondary metabolites. *Nord. J. Bot.* **2012**, *30*, 470–481. [CrossRef]
2. Szmidt, A.E.; Nilsson, M.-C.; Briceño, E.; Zackrisson, O.; Wang., X.-R. Establishment and genetic structure of *Empetrum hermaphroditum* populations in northern Sweden. *J. Veg. Sci.* **2002**, *13*, 627–634. [CrossRef]
3. Hagerup, O. Studies on the *Empetraceae*. *Biol. Meddr.* **1946**, *20*, 1–49.
4. Lorion, J.; Small, E. Crowberry (*Empetrum*): A chief arctic traditional indigenous fruit in need of economic and ecological management. *Bot. Rev.* **2021**, *87*, 259–310. [CrossRef]
5. Mazurenko, M.T. *Heather Shrubs of the Far East (Structure and Morphogenesis)*; Khokhryakov, A.P., Ed.; Nauka: Moscow, Russia, 1982; p. 184. (In Russian)
6. Sulavik, J.; Auestad, I.; Boudreau, S.; Halvorsen, R.; Rydgren, K. Population re-establishment and spatial dynamics of crowberry (*Empetrum nigrum* ssp. *hermaphroditum*), a foundation species in restored alpine ecosystems. *Ecol. Evol.* **2024**, *14*, e70242. [CrossRef]
7. Zverev, V.E.; Zvereva, E.L.; Kozlov, M.V. Slow growth of *Empetrum nigrum* in industrial barrens: Combined effect of pollution and age of explant plants. *Environ. Pollut.* **2008**, *156*, 454–460. [CrossRef]
8. Tuomi, M.W.; Utsi, T.A.; Yoccoz, N.G.; Armstrong, C.W.; Gonzalez, V.; Hagen, S.B.; Jónsdóttir, I.S.; Pugnaire, F.I.; Shea, K.; Wardle, D.A.; et al. The increase of an allelopathic and unpalatable plant undermines reindeer pasture quality and current management in the Norwegian tundra. *Commun. Earth Environ.* **2024**, *5*, 414. [CrossRef]
9. Tybirk, K.; Nilsson, M.-C.; Michelsen, A.; Kristensen, H.L.; Shevtsova, A.; Strandberg, M.T.; Johansson, M.; Nielsen, K.E.; Riis-Nielsen, T.; Strandberg, B.; et al. Nordic *Empetrum* dominated ecosystems: Function and susceptibility to environmental changes. *AMBIO A J. Hum. Environ.* **2000**, *29*, 90–97. [CrossRef]
10. Bråthen, K.A.; Fodstad, C.H.; Gallet, C. Ecosystem disturbance reduces the allelopathic effects of *Empetrum hermaphroditum* humus on tundra plants. *J. Veg. Sci.* **2010**, *21*, 786–795. [CrossRef]
11. Parkinson, L.V.; Mulder, C.P.H.; Putman, M.; Ruggles, A.; Sousa, E.E.; Spellman, K.V. Crowberry in a changing climate: Threats and opportunities. In *Berries in Alaska's Changing Environment Series: Empetrum nigrum*; Institute of Arctic Biology and International Arctic Research Center, University of Alaska Fairbanks: Fairbanks, AK, USA; 2024; pp. 1–19.
12. Ruotsalainen, A.L.; Markkola, A.M.; Kozlov, M.V. Birch effect on root fungal colonisation of crowberry are uniform along different environmental gradients. *Basic Appl. Ecol.* **2010**, *11*, 459–467. [CrossRef]
13. Wollenweber, E.; Dörr, M.; Stelzer, R.; Arriaga-Giner, F.A. Lepophilic phenolics from the leaves of *Empetrum nigrum*—Chemical structures and exudate localization. *Bot. Acta* **1992**, *105*, 300–305. [CrossRef]
14. Kochkin, R.A.; Lobanov, A.A.; Andronov, S.V.; Kostricin, V.V.; Popov, A.A.; Lobanova, L.P.; Kobelkova, I.V.; Kambarov, A.O. Efficiency of black crowberry (*Empetrum nigrum* L.) in correction of cold stress. *J. New Med. Technol.* **2017**, *24*, 66–72. (In Russian) [CrossRef]
15. Seeram, N.P. Berry fruits: Compositional elements, biochemical activities, and the impact of their intake on human health, performance, and disease. *J. Agric. Food Chem.* **2008**, *56*, 627–629. [CrossRef]
16. Laaksonen, O.; Sandell, M.; Järvinen, R.; Kallio, H. Orosensory contributing compounds in crowberry (*Empetrum nigrum*) press-byproducts. *Food Chem.* **2011**, *124*, 1514–1524. [CrossRef]
17. Svanberg, I.; Ægisson, S. Edible wild plant use in the Faroe Islands and Iceland. *Acta Soc. Bot. Pol.* **2012**, *81*, 233–238. [CrossRef]
18. Manninen, O.H.; Peltola, R. Effects of picking methods on the berry production of bilberry (*Vaccinium myrtillus*), lingonberry (*V. vitis-idaea*) and crowberry (*Empetrum nigrum* ssp. *hermaphroditum*) in Northern Finland. *Silva Fenn.* **2013**, *47*, 972. [CrossRef]
19. Jurikova, T.; Mlcek, J.; Skrovankova, S.; Balla, S.; Sochor, J.; Baron, M.; Sumczynski, D. Black Crowberry (*Empetrum nigrum* L.) Flavonoids and Their Health Promoting Activity. *Molecules* **2016**, *21*, 1685. [CrossRef] [PubMed]
20. Isaeva, M.A. Development of technology for dry extract of the black crowberry herb. In Proceedings of the Young Pharmacy—Potential of the Future 2022, St. Petersburg, Russia, 5–6 May 2022; Conference Abstract. pp. 709–712. (In Russian)

21. Bezverkhniaia, E.A.; Povet'eva, T.N.; Kadyrova, T.V.; Suslov, N.I.; Nesterova, Y.V.; Afanas'eva, O.G.; Kul'pin, P.V.; Yusova, Y.G.; Ermilova, E.V.; Miroshnichenko, A.G.; et al. Screening Study for Anticonvulsive Activity of Lipophilic Fractions from *Empetrum nigrum* L. *Res. Results Pharmacol.* **2020**, *6*, 67–73. [CrossRef]
22. Hagen, D. Propagation of native Arctic and alpine species with a restoration potential. *Polar Res.* **2002**, *21*, 37–47. [CrossRef]
23. Bell, J.N.; Tallis, J.H. *Empetrum nigrum* L. *J. Ecol.* **1973**, *61*, 289–305. [CrossRef]
24. Baskin, C.C.; Zackrisson, O.; Baskin, J.M. Role of warm stratification in promoting germination of seeds of *Empetrum hermaphroditum* (*Empetraceae*), a circumboreal species with a stony endocarp. *Am. J. Bot.* **2002**, *89*, 486–493. [CrossRef]
25. Qarachoboogh, A.F.; Alijanpour, A.; Hosseini, B.; Shafiei, A.B. Efficient and reliable propagation and rooting of foetid juniper (*Juniperus foetidissima* Willd.), as an endangered plant under in vitro condition. *In Vitro Cell. Dev. Biol.-Plant* **2022**, *58*, 399–406. [CrossRef]
26. Wang, Y.; Zhang, X.; Jiang, Z.; Yang, X.; Liu, X.; Ou, X.; Su, W.; Chen, R. Establishment and Optimization of Micropropagation System for Southern Highbush Blueberry. *Horticulturae* **2023**, *9*, 893. [CrossRef]
27. Hanus-Fajerska, E.; Wiszniewska, A.; Czaicki, P. Effectiveness of *Daphne* L. (*Thymelaeaceae*) in vitro propagation, rooting of microshoots and acclimatization of plants. *Acta Agrobot.* **2012**, *65*, 21–28. [CrossRef]
28. Nechiporenko, I.; Kazakov, P.; Palamarchuk, D. Prospects for obtaining planting material of a less widespread berry crop—Black crowberry (*Empetrum nigrum* L.) in vitro conditions. *BIO Web Conf.* **2024**, *139*, 01008. [CrossRef]
29. Park, S.Y.; Kim, Y.-W.; Moon, H.-K. Practical factors controlling in vitro multiplication and rooting in *Empetrum nigrum* var. *japonicum*, an endangered woody species. *Korean J. Plant Res.* **2012**, *25*, 739–744. [CrossRef]
30. Wei, X.; Chen, J.; Zhang, C.; Wang, Z. In vitro shoot culture of *Rhododendron fortunei*: An important plant for bioactive phytochemicals. *Ind. Crops Prod.* **2018**, *126*, 456–465. [CrossRef]
31. Tseplyaev, A.N.; Treshchevskaya, E.I.; Turtanova, E.N. Experience of growth in containers of the planting material received by the method of clonal propagation in vitro. *For. Eng. J.* **2018**, *8*, 124–130. (In Russian) [CrossRef]
32. Chizhik, O.V.; Filipenia, V.L.; Gorbatsevich, V.I. Development of technology for growing clonal material of promising species of woody shrub species of the *Vaccinium* genus using microbial preparations. In *Biotechnological Methods in Conservation of Biodiversity and Plant Breeding Proceeding of the International Scientific Conference Minsk, Minsk, Belarus, 18–20 August 2014*; Reshetnikov, V.N., Titok, V.V., Spiridovich, E.V., Fomenko, T.I., Kuzovkova, A.A., Eds.; National Academy of Sciences of Belarus The Central Botanical Gardens: Minsk, Belarus, 2014; pp. 250–252. (In Russian)
33. Clapa, D.; Fira, A.; Vescan, L.A. Aspects regarding the in vitro culture and ex vitro rooting in *Vaccinium macrocarpon* cultivar 'Piligrim'. *Bull. UASVM Anim. Sci. Biotechnol.* **2012**, *69*, 226–234. [CrossRef]
34. Grzelak, M.; Pacholczak, A.; Nowakowska, K. Challenges and insights in the acclimatization step of micropropagated woody plants. *Plant Cell Tiss. Organ Cult.* **2024**, *159*, 72. [CrossRef]
35. Chandra, S.; Bandopadhyay, R.; Kumar, V.; Chandra, R. Acclimatization of tissue cultured plantlets: From laboratory to land. *Biotechnol. Lett.* **2010**, *32*, 1199–1205. [CrossRef] [PubMed]
36. Wei, X.; Chen, J.; Zhang, C.; Liu, H.; Zheng, X.; Mu, J. Ericoid mycorrhizal fungus enhances microcutting rooting of *Rhododendron fortunei* and subsequent growth. *Hortic. Res.* **2020**, *7*, 140. [CrossRef]
37. Sergeeva, K.; Nechiporenko, I.; Popletaeva, S. Effect of MF3 (peptidyl-prolyl cis/trans isomerase) protein from *Pseudomonas fluorescens* on ex vitro adaptation and post-adaptation of hardy kiwi (*Actinidia arguta* Planch. ex Miq.) plants. *BIO Web Conf.* **2024**, *139*, 01010. [CrossRef]
38. Song, G.Q. Blueberry (*Vaccinium corymbosum* L.). In *Agrobacterium Protocols: Methods in Molecular Biology, Volume 1224*; Wang, K., Ed.; Springer: New York, NY, USA, 2015; pp. 121–132. [CrossRef]
39. Shiwani, K.; Sharma, D.; Kumar. Improvement of plant survival and expediting acclimatization process. In *Commercial Scale Tissue Culture for Horticulture and Plantation Crops*; Gupta, S., Chaturvedi, P., Eds.; Springer: Singapore, 2022; pp. 227–291. [CrossRef]
40. Kosolapov, V.M.; Cherniavskih, V.I.; Dumacheva, E.V.; Sajfutdinova, L.D.; Zhuchenko, A.A.; Glinushkin, A.P.; Grishina, H.V.; Kalinitchenko, V.P.; Akimova, S.V.; Semenova, N.A.; et al. Using microorganismal consortium and bioactive substances to treat seeds of two scots pine ecotypes as a technique to increase re-afforestation efficiency on chalk outcrops. *Forest* **2023**, *14*, 1093. [CrossRef]
41. Dewir, Y.H.; Habib, M.M.; Akqarawi, A.A.; Alshahrani, T.S.; Alaizari, A.A.; Malik, J.A.; Alwahibi, M.S.; Murthy, H.N. Mycorrhization enhances vegetative growth, leaf gas exchange, and root development of micropropagated *Philodendron bipinnatifidum* Schott ex Endl. plantlets during acclimatization. *Horticulturae* **2023**, *9*, 276. [CrossRef]
42. Vohník, M. Ericoid mycorrhizal symbiosis: Theoretical background and methods for its comprehensive investigation. *Mycorrhiza* **2020**, *30*, 671–695. [CrossRef] [PubMed]
43. Baba, T.; Hirose, D.; Sasaki, N.; Watanabe, N.; Kobayashi, N.; Kurashige, Y.; Karimi, F.; Ban, T. Mycorrhizal formation and diversity of endophytic fungi in hair roots of *Vaccinium oldhamii* Miq. in Japan. *Microbes Environ.* **2016**, *31*, 186–189. [CrossRef]
44. Martin, F.M.; van der Heijden, M.G.A. The mycorrhizal symbiosis: Research frontiers in genomics, ecology, and agricultural application. *New Phytol.* **2024**, *242*, 1486–1506. [CrossRef]

45. Olsrud, M.; Michelsen, A.; Wallander, H. Ergosterol content in ericaceous hair roots correlates with dark septate endophytes but not with ericoid mycorrhizal colonization. *Soil Biol. Biochem.* **2007**, *39*, 1218–1221. [CrossRef]
46. Walker, J.F.; Aldrich-Wolfe, L.; Riffel, A.; Barbare, H.; Simpson, N.B.; Trowbridge, J.; Jumpponen, A. Diverse *Helotiales* associated with the roots of three species of arctic *Ericaceae* provide no evidence for host specificity. *New Phytol.* **2011**, *191*, 515–527. [CrossRef] [PubMed]
47. Nordin, A.; Näsholm, T. Nitrogen storage forms in nine boreal understory plant species. *Oecologia* **1997**, *110*, 487–492. [CrossRef]
48. Read, D.J.; Perez-Moreno, J. Mycorrhizas and nutrient cycling in ecosystems—A journey toward relevance? *New Phytol.* **2003**, *157*, 475–492. [CrossRef]
49. Dunleavy, H.R.; Mack, M.C. Nonlinear responses of ericaceous and ectomycorrhizal Arctic shrubs across a long-term experimental nutrient gradient. *Ecosphere* **2024**, *15*, e4888. [CrossRef]
50. Hupperts, S.F.; Islam, K.S.; Gundale, M.J.; Kardol, P.; Sundqvist, M.K. Warming influences carbon and nitrogen assimilation between a widespread ericaceous shrub and root-associated fungi. *New Phytol.* **2024**, *241*, 1062–1073. [CrossRef]
51. Monaci, F.; Leidi, E.O.; Mingorance, M.D.; Valdés, B.; Oliva, S.R.; Bargagli, R. Selective uptake of major and trace elements in *Erica andevalensis*, an endemic species to extreme habitats in the Iberian Pyrite Belt. *J. Environ. Sci.* **2011**, *23*, 444–452. [CrossRef]
52. Geel, M.V.; Jacquemyn, H.; Peeters, G.; van Asker, K.; Honnay, O.; Ceulemans, T. Diversity and community structure of ericoid mycorrhizal fungi in European bogs and heathlands across a gradient of nitrogen deposition. *New Phytol.* **2020**, *228*, 1640–1651. [CrossRef]
53. Cairney, J.W.G.; Ashfold, A.E. Biology of mycorrhizal associations of epacrids (*Ericaceae*). *New Phytol.* **2002**, *154*, 305–326. [CrossRef] [PubMed]
54. Watkinson, S.C. Mutualistic symbiosis between fungi and autotrophs. In *The Fungi*, 3rd ed.; Watkinson, S.C., Boddy, L., Money, N.P., Eds.; Elsevier Science & Technology: Amsterdam, The Netherlands, 2015; pp. 205–243. [CrossRef]
55. Grelet, G.-A.; Johnson, D.; Vrålstad, T.; Alexander, I.J.; Anderson, I.C. New insights into the mycorrhizal *Rhizoscyphus ericae* aggregate: Spatial structure and co-colonization of ectomycorrhizal and ericoid roots. *New Phytol.* **2010**, *188*, 210–222. [CrossRef] [PubMed]
56. Wei, X.; Zhang, W.; Zulfiqar, F.; Zhang, C.; Chen, J. Ericoid mycorrhizal fungi as biostimulants for improving propagation and production of ericaceous plants. *Front. Plant Sci.* **2022**, *13*, 1027390. [CrossRef] [PubMed]
57. Wang, B.; Qiu, Y.L. Phylogenetic distribution and evolution of mycorrhizas in land plants. *Mycorrhiza* **2006**, *16*, 299–363. [CrossRef]
58. Ouhaddou, R.; Anli, M.; Ben-Laouane, R.; Boutasknit, A.; Baslam, M.; Meddich, A. The Importance of the Glomus Genus as a Potential Candidate for Sustainable Agriculture Under Arid Environments: A Review. *Int. J. Plant Biol.* **2025**, *16*, 32. [CrossRef]
59. Kokkoris, V.; Banchini, C.; Paré, L.; Abdellatif, L.; Séguin, S.; Hubberd, K.; Findlay, W.; Dalpé, Y.; Dettman, J.; Corradi, N.; et al. *Rhizophagus irregularis*, the model fungus in arbuscular mycorrhiza research, forms dimorphic spores. *New Phytol.* **2024**, *242*, 1771–1784. [CrossRef] [PubMed]
60. Sheikh-Assadi, M.; Khandan-Mirkohi, A.; Taheri, M.R.; Babalar, M.; Sheikhi, H.; Nicola, S. Arbuscular mycorrhizae contribute to growth, nutrient uptake, and ornamental characteristics of statice (*Limonium sinuatum* [L.] Mill.) subject to appropriate inoculum and optimal phosphorus. *Horticulturae* **2023**, *9*, 564. [CrossRef]
61. Hodge, A.; Storer, K. Arbuscular mycorrhiza and nitrogen: Implications for individual plants through to ecosystems. *Plant Soil* **2015**, *386*, 1–19. [CrossRef]
62. Toljander, J.F.; Lindahl, B.D.; Paul, L.R.; Elfstrand, M.; Finlay, R.D. Influence of arbuscular mycorrhizal mycelial exudates on soil bacterial growth and community structure. *FEMS Microbiol. Ecol.* **2007**, *61*, 295–304. [CrossRef]
63. Bortolot, M.; Buffoni, B.; Mazzarino, S.; Hoff, G.; Martino, E.; Fiorilli, V.; Salvioli Di Fossalunga, A. The Importance of Mycorrhizal Fungi and Their Associated Bacteria in Promoting Crops' Performance: An Applicative Perspective. *Horticulturae* **2024**, *10*, 1326. [CrossRef]
64. Eccher, T.; Martinelli, M. Inoculation of *Rhododendron* cultivars in vitro with different strains of ericoid endomycorrhizae. *Acta Hortic.* **2010**, *865*, 327–332. [CrossRef]
65. Corkidi, L.; Evans, M.; Bohn, J. Infectivity and effectiveness of arbuscular mycorrhizal fungi in horticultural practices. *Comb. Proc.-Int. Plant Propag. Soc.* **2008**, *58*, 241–244.
66. Bubier, J.L.; Smith, R.; Juutinen, S.; Moore, T.R.; Minocha, R.; Long, S.; Minocha, S. Effect of nutrient addition on leaf chemistry, morphology, and photosynthetic capacity of three bog shrubs. *Oecologia* **2011**, *167*, 355–368. [CrossRef] [PubMed]
67. Marty, C.; Lévesque, J.-A.; Bradley, R.L.; Lafond, J.; Paré, M.C. Lowbush blueberry fruit yield and growth response to inorganic and organic N-fertilization when competing with two common weed species. *PLoS ONE* **2019**, *14*, e0226619. [CrossRef]
68. Hazard, C.; Gosling, P.; Mitchell, D.K.; Doohan, F.M.; Bending, G.D. Diversity of fungi associated with hair roots of ericaceous plants is affected by land use. *FEMS Microbiol. Ecol.* **2014**, *87*, 586–600. [CrossRef]
69. Vohník, M.; Réblová, M. Fungi in hair roots of *Vaccinium* spp. (*Ericaceae*) growing on decomposing wood: Colonization patterns, identity, and in vitro symbiotic potential. *Mycorrhiza* **2023**, *33*, 69–86. [CrossRef]

70. Chaudhary, A.; Poudyal, S.; Kaundal, A. Role of arbuscular mycorrhizal fungi in maintaining sustainable agroecosystems. *Appl. Microbiol.* **2025**, *5*, 6. [CrossRef]
71. Jarosz, Z.; Michałojć, Z.; Pitura, K.; Dzida, K.; Koter, M. Influence of fertilization and mycorrhizae on the nutritional status of rhododendron (*Rhododendron hybridum*) in a nursery. *Agriculture* **2021**, *11*, 538. [CrossRef]
72. Williams, A. Current nursery practice with regard to mycorrhizae and the propagation of New Zealand's native plants. *Ecol. Manag. Restor.* **2010**, *11*, 220–223. [CrossRef]
73. Lloyd, G.; McCown, B. Commercially-feasible micropropagation of mountain laurel, *Kalmia latifolia*, by use of shoot-tip culture. *Comb. Proc.-Int. Plant Propag. Soc.* **1980**, *30*, 421–426.
74. Cabrera-Ariza, A.M.; Silva-Flores, P.; González-Ortega, M.; Acevedo-Tapia, M.; Cartes-Rodríguez, E.; Palfner, G.; Ramos, P.; Santelices-Moya, R.E. Early effects of mycorrhizal fungal inoculum and fertilizer on morphological and physiological variables of nursery-grown *Nothofagus alessandrii* plants. *Plants* **2023**, *12*, 1521. [CrossRef]
75. Trejo, D.; Sangabriel-Conde, W.; Gavito-Pardo, M.E.; Banuelos, J. Mycorrhizal inoculation and chemical fertilizer interactions in pineapple under field conditions. *Agriculture* **2021**, *11*, 934. [CrossRef]
76. Michałojć, Z.; Koter, M. Effect of fertilization and mycorrhization on growth and nutritional status of cranberry (*Vaccinium macrocarpon* Ait.) in the nursery. *J. Hortic. Res.* **2014**, *23*, 49–56. [CrossRef]
77. Zinati, G.M.; Dighton, J.; Both, A.-J. Fertilizer, irrigation, and natural ericaceous root and soil inoculum (NERS): Effect on container-grown ericaceous nursery crop biomass, tissue nutrient concentration, and leachate nutrient quality. *HortScience* **2011**, *46*, 799–807. [CrossRef]
78. Koizumi, T.; Nara, K. Communities of putative ericoid mycorrhizal fungi isolated from alpine dwarf shrubs in Japan: Effect of host identity and microhabitat. *Microbes Environ.* **2017**, *32*, 147–153. [CrossRef]
79. Kjøller, R.; Olsrud, M.; Michelsen, A. Co-existing ericaceous plant species in a subarctic mire community share fungal root endophytes. *Fungal Ecol.* **2010**, *3*, 205–214. [CrossRef]
80. Lareau, M.J. Rooting and establishment of in vitro blueberry plantlets in the presence of mycorrhizal fungi. *Acta Hort.* **1985**, *165*, 197–202. [CrossRef]
81. Nieuwdorp, P.J. Some investigations on the mycorrhiza of *Calluna, Erica* and *Vaccinium*. *Acta Bot. Neerl.* **1969**, *18*, 180–196. [CrossRef]
82. Hagerup, O. Studies on the significance of polyploidy (IV. *Oxycoccus*). *Hereditas* **1940**, *26*, 399–410. [CrossRef]
83. Sidorovich, E.A.; Kudinov, M.A.; Ruban, N.N.; Sherstenikina, A.V.; Rupasova, Z.A.; Shapiro, D.K.; Gorbenko, S.V. *Large-Fruited Cranberries in Belarus*; Nauka i technika: Minsk, Belarus, 1987; p. 238. (In Russian)
84. Darnell, R.L.; Stutte, G.W.; Martin, G.C.; Lang, G.A.; Early, J.D. Development physiology of rabbiteye blueberry. In *Horticultural Reviews, Volume 13*; Janick, J., Ed.; John Wiley & Sons, Inc.: Hoboken, NJ, USA, 1992; pp. 339–406. [CrossRef]
85. Medappa, K.C.; Dana, M.N. Influence of pH, calcium, iron, and aluminum on the uptake of radiophosphorus by cranberry plant. *Soil Sci. Soc. Amer. Proc.* **1968**, *32*, 381–383. [CrossRef]
86. Kumar, S.; Kumar, S.; Mohapatra, T. Interaction between macro- and micro-nutrients in plants. *Front. Plant. Sci.* **2021**, *12*, 1–9. [CrossRef]
87. Esposti, M.D.D.; de Siqueira, D.L.; Pereira, P.R.G.; Venegas, V.H.A.; Salomão, L.C.C.; Filho, J.A.M. Assessment of nitrogenized nutrition of citrus rootstock using chlorophyll concentration in the leaf. *J. Plant Nutr.* **2003**, *26*, 1287–1299. [CrossRef]
88. Boorboori, M.R.; Lackóová, L. Arbuscular mycorrhizal fungi and salinity stress mitigation in plants. *Front. Plant Sci.* **2025**, *15*, 1504970. [CrossRef]
89. Maffia, A.; Oliva, M.; Marra, F.; Mallamaci, C.; Nardi, S.; Muscolo, A. Humic substances: Bridging ecology and agriculture for a greener future. *Agronomy* **2025**, *15*, 410. [CrossRef]
90. Fuentes-Quiroz, A.; Herrera, H.; Alvarado, R.; Sagredo-Saez, C.; Isabel-Mujica, M.; Vohník, M.; Rolli, E. Cultivable root-symbiotic bacteria of a pioneer ericaceous dwarf shrub colonizing volcanic deposits and their potential to promote host fitness. *J. Soil Sci. Plant Nutr.* **2024**, *24*, 3355–3363. [CrossRef]
91. Pani, S.; Kumar, A.; Sharma, A. *Trichoderma harzianum*: An overview. *Bull. Environ. Pharmacol. Life Sci.* **2021**, *10*, 32–39.
92. Yao, X.; Guo, H.; Zhang, K.; Zhao, M.; Ruan, J.; Chen, J. *Trichoderma* and its role in biological control of plant fungal and nematode disease. *Front. Microbiol.* **2023**, *14*, 1160551. [CrossRef]
93. Sagar, A.; Yadav, S.S.; Sayyed, R.Z.; Sharma, S.; Ramteke, P.W. *Bacillus subtilis*: A multifarious plant growth promoter, biocontrol agent, and bioalleviator of abiotic stress. In *Bacilli in Climate Resilient Agriculture and Bioprospecting*; Islam, M.T., Rahman, M., Pandey, P., Eds.; Springer: Cham, Switzerland, 2022; pp. 561–580. [CrossRef]
94. Ortiz, A.; Sansinenea, E. 5—*Bacillus* sp. as biofertilizers applied in horticultural crops. In *Bio-Inoculants in Horticultural Crops. Advances in Bio-Inoculant Sciences, Volume 3*; Rakshit, A., Meena, V.S., Fraceto, L.F., Parihar, M., Mendon, A.B., Singh, H.B., Eds.; Elsevier Inc.: Amsterdam, The Netherlands, 2024; pp. 97–108. [CrossRef]
95. Silva, L.I.; Pereira, M.C.; Carvalho, A.M.X.; Buttrós, V.H.; Pasqual, M.; Dória, J. Phosphorus-solubilizing microorganisms: A key to sustainable agriculture. *Agriculture* **2023**, *13*, 462. [CrossRef]

96. Taylor, T.B.; Silby, M.W.; Jackson, R.W. *Pseudomonas fluorescens*. *Trends Microbiol.* **2025**, *33*, 250–251. [CrossRef]
97. Mohan, V.; Wibisono, R.; Chalke, S.; Fletcher, G.; Leroi, F. The anti-listeria Activity of *Pseudomonas fluorescens* isolated from the horticultural environment in New Zealand. *Pathogens* **2023**, *12*, 349. [CrossRef] [PubMed]
98. Mukhina, M.T.; Borovik, R.A.; Korshunov, A.A. Slow-and control-release fertilizers: Brief history and modern ternts. *Plodorodie* **2021**, 77–82. (In Russian) [CrossRef]
99. Kassem, I.; Ablouh, E.-H.; Bouchtaoui, F.-Z.E.; Jaouahar, M.; Achaby, M.E. Polymer coated slow/ controlled release granular fertilizers: Fundamentals and research trends. *Prog. Mater. Sci.* **2024**, *144*, 101269. [CrossRef]
100. Oertli, J.J.; Lunt, O.R. Controlled release of fertilizer minerals by incapsulating membranes: I. factors influencing the rate of release. *Soil Sci. Soc. Am. J.* **1962**, *26*, 579–583. [CrossRef]
101. Christianson, C.B. Factors affecting N release of urea from reactive layer coated urea. *Fertil Res.* **1988**, *16*, 273–284. [CrossRef]
102. Wardle, D.A.; Nilsson, M.-C.; Gallet, C.; Zackrisson, O. An ecosystem-level perspective of allelopathy. *Biol. Rev.* **1998**, *73*, 305–319. [CrossRef]
103. Nilsson, M.-C.; Gallet, C.; Wallstedt, A. Temporal variability of phenolics and batatasin-Ill in *Empetrum hermaphroditum* leaves over an eight-year period: Interpretations of ecological function. *OIKOS* **1998**, *81*, 6–16. [CrossRef]
104. Lawrencia, D.; Wong, S.K.; Low, D.Y.S.; Goh, B.H.; Goh, J.K.; Ruktanonchai, U.R.; Soottitantawat, A.; Lee, L.H.; Tang, S.Y. Controlled release fertilizers: A review on coating materials and mechanism of release. *Plants* **2021**, *10*, 238. [CrossRef] [PubMed]
105. Dai, S.; Ravi, P.; Tam, K.C. PH-responsive polymers: Synthesis, properties and applications. *Soft Matter* **2008**, *4*, 435–449. [CrossRef] [PubMed]
106. Kocak, G.; Tuncer, C.; Bütün, V. PH-responsive polymers. *Polym. Chem.* **2017**, *8*, 144–176. [CrossRef]
107. Majeed, Z.; Ramli, N.K.; Mansor, N.; Man, Z. A comprehensive review on biodegradable polymers and their blends used in controlled- release fertilizer processes. *Rev. Chem. Eng.* **2015**, *31*, 69–95. [CrossRef]
108. Tanan, W.; Panichpakdee, J.; Suwanakood, P.; Saengsuwan, S. Biodegradable hydrogels of cassava starch-g-polyacrylic acid/natural rubber/polyvinyl alcohol as environmentally friendly and highly efficient coating material for slow-release urea fertilizers. *J. Ind. Eng. Chem.* **2021**, *101*, 237–252. [CrossRef]
109. Januszkiewicz, R.; Kulczycki, G.; Samoraj, M. Foliar fertilization of crop plants in Polish agriculture. *Agriculture* **2023**, *13*, 1715. [CrossRef]
110. Zydlik, Z.; Zydlik, P.; Kafkas, N.E.; Yesil, B.; Cieśliński, S. Foliar application of some macronutrients and micronutrients improves yield and fruit quality of highbush blueberry (*Vaccinium corymbosum* L.). *Horticulturae* **2022**, *8*, 664. [CrossRef]
111. Gahan, J.; Schmalenberger, A. The role of bacteria and mycorrhiza on plant sulfur supply. *Front. Plant Sci.* **2014**, *5*. [CrossRef]
112. Ma, W.; Tang, S.; Dengzeng, Z.; Zhang, D.; Zhang, T.; Ma, X. Root exudates contribute to belowground ecosystem hotspots: A review. *Front. Microbiol.* **2022**, *13*, 937940. [CrossRef] [PubMed]
113. Zhao, F.-J.; Tausz, M.; De Kok, L.J. Role of sulfur for plant production in agricultural and natural ecosystems. In *Sulfur Metabolism in Phototrophic Organisms. Advances in Photosynthesis and Respiration, Volume 27*; Hell, R., Dahl, C., Knaff, D., Leustek, T., Eds.; Springer: Dordrecht, The Netherlands, 2008; pp. 417–435. [CrossRef]
114. Cooper, R.M.; Williams, J.S. Elemental sulphur as an induced antifungal substance in plant defence. *J. Exp. Bot.* **2004**, *55*, 1947–1953. [CrossRef] [PubMed]
115. Zenda, T.; Liu, S.; Dong, A.; Duan, H. Revisiting sulphur-the once neglected nutrient: It's roles in plant growth, metabolism, stress tolerance and crop production. *Agriculture* **2021**, *11*, 626. [CrossRef]
116. Zhou, J.; Zhang, H.; Huang, Y.; Jiao, S.; Zheng, X.; Lu, W.; Jiang, W.; Bai, X. Impact of sulfur deficiency and excess on the growth and development of soybean seedlings. *Int. J. Mol. Sci.* **2024**, *25*, 11253. [CrossRef]
117. Alhendawi, R.A.; Kirkby, E.A.; Pilbeam, D.J. Evidence that sulfur deficiency enhances molybdenum transport in xylem sap of tomato plants. *J. Plant Nutr.* **2005**, *28*, 1347–1353. [CrossRef]
118. Astolfi, S.; Zuchi, S.; Hubberten, H.M.; Pinton, R.; Hoefgen, R. Supply of sulphur to S-deficient young barley seedlings restores their capability to cope with iron shortage. *J. Exp. Bot.* **2010**, *61*, 799–806. [CrossRef] [PubMed]

MDPI AG
Grosspeteranlage 5
4052 Basel
Switzerland
Tel.: +41 61 683 77 34

Horticulturae Editorial Office
E-mail: horticulturae@mdpi.com
www.mdpi.com/journal/horticulturae

www.ingramcontent.com/pod-product-compliance
Lightning Source LLC
Chambersburg PA
CBHW061500210926
51850CB00008B/1772
9783725855476